SolidWorks Flow Simulation 2021 Black Book (Colored)

By
Gaurav Verma
Matt Weber
(CADCAMCAE Works)

Edited by
Kristen

ISBN # 978-1-77459-008-9

DEDICATION

To teachers, who make it possible to disseminate knowledge
to enlighten the young and curious minds
of our future generations

To students, who are the future of the world

THANKS

To my friends and colleagues

To my family for their love and support

Table of Contents

Chapter 3 : Analyzing and Generating Results of Analysis

Chapter 4 : Practical and Practice

Chapter 5 : Advanced Boundary Conditions

Chapter 6 : Basics of CFD

Chapter 7 : Practical and Practice

Preface

SolidWorks Flow Simulation 2021 is an Add-In for SolidWorks used to perform computational fluid dynamics related analysis. SOLIDWORKS Flow Simulation can perform complex calculations of computational fluid dynamics and can quickly and easily simulate fluid flow, heat transfer, and fluid forces that are critical to the success of your design.

The **SolidWorks Flow Simulation 2021 Black Book** is the 4th edition of our series on SolidWorks Flow Simulation. The book is targeted for beginners of SolidWorks Flow Simulation. This book covers the basic equations and terms of Fluid Dynamics theory. The book covers all the major tools of Flow Simulation modules like Fluid Flow, Thermal Fluid Flow, and Electronic Cooling modules. A chapter on basic concepts of CFD has been added discuss behind the scene calculations of SolidWorks CFD software. This book can be used as supplement to Fluid Dynamics course if your subject requires the application of Software for solving real-world problems. Some of the salient features of this book are :

In-Depth explanation of concepts

Every new topic of this book starts with the explanation of the basic concepts. In this way, the user becomes capable of relating the things with real world.

Topics Covered

Every chapter starts with a list of topics being covered in that chapter. In this way, the user can easy find the topic of his/her interest easily.

Instruction through illustration

The instructions to perform any action are provided by maximum number of illustrations so that the user can perform the actions discussed in the book easily and effectively. There are about 500 illustrations that make the learning process effective.

Tutorial point of view

At the end of concept's explanation, the tutorial make the understanding of users firm and long lasting. Almost each chapter of the book has tutorials that are real world projects. Moreover, most of the tools in this book are discussed in the form of tutorials.

Project

Free projects and exercises are provided to students for practicing.

For Faculty

If you are a faculty member, then you can ask for video tutorials on any of the topic, exercise, tutorial, or concept.

Formatting Conventions Used in the Text

All the key terms like name of button, tool, drop-down etc. are kept bold.

Free Resources

Link to the resources used in this book are provided to the users via email. To get the resources, mail us at ***cadcamcaeworks@gmail.com*** with your contact information. With your contact record with us, you will be provided latest updates and informations regarding various technologies. The format to write us mail for resources is as follows:

Subject of E-mail as ***Application for resources of............book***.
Also, given your information like
Name:
Course pursuing/Profession:
Contact Address:
E-mail ID:

Note: We respect your privacy and value it. If you do not want to give your personal informations then you can ask for resources without giving your information.

About Author

The author of this book, Gaurav Verma, has written many books on CAD/CAM/CAE available already in market. He has written and assisted in more than 15 top selling titles in CAD/CAM/CAE. He has authored **AutoCAD Electrical Black Books** which are available in both **English** and **Russian** language. He has provided consultant services to many industries in US, Greece, Canada, and UK. **SolidWorks Simulation Black Books** are one of the most selling books in SolidWorks Simulation field. The author has hands on experience on almost all the CAD/CAM/CAE packages. If you have any query/doubt in any CAD/CAM/CAE package, then you can contact the author by writing at cadcamcaeworks@gmail.com

For Any query or suggestion

If you have any query or suggestion, please let us know by mailing us on *cadcamcaeworks@gmail.com*. Your valuable constructive suggestions will be incorporated in our books and your name will be addressed in special thanks area of our books on your confirmation.

This page is left blank intentionally

Chapter 1

Starting with Computational Fluid Dynamics

Topics Covered

The major topics covered in this chapter are:

- ***Introduction to CFD***
- ***Starting SolidWorks Flow Simulation***
- ***Methods of Creating Lids***
- ***Check Geometry***
- ***Leak Tracking Technique***
- ***Engineering Database***

INTRODUCTION OF FLUID MECHANICS

During the course, you will know various aspects of SolidWorks Flow Simulation for various practical problems. But, keep in mind that all computer software work on same concept of GIGO which means Garbage In - Garbage Out. So, if you have specified any wrong parameter while defining properties of analysis then you will not get the correct results. This problem demands a good knowledge of Fluid Mechanics so that you are well conversant with the terms of classical fluid mechanics and can related the results to the theoretical concepts. In this chapter, we will discuss the basics of Fluid Mechanics and we will try to related them with analysis wherever possible.

BASIC PROPERTIES OF FLUIDS

There are various basic properties required while performing analysis on fluid. These properties are collected by performing experiments in labs. Most of these properties are available in the form of tables in Steam Tables or Design Data books. These properties are explained next.

Mass Density, Weight Density, and Specific Gravity

Density or Mass Density is the mass of fluid per unit volume. In SI units, mass is given by kg and volume is given by m³. So, mathematically we can say,

$$\text{Density (or Mass Density) } \rho = \frac{Mass\ of Fluid}{Volume\ Occupied\ by\ Fluid}\ \text{kg/m3}$$

If you are asked for weight density then multiply mass by gravity coefficient. Mathematically it can be expressed as:

$$\text{Weight Density w} = \frac{Mass\ of Fluid \times Gravity\ Coefficient}{Volume\ Occupied\ by\ Fluid}\ \text{N/m3}$$

Most of the time, fluid density is available as **Specific Gravity**. Specific gravity is the ratio of weight density of fluid to weight density of water in case of liquid. In case of gases, it is the ratio of weight density of fluid to weight density of air. Note that weight density of water is 1000 kg/m^3 at 4 °C and weight density of air is 1.225 kg/m^3 at 15 °C.

Note that as the temperature of liquid rises, its density is reduced and vice-versa. The same is true for gases as well.

Viscosity

Viscosity is the coefficient of friction between different layers of fluid. In other terms, it is the shear stress required to produce unit rate of shear strain in one layer of fluid. Mathematically it can be expressed as:

$$\mu = \frac{\tau}{(\frac{dx}{dy})}\ \text{N.s/m}^2 \text{ or Pa.s}$$

where μ is viscosity, τ is shear stress (or force applied tangentially to the layer of fluid) and (dx/dy) is the shear strain.

As the density of fluid changes with temperature so does the viscosity with temperature. The formula for viscosity of fluid at different temperature is given next.

For Liquids, $\mu = \mu_0 \frac{1}{1 + \alpha t + \beta t^2}$

For Gases, $\mu = \mu_0 + \alpha t - \beta t^2$

here, μ_0 is viscosity at 0 °C
α and β are constants for fluid (for water α is 0.03368 and β is 0.000221)
(for air α is 5.6x10-8 and β is 1.189x10-10)
t is the temperature

Problem on Viscosity

Dynamic viscosity of lubricant oil used between shaft and sleeve is 8 poise. The shaft has a diameter of 0.4 m and rotates at 250 r.p.m. Find out the power lost due to viscosity of fluid if length of sleeve is 100 mm and thickness of oil film is 1.5 mm; refer to Figure-2.

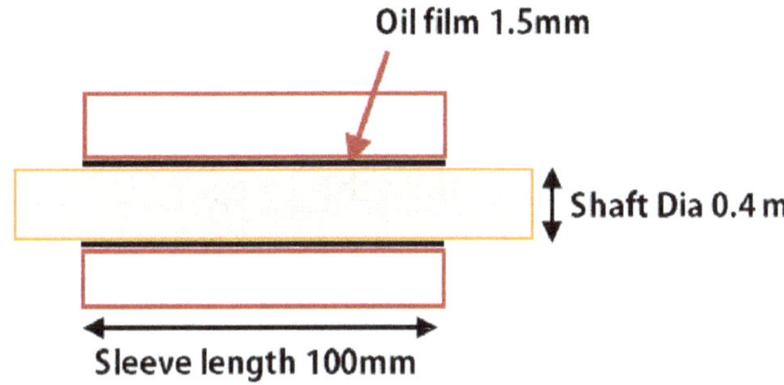

Figure-1. Viscosity problem

Solution:

Viscosity μ = 8 poise = 8/10 N.s/m² =0.8 N.s/m²

Tangential velocity of shaft $u = \frac{\pi \times D \times N}{60} = \frac{\pi \times 0.4 \times 250}{60}$=5.236 m/s

Using the relation, $\tau = \mu \frac{dx}{dy}$

where dx is 5.236
and dy is 1.5x10-3

$$\tau = 0.8 \times \frac{5.236}{1.5 \times 10^{-3}} = 2792.53 \text{ N/m2}$$

Shear force F = τ x Area

$F = \tau \times \pi D \times L = 2792.53 \times \pi \times 0.4 \times 100 \times 10^{-3} = 350.92$ N

Torque (T) = Force x Radius = 350.92 x 0.2 = 70.184 N.m

Power = 2 π.N.T/60 = (2 π x 250 x 70.184)/60=1837.41 W Ans.

Now, you may ask how this problem relates with CFD. As discussed earlier, the viscosity changes with temperature and as fluid flows through pipe or comes in contact with rolling shaft, its temperature rises. In such cases, CFD gives the approximate viscosity and temperature of fluids in the system at different locations. This data later can be used to find solution for other engineering problems.

TYPES OF FLUIDS

There are mainly 5 types of fluids:

Ideal Fluids: These fluids are incompressible and have no viscosity which means they flow freely without any resistance. This category of fluid is imaginary and used in some cases of calculations.

Real Fluids: These are the fluids found in real world. These fluids have viscosity values as per their nature and can be compressible in some cases.

Newtonian Fluids: Newtonian fluids are those in which shear stress is directly proportional to shear strain. In a specific temperature range, water, gasoline, alcohol etc. can be Newtonian fluids.

Non-Newtonian Fluids: Those fluids in which shear stress is not directly proportional to shear strain. Most of the time Real Fluids fall in this category.

Ideal Plastic Fluids: Those fluids in which shear stress is more than yield value and they deform plastically. The shear stress in these fluids is directly proportional to shear strain.

THERMODYNAMIC PROPERTIES OF FLUID

Most of the liquids are not considered as compressible in general applications as their molecules are already bound closely to each other. But, Gas have large gap between their molecules and can be compressed easily relative to liquids. As we pick pressure to compress the gas, other thermodynamic properties also come into play. The relationship between Pressure, Temperature, and specific Volume is given by;

$$P.V = RT$$

here,

P = Absolute pressure of a gas in N/m^2
V = Specific Volume = $1/\rho$
R = Gas Constant (for Air is 287 J/Kg-K
T = Absolute Temperature
ρ = Density of gas

If the density of gas changes with constant temperature then the process is called Isothermal process and if density changes with no heat transfer then the process is called Adiabatic process.

For Isothermal process, $p/\rho = \text{Constant}$

For Adiabatic process, $p/\rho k = \text{Constant}$

here, k is Ratio of specific heat of a gas at constant pressure and constant volume (1.4 for air).

Universal Gas Constant

By Pressure, Temperature, volume equation,

p.V = nMRT

Here,
p = Absolute pressure of a gas in N/m²
V = Specific Volume = 1/ ϱ
n = Number of moles in a Volume of gas
M = Mass of gas molecules/ Mass of Hydrogen atom = n x m (m is mass gas in kg)
R = Gas Constant (for Air is 287 J/Kg-K)
T = Absolute Temperature

MxR is called Universal Gas constant and is equal to 8314 J/kg-mole K for water.

Compressibility of Gases

Compressibility is reciprocal of bulk modulus of elasticity K, which is defined as ratio of Compressive stress to volumetric strain.

Bulk Modulus = Increase in pressure/ Volumetric strain

K = -(dp/dV)xV

Vapour Pressure and Cavitation

When a liquid converts into vapour due to high temperature in a vessel then vapours exert pressure on the walls of vessel. This pressure is called **Vapour pressure**.

When a liquid flows through pipe, sometimes bubbles are formed in the flow. When these bubbles collapse at the adjoining boundaries then they erode the surface of tube due to high pressure burst of bubble. This erosion is in the form of cavities at the surface of tube and the phenomena is called **Cavitation**.

PASCAL'S LAW

Pascal's Law states that pressure at a point in static fluid is same in all directions. In mathematical form px=py=pz in case of static fluids.

FLUID DYNAMICS

Up to this point, the rules stated in this chapter were for static fluid that is fluid at rest. Now, we will discuss the rules for flowing fluid.

Bernoulli's Incompressible Fluid Equation

Bernoulli's equation states that the total energy stored in fluid is always same in a closed system. In the language of mathematics,

$$p + \frac{1}{2}\rho V^2 + \rho gh = constant$$

Here, p is pressure
ϱ is density
V is velocity of fluid
h is height of fluid
g is gravitational acceleration

Eulerian and Lagrangian Method of Analysis

There are two different points of view in analyzing problems in fluid mechanics. The first view, appropriate to fluid mechanics, is concerned with the field of flow and is called the eulerian method of description. In the eulerian method, we compute the pressure field p(x, y, z, t) of the flow pattern, not the pressure changes p(t) that a particle experiences as it moves through the field.

The second method, which follows an individual particle moving through the flow, is called the lagrangian method. The lagrangian approach, which is more appropriate to solid mechanics, will not be treated in this book. However, certain numerical analyses of sharply bounded fluid flows, such as the motion of isolated fluid droplets, are very conveniently computed in lagrangian coordinates.

Fluid dynamic measurements are also suited to the eulerian system. For example, when a pressure probe is introduced into a laboratory flow, it is fixed at a specific position (x, y, z). Its output thus contributes to the description of the eulerian pressure field p(x, y, z, t). To simulate a lagrangian measurement, the probe would have to move downstream at the fluid particle speeds; this is sometimes done in oceanographic measurements, where flow meters drift along with the prevailing currents.

Now, we know the two methods of analyzing fluid mechanics problem. But, there are further three approaches for these two methods by which problems are derived to solution. These approaches are:

Control Volume
Differential
Experimental

Control volume analysis, is accurate for any flow distribution but is often based on average or "one dimensional" property values at the boundaries. It always gives useful "engineering" estimates. In principle, the differential equation approach can be applied to any problem. Only a few problems, such as straight pipe flow, yield to exact analytical solutions. But the differential equations can be modeled numerically, and the flourishing field of computational fluid dynamics (CFD) can now be used to give good estimates for almost any geometry. Finally, the dimensional analysis applies to any problem, whether analytical, numerical, or experimental. It is particularly useful to reduce the cost of experimentation.

Since the Differential Equation approach is more concerned to CFD so we will discuss this approach a little deeper.

DIFFERENTIAL APPROACH OF FLUID FLOW ANALYSIS

As discussed earlier, in this approach, the fluid is divided in to very small finite number of elements via a computer process called meshing. Various equations for different properties of a fluid are given next.

Acceleration

Acceleration **a** can be given as:

$$\mathbf{a} = \frac{d\mathbf{V}}{dt} = \mathbf{i}\frac{du}{dt} + \mathbf{j}\frac{dv}{dt} + \mathbf{k}\frac{dw}{dt}$$

Various components of acceleration **a** are:

$$a_x = \frac{du}{dt} = \frac{\partial u}{\partial t} + u\frac{\partial u}{\partial x} + v\frac{\partial u}{\partial y} + w\frac{\partial u}{\partial z} = \frac{\partial u}{\partial t} + (\mathbf{V}\cdot\nabla)u$$

$$a_y = \frac{dv}{dt} = \frac{\partial v}{\partial t} + u\frac{\partial v}{\partial x} + v\frac{\partial v}{\partial y} + w\frac{\partial v}{\partial z} = \frac{\partial v}{\partial t} + (\mathbf{V}\cdot\nabla)v$$

$$a_z = \frac{dw}{dt} = \frac{\partial w}{\partial t} + u\frac{\partial w}{\partial x} + v\frac{\partial w}{\partial y} + w\frac{\partial w}{\partial z} = \frac{\partial w}{\partial t} + (\mathbf{V}\cdot\nabla)w$$

Summing these into a vector, we obtain the total acceleration as

$$a = \frac{dV}{dt} = \frac{\partial V}{\partial t} + \left(u\frac{\partial V}{\partial x} + v\frac{\partial V}{\partial y} + w\frac{\partial V}{\partial z}\right) = \frac{\partial V}{\partial t} + (V.\nabla)V$$

Similarly, you can divide other parameters as vector like Force, Pressure, Temperature, and so on.

INTRODUCTION TO CFD

The CFD stands for Computational Fluid Dynamics. The SolidWorks CFD Black Book is truly for beginners. If you never studied the CFD before, if you have never worked in the area, and if you have no real idea as to what the CFD is all about, then this book is for you. Absolutely no prior knowledge of CFD is assumed on your part, only your desire to learn something about the subject is taken for granted.

Computational Fluid Dynamics constitutes a new "third approach" in the philosophical study and development of the whole discipline of fluid dynamics. In the seventeenth century, the foundations for experimental fluid dynamics were laid in France and England. The eighteenth and nineteenth centuries saw the gradual development of theoretical fluid dynamics, again primarily in Europe. As a result, throughout most of the twentieth century the study and practice of fluid dynamics involved the use of theory on the one hand and pure experiment on the other hand.

However, to keep things in context, CFD provides a new third approach-but nothing more than that. It nicely and synergistically complements the other two approaches of pure theory and pure experiment, but it will never replace either of these approaches. There will always be a need for theory and experiment. The future advancement of fluid dynamics will rest upon a balance of all three approaches, with computational fluid dynamics helping to interpret and understand the results of theory, experiment, and vice-versa; refer to Figure-2.

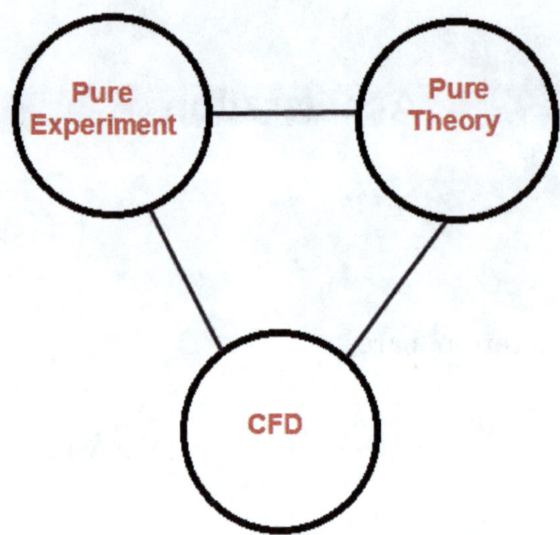

Figure-2. The three dimensions of fluid dynamics

Finally, we note that computational fluid dynamics is commonplace enough today that the composition CFD is universally accepted for the phrase "Computational Fluid Dynamics". We will use this composition throughout the book.

Computational fluid dynamics (CFD) is the use of applied math, physics, and computational software package to examine how a gas or liquid flows, also as how the gas or liquid affects objects as it flows past. Computational fluid dynamics is based on the Navier-Stokes equations. These equations describe how the velocity, pressure, temperature, and density of a moving fluid are related.

Navier-Stokes equation

The Navier-Stokes equations are the elemental partial differentials equations that describe the flow of incompressible fluids (Newtonian Fluids). Using the rate of stress and rate of strain, it can be shown that the parts of a viscous force F during a non rotating frame are given by-

$$\begin{aligned}\frac{F_i}{V} &= \frac{\partial}{\partial x_j}\left[\eta\left(\frac{\partial u_i}{\partial x_j}+\frac{\partial u_j}{\partial x_i}\right)+\lambda\delta_{ij}\nabla\cdot\mathbf{u}\right]\\ &= \frac{\partial}{\partial x_j}\left[\eta\left(\frac{\partial u_i}{\partial x_j}+\frac{\partial u_j}{\partial x_i}-\tfrac{2}{3}\delta_{ij}\nabla\cdot\mathbf{u}\right)+\mu_B\delta_{ij}\nabla\cdot\mathbf{u}\right],\end{aligned}$$

Where η is the dynamic viscosity, λ is the second viscosity coefficient, δ_{ij} is the Kronecker delta, $\nabla.u$ is the divergence, μ_B is the bulk viscosity and Einstein summation has been used to sum over j=1,2,and 3.

Dynamic Viscosity

Dynamic viscosity is the force required by a fluid to overcome its own internal molecular friction so that the fluid can flow. In other words, dynamic viscosity is defined as the tangential

force per unit area required to move the fluid in one horizontal plane with reference to other plane with a unit velocity whereas the fluid's molecules maintain a unit distance apart.

A parameter η is defined as

[Shear Stress] = η [Strain Rate]

Written explicitly.

$$\sigma = \eta \dot{e} = \eta \frac{1}{l} \frac{dl}{dt} = \eta \frac{u}{l},$$

Where l is the length scale and u is the velocity scale. In cgs η has units of g cm-1 s-1. Dynamic viscosity is related to kinematic viscosity ν by

$$\eta = \rho \nu$$

Where ρ is the density.

Second Viscosity Coefficient

For a compressible fluid, i.e. one for which $\nabla.u = 0$, where $\nabla.u$ is the divergence Σ of the velocity field, the stress tensor of the fluid can be written

$$S_{ij} = \eta \left(\frac{\partial u_i}{\partial x_j} + \frac{\partial u_j}{\partial x_i} \right) + \lambda \delta_{ij} \nabla \cdot \mathbf{u},$$

Where δ_{ij} is the Kronecker delta, η is the dynamic viscosity, and λ is the second coefficient of viscosity. λ is analogous to the first Lame constant. For an incompressible fluid, the term involving λ drops out from the equation, so λ can be ignored.

Kronecker Delta

The simplest interpretation of the Kronecker delta is as the discrete version of the delta function defined by

$$\delta_{ij} = \begin{cases} 1 & i = j \\ 0 & i \neq j \end{cases}$$

The Kronecker delta is implemented in the Wolfram language as KroneckerDelta[i, j], as well as in a generalized form KroneckerDelta[i, j, ...] that returns 1 if all arguments are equal and 0 otherwise.

It has the contour integral representation

$$\delta_{mn} = \frac{1}{2\pi i} \oint_{\gamma} z^{m-n-1} \, dz,$$

Where γ is a contour corresponding to the unit circle and m and n are integers.

In three space, the Kronecker delta satisfies the identities

$$\delta_{ii} = 3$$
$$\delta_{ij}\epsilon_{ijk} = 0$$
$$\epsilon_{ipq}\epsilon_{jpq} = 2\delta_{ij}$$
$$\epsilon_{ijk}\epsilon_{pqk} = \delta_{ip}\delta_{jq} - \delta_{iq}\delta_{jp},$$

where Einstein summation is implicitly assumed. i,j=1,2,3,and ϵ_{ijk} is the permutation symbol.

Technically, the Kronecker delta is a tensor defined by the relationship

$$\delta_l^k \frac{\partial x_i'}{\partial x_k}\frac{\partial x_l}{\partial x_j'} = \frac{\partial x_i'}{\partial x_k}\frac{\partial x_k}{\partial x_j'} = \frac{\partial x_i'}{\partial x_j'}.$$

Since, by definition, the coordinates xi and xj are independent for i=j,

$$\frac{\partial x_i'}{\partial x_j'} = \delta'^i_j,$$

So,

$$\delta'^i_j = \frac{\partial x_i'}{\partial x_k}\frac{\partial x_l}{\partial x_j'}\delta_l^k,$$

and δ'^i_j is really a mixed second-rank tensor. It satisfies

$$\delta_{ab}{}^{jk} = \epsilon_{abi}\epsilon^{jki}$$
$$= \delta_a^j\delta_b^k - \delta_a^k\delta_b^j$$
$$\delta_{abjk} = g_{aj}g_{bk} - g_{ak}g_{bj}$$
$$\epsilon_{aij}\epsilon^{bij} = \delta_{ai}{}^{bi}$$
$$= 2\delta_a^b.$$

Divergence

The divergence of a vector field **F**, denoted div(**F**) or ∇.**F** (the notation used in this work), is defined by a limit of the surface integral

$$\nabla\cdot\mathbf{F} \equiv \lim_{V\to 0}\frac{\oint_S \mathbf{F}\cdot d\mathbf{a}}{V}$$

Where the surface integral provides the value of **F** integrated over a closed small boundary surface S=∂V surrounding a volume component V that is taken to size zero using a limiting method. The divergence of vector field is thus a scalar field. If ∇.**F**=0, then the filed is alleged to be a divergence less field. The symbol∇. is referred to as nabla or del.

Bulk Viscosity

The Bulk viscosity μ_B of a fluid is defined as

$$\mu_B = \lambda + \tfrac{2}{3}\mu,$$

Where λ is the second viscosity coefficient and μ is the shear viscosity.

Reynolds Number

The Reynolds number for a flow through a pipe is defined as

$$\mathrm{Re} \equiv \frac{\rho \bar{u} d}{h} = \frac{\bar{u} d}{\nu},$$

Where ρ is the density of the fluid, **u** is the velocity scale, d is the pipe diameter and v is the kinematic viscosity of the fluid. Poiseuille (laminar) flow is experimentally found to occur for **Re<30**. At larger
Reynolds numbers flow become turbulent.

Density

A measure of a substance's mass per unit of volume. For a substance with mass **m** and volume **V**,

$$\rho \equiv \frac{m}{V}.$$

For a body weight wa placed in a fluid of weight ww,

$$\rho = G_s \rho_w = \frac{w_a}{w_a - w_w} \rho_w,$$

Where Gs is the specific gravity for an ideal gas,

$$\rho = \frac{mP}{kT}.$$

Kinematic Viscosity

A coefficient which describes the diffusion of momentum. Let η be the dynamic viscosity, then

$$\nu \equiv \frac{\eta}{\rho}.$$

The unit of kinematic viscosity is Stoke, equal to 1 cm^2 per second

V_{water} = 1.0 X 10-6 m^2 per second =0.010 cm^2 per second
V_{air} = 1.5 X 10-5 m^2 per second = 0.15 cm^2 per second

INTRODUCTION TO SOLIDWORKS FLOW SIMULATION

The Flow Simulation of SOLIDWORKS software provides design engineers access to powerful CFD analysis capabilities that facilitate to fasten up product innovation. Leveraging the acquainted SOLIDWORKS 3D CAD atmosphere, this extensive technology is not regarding ensuring your product works but it is regarding understanding how your product will behave within the real world.

Built to tackle CFD engineering challenges, SOLIDWORKS Flow Simulation permits engineers to take advantage of CAD integration, advanced geometry meshing capabilities, powerful answer convergence, and automatic flow regime determination while not sacrificing simple use or accuracy.

Product engineers and CFD specialists alike, armed with the ability of SOLIDWORKS Flow Simulation, can predict flow fields, combination processes, heat transfer, and directly verify pressure drop, comfort parameters, fluid forces, and fluid structure interaction throughout design. SOLIDWORKS Flow Simulation allows true coincidental CFD, without the requirement for advanced CFD expertise. SOLIDWORKS Flow Simulation software package takes the complexity out of flow analysis and allows engineers to simply simulate fluid flow, heat transfer, and fluid forces thus engineers can investigate the impact of a liquid or gas flow on product performance. Following are some of the tasks that can be performed by using SolidWorks Flow Simulation:

Evaluate and optimize complex flows:

- Examine complicated flows through and around your parts with parametric analysis.
- Understand the flow of non-Newtonian liquids, like blood and liquid plastic.
- Assess the impact of various impellers and fans on your design.
- Align your model with flow conditions, like pressure drop, to satisfy design goals.
- Detect turbulences and recirculation problems with animated flow trajectories

Reduce the risk of overheating in your design:

- Visualize and perceive temperature distribution in and around your product.
- Couple flow with thermal analysis, simulating convection,conduction, and radiation effects.
- Simulate advanced radiation with clear material and wavelength-dependent radiative properties with the HVAC module.
- Apply time and coordinate dependent boundary conditions and warmth sources.
- Find the most effective dimensions to satisfy your design goals, like heat exchanger efficiency
- Optimize the thermal performance of your PCBs and electronics components

You can perform thermal analysis on designs incorporating printed circuit boards (PCBs) and electronics with SOLIDWORKS Flow Simulation and also the Electronic Cooling Module. The Electronic Cooling Module features a comprehensive set of intelligent models additionally to the core SOLIDWORKS Flow Simulation models to modify a broad range of electronic cooling applications to be designed quickly and accurately. The models enclosed for electronic thermal simulations are:

- Fans
- PCB generator tool
- Electrical contact condition
- Heat Pipe Compact model
- Thermoelectric cooler (TEC)
- Heat sink simulation
- Joule Heating calculation
- Extensive library of electronic models
- Two-Resistor Component Compact Model (JEDEC standard)

Predict and achieve airflow and comfort parameters in working and living environments

You can perceive and measure thermal comfort levels for multiple environments using thermal comfort factor analysis with SOLIDWORKS Flow Simulation and also the HVAC Application Module. Assessment of the thermal atmosphere within the occupied zone needs knowing the Thermal Comfort Parameters as well as factors which give information regarding air quality, calculated with the HVAC module, including:

- Predicted Mean Vote (PMV)
- Contaminant Removal Effectiveness (CRE)
- Draft Temperature
- Operative Temperature
- Air Diffusion Performance Index (ADPI)
- Local Air Quality Index (LAQI)
- Predicted Percent Dissatisfied (PPD)

Gain valuable insights with powerful and intuitive results visualization tools:

- Utilize Section or Surface plots to check the distribution of resultant values, as well as velocity, pressure, vorticity, temperature, and mass fraction
- Compare the Fluid Flow results for varied configurations with the Compare Mode
- Measure results at any location with the point, Surface, and Volume Parameter tool
- Graph results variation on any SOLIDWORKS sketch
- List results and automatically export information to Microsoft® Excel®
- Communicate your CFD results in 3D with SOLIDWORKS eDrawings®

STARTING SOLIDWORKS FLOW SIMULATION

If you want to open the existing file of SolidWorks software for simulation process then steps are as follow:

- Double-click on the file from the respective folder. The file will be opened in the respective tab (**Part** or **Assembly**) of SolidWorks window; refer to Figure-3.

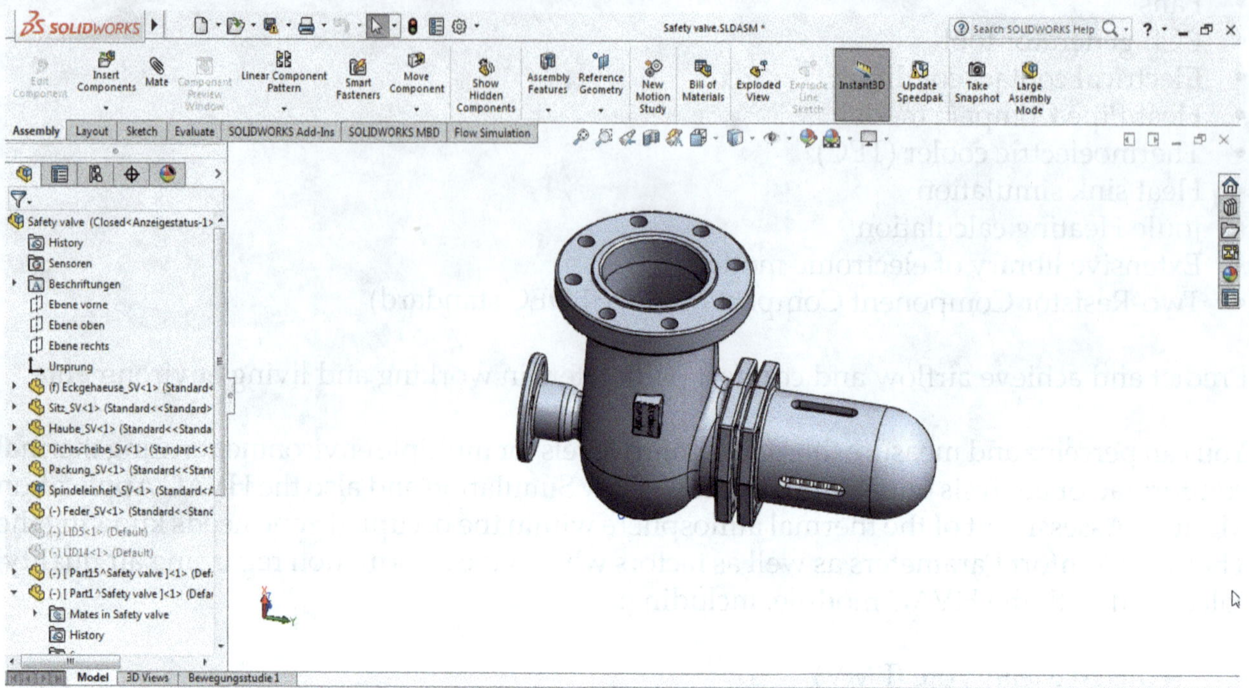

Figure-3. Opened existing file of SolidWorks

- Click on the **SOLIDWORKS Flow Simulation** tool from the **SOLIDWORKS Add-Ins CommandManager** in the **Ribbon**. The **Flow Simulation CommandManager** will be added in the **Ribbon** and the Flow Simulation environment will become active.
- Click on the **Flow Simulation CommandManager** from **Ribbon**. The tools of **Flow Simulation** will be displayed; refer to Figure-4.

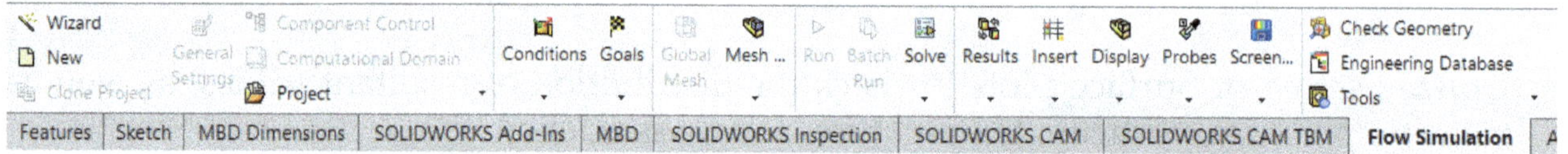

Figure-4. Tools of Flow Simulation

Rest of the process of simulation of a part or component will be discussed later.

If you do not have a part for simulation and want to create a new component for simulation process then double-click on the **SolidWorks** icon from desktop or start SolidWorks using Start menu. The SolidWorks application window will be displayed; refer to Figure-5.

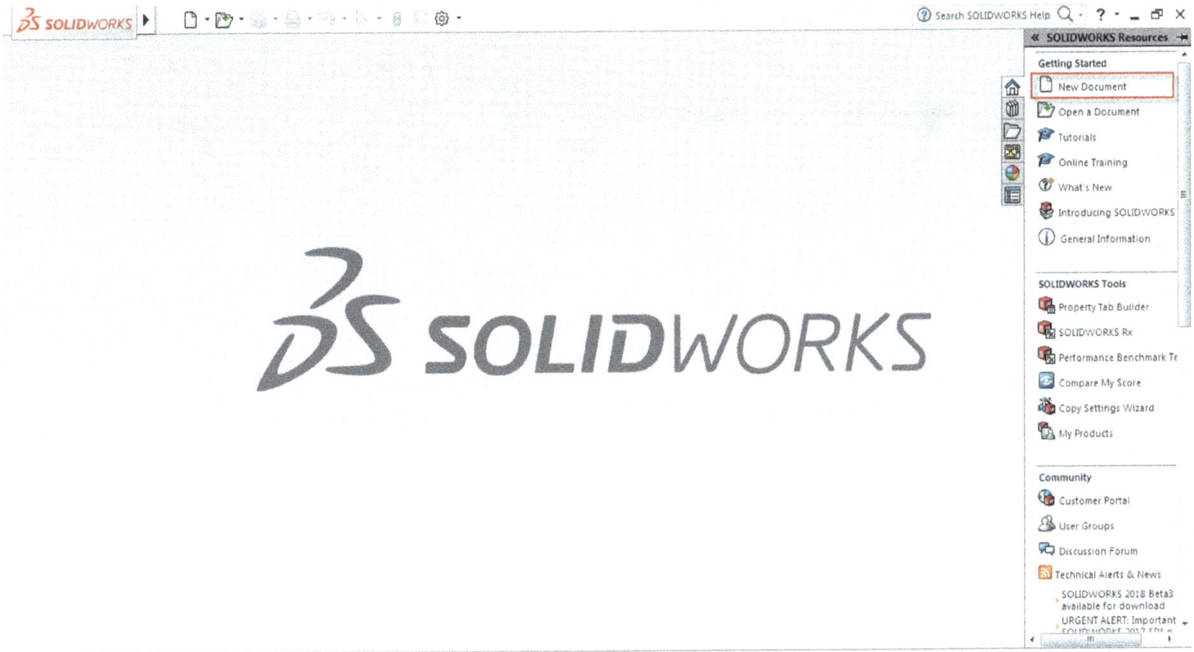

Figure-5. SolidWorks starting window

- Click on the **New Document** button from **SOLIDWORKS Resources** box. The **New SOLIDWORKS Document** dialog box will be displayed; refer to Figure-6.

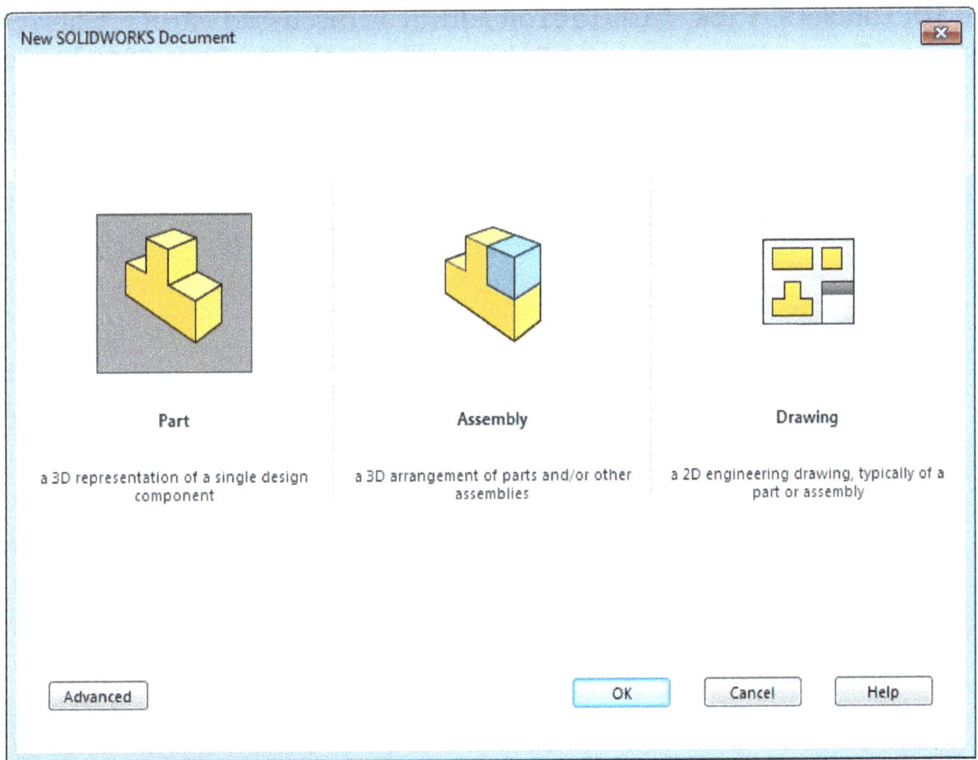

Figure-6. New SOLIDWORKS Document dialog box

- Select the **Part** button and click on the **OK** button. The SolidWorks working environment will be displayed; refer to Figure-7.

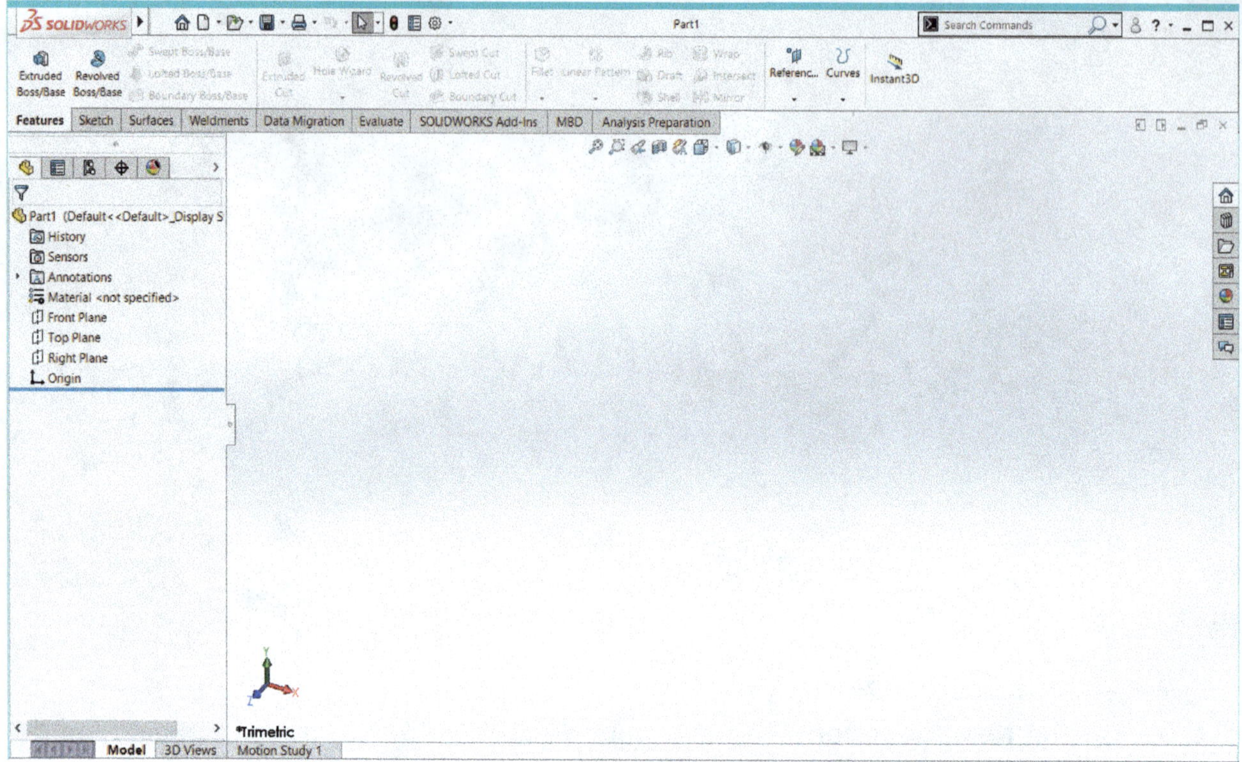

Figure-7. SolidWorks working Window

- Activate **SOLIDWORKS Flow Simulation** add-in as discussed earlier. Click on the **Flow Simulation CommandManager** from **Ribbon**. The tools of **Flow Simulation** will be displayed.
- You can also open an existing SolidWorks file by clicking on **Open a Document** button from **SOLIDWORKS Resources** box. The **Open** dialog box will be displayed; refer to Figure-8.

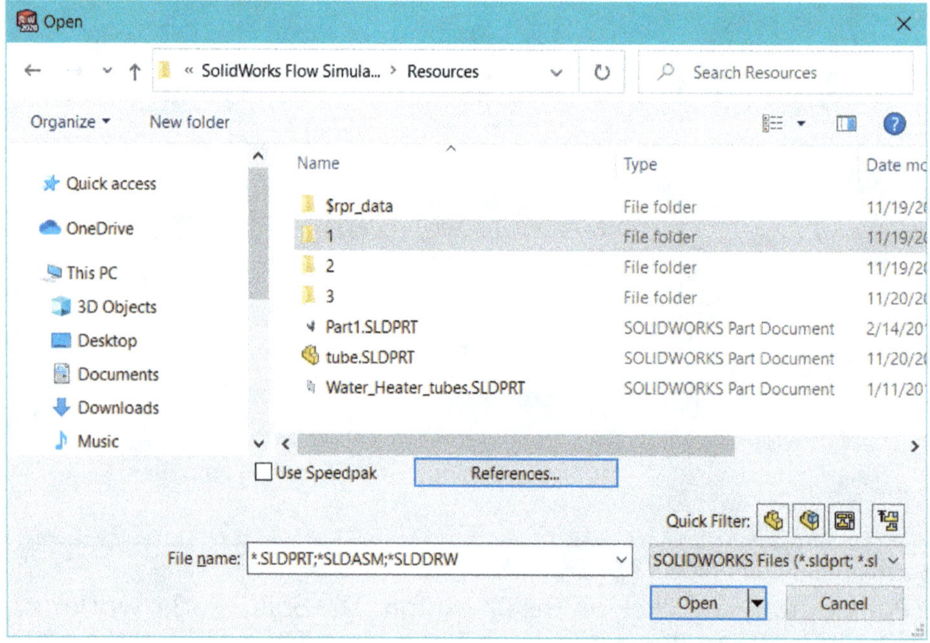

Figure-8. The Open dialog box

- Select the desired file and click on the **Open** button from the **Open** dialog box. The file will open in the SolidWorks window. Note that only the files of format selected in the **Files Type** drop-down will be displayed.

PREPARING COMPONENT FOR SIMULATION

Open the part on which you want to perform the analysis; refer to Figure-9.

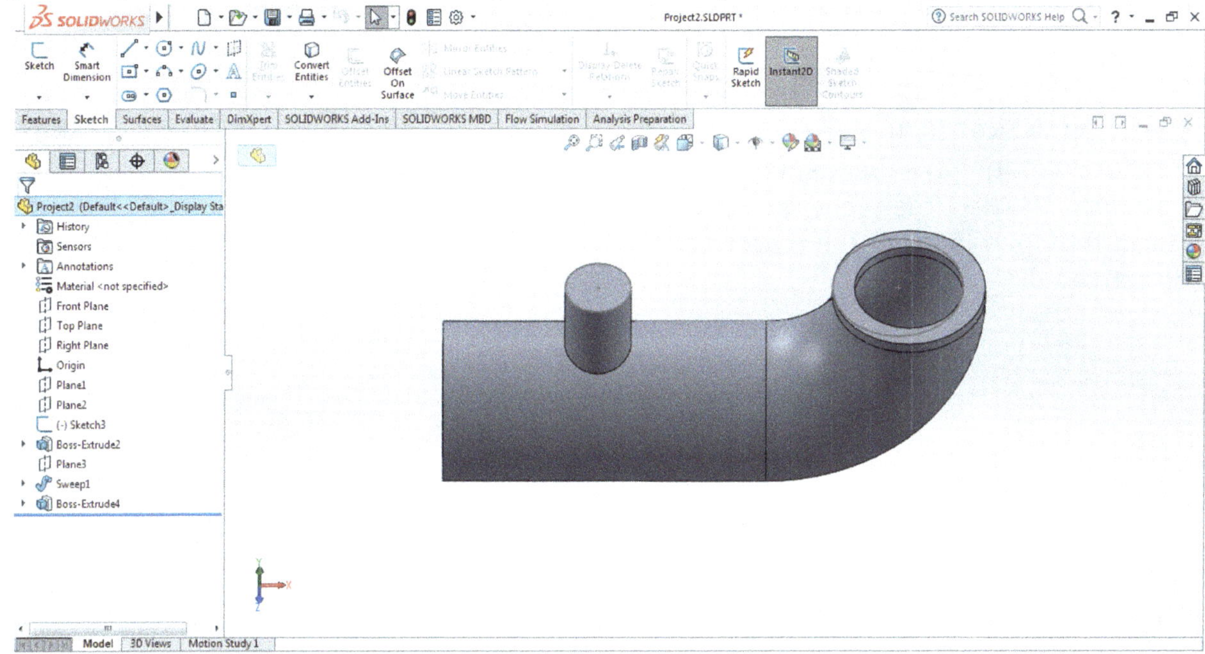

Figure-9. Created component for simulation

The tools and procedure used to create a component in SolidWorks are discussed in our another book SolidWorks 2021 Black Book. You can read this book to clear all the queries regarding model creation in SolidWorks.

- After creating/opening the component, click on the **Flow Simulation CommandManager**. The tools in the **Flow Simulation CommandManager** will be displayed along with the model; refer to Figure-10.

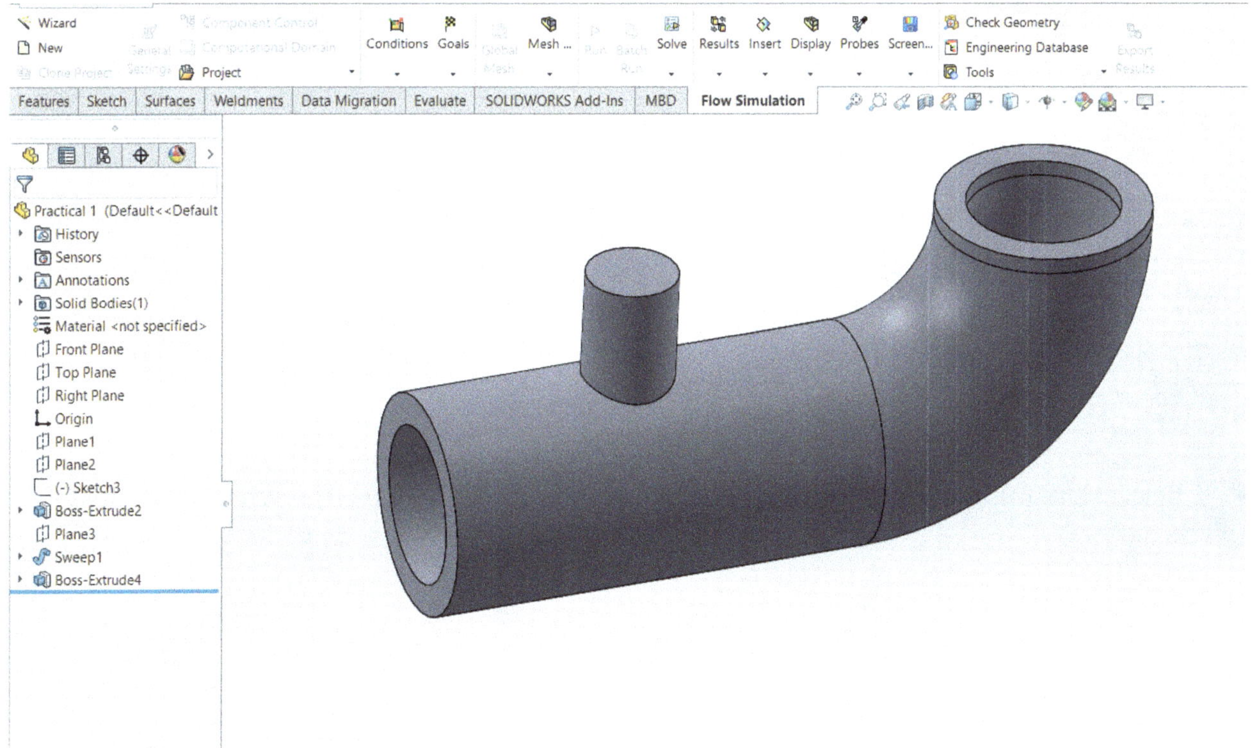

Figure-10. Flow Simulation tab along with component

The openings and closing of the displayed components for fluid flow are open. To apply the simulation, we need to close these openings so that we can define inlet and outlet faces for fluid. There are two methods to close these openings which are discussed next.

Creating lids by Solid Modeling (First Method)

- Click on the **Sketch** tool from **Sketch** drop-down; refer to Figure-11. The **Edit Sketch FeatureManager** will be displayed at the left of the screen.
- You need to select the plane on which you want to create the first lid; refer to Figure-12. Select the open face of tube where you want to place the lid, the Sketch workbench will be activated.

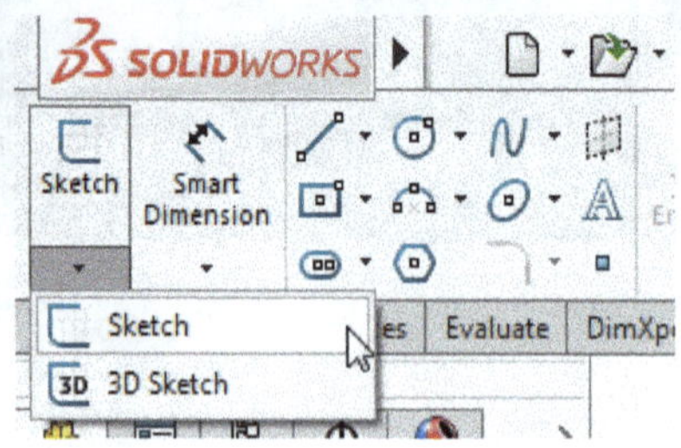

Figure-11. Sketch button

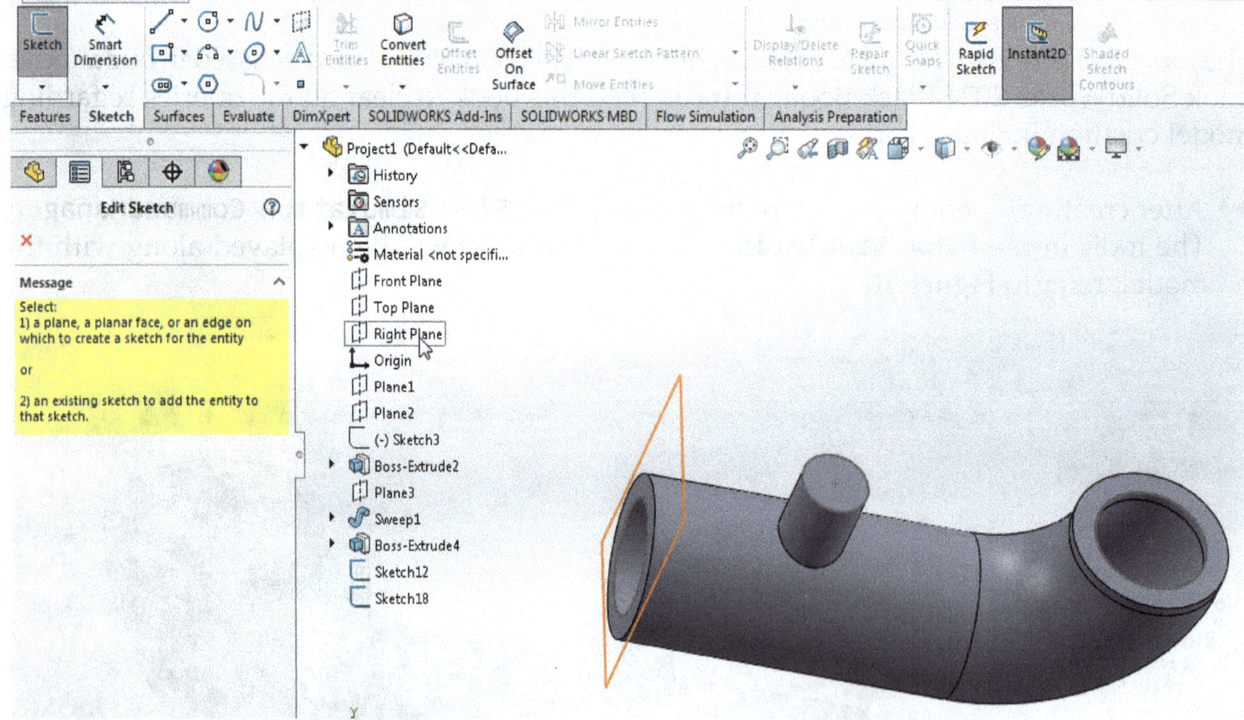

Figure-12. Selecting plane for creating lid

- Click on the inner edge of the opening; refer to Figure-13. The edge will be selected.

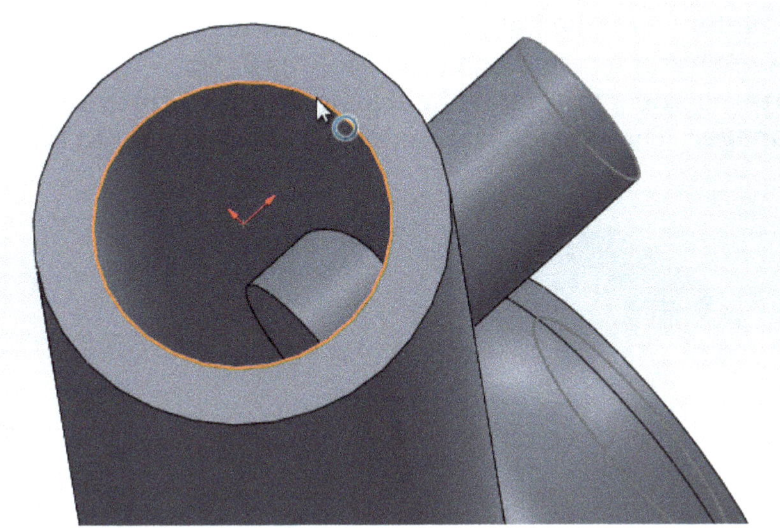

Figure-13. Selecting inner edge of opening

- Click on the **Convert Entities** tool from the **Sketch CommandManager** in the **Ribbon**; refer to Figure-14. The selected edge will be projected on the sketching plane.

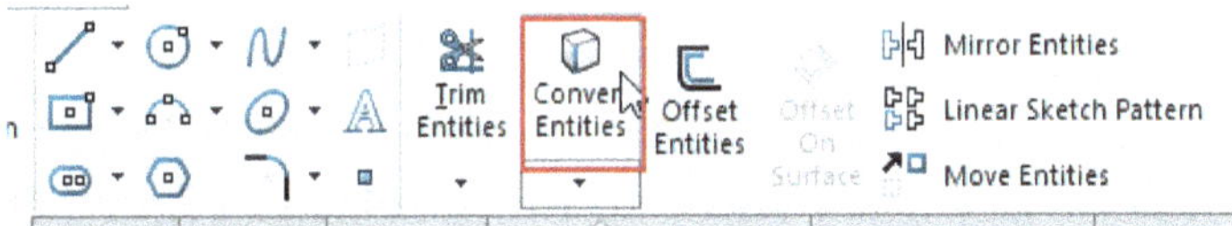

Figure-14. Convert Offset button

- Click on the **Extruded Boss/Base** tool from **Features CommandManager** of the **Ribbon**; refer to Figure-15. The **Boss-Extrude PropertyManager** will be displayed on the left in the screen; refer to Figure-16.

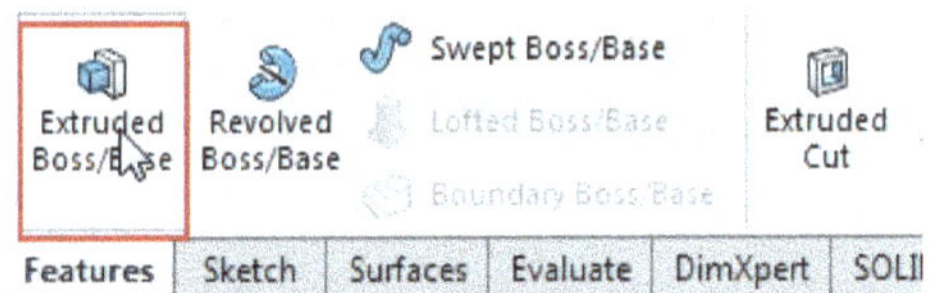

Figure-15. Extruded Boss/Base tool

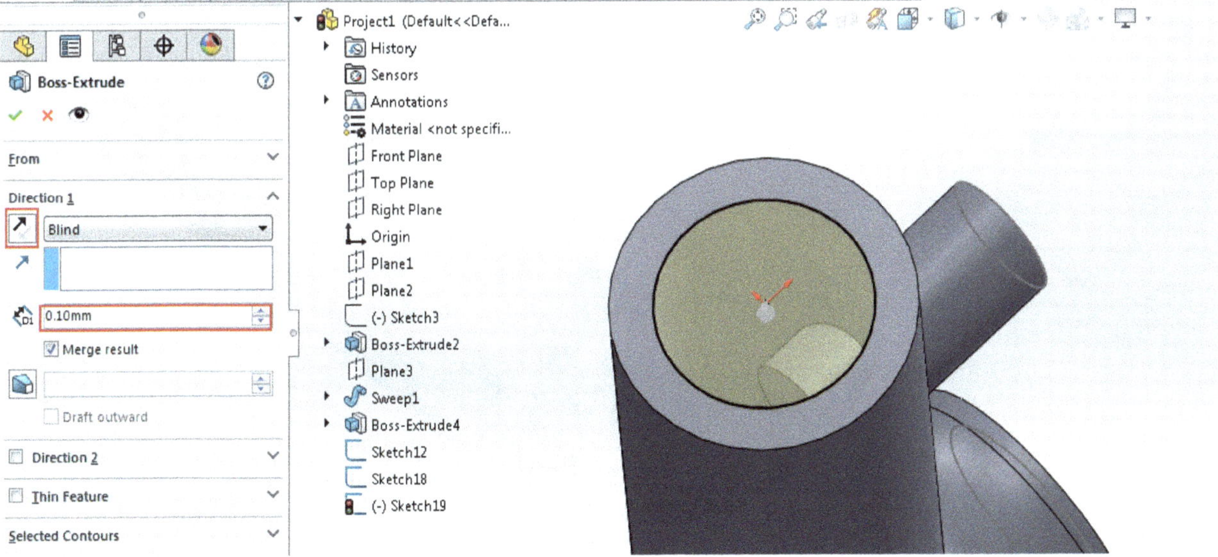

Figure-16. The Boss Extrude feature

The recently projected sketch will be selected automatically for extrude. If not selected by default then select it from drawing area.

- Click on the **Blind** option from **Direction 1** drop-down and click on the **Reverse Direction** button to fill the hole by extrude feature.
- Click in the **Depth** edit box and enter the value as **0.50**.
- After specifying the parameters, click on the **OK** button from **Boss-Extrude PropertyManager**. The extrude feature will be created; refer to Figure-17.

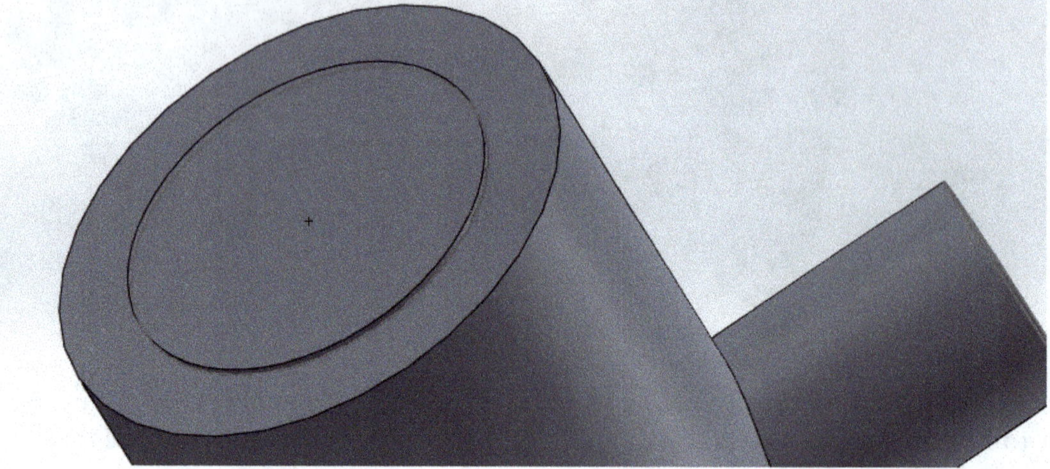

Figure-17. Applied extrude

- Click twice with a pause on the **Boss-Extrude** feature in the **FeatureManager Design Tree** to rename the Boss Extrude feature to LID1; refer to Figure-18. This renaming will help us to identify the lids.

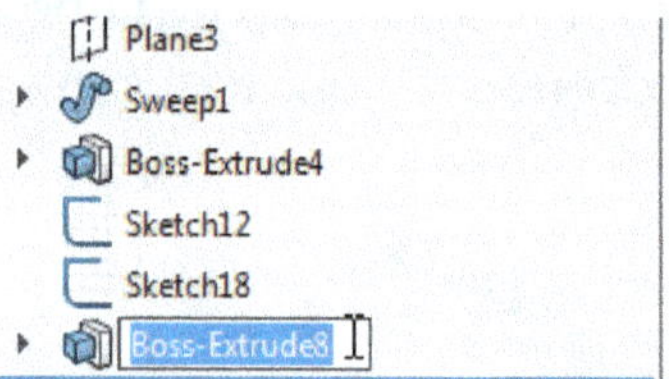

Figure-18. Changing name

- Similarly, convert the inner edge of the other opening to the entity; refer to Figure-19.

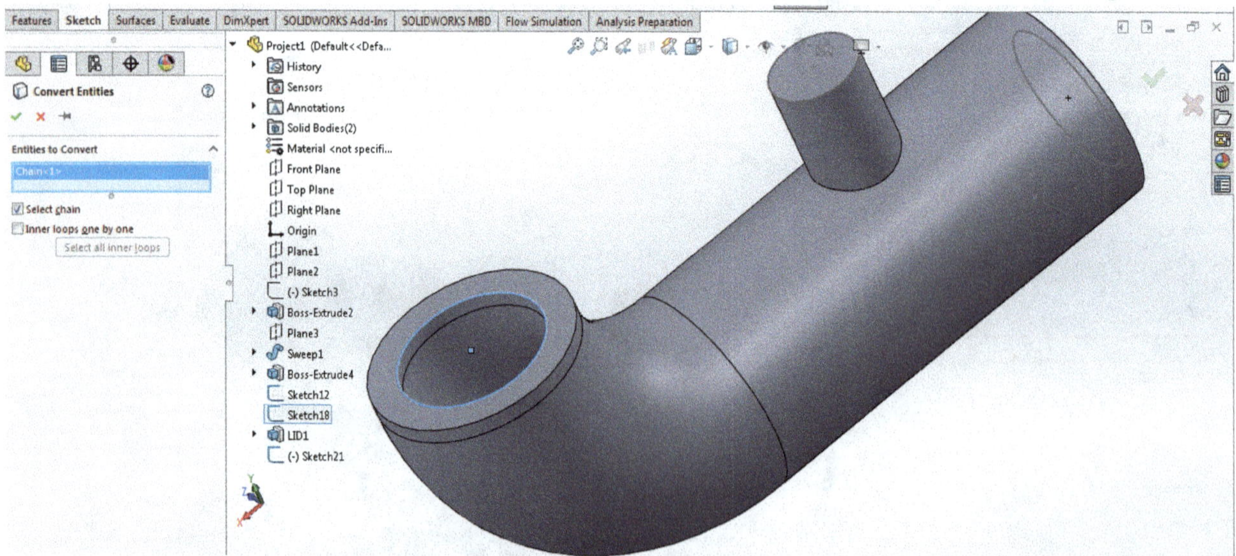

Figure-19. Selection for converting entity

- After selection of edge, click on the **OK** button from the **Convert Entities PropertyManager**. The entity will be converted to sketch.

- Extrude the converted entity to 0.5 mm as discussed earlier. After extruding, the model will be displayed; refer to Figure-20.

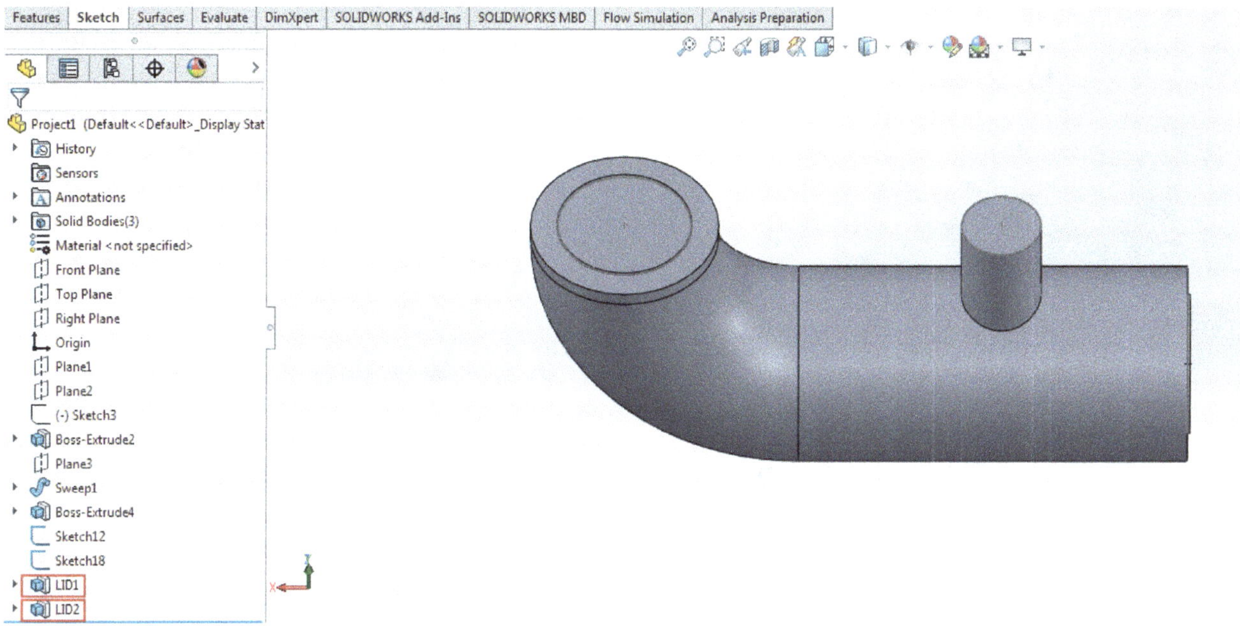

Figure-20. Created other lid

- Rename the recently created extrude feature to LID2.

Creating Lids using the Create Lids tool (Second Method)

- Click on the **Create Lids** tool from the **Tools > Flow Simulation > Tools >** cascading menu in the Menu bar; refer to Figure-21. The **Create Lids PropertyManager** will be displayed at the left in the SolidWorks screen; refer to Figure-22.

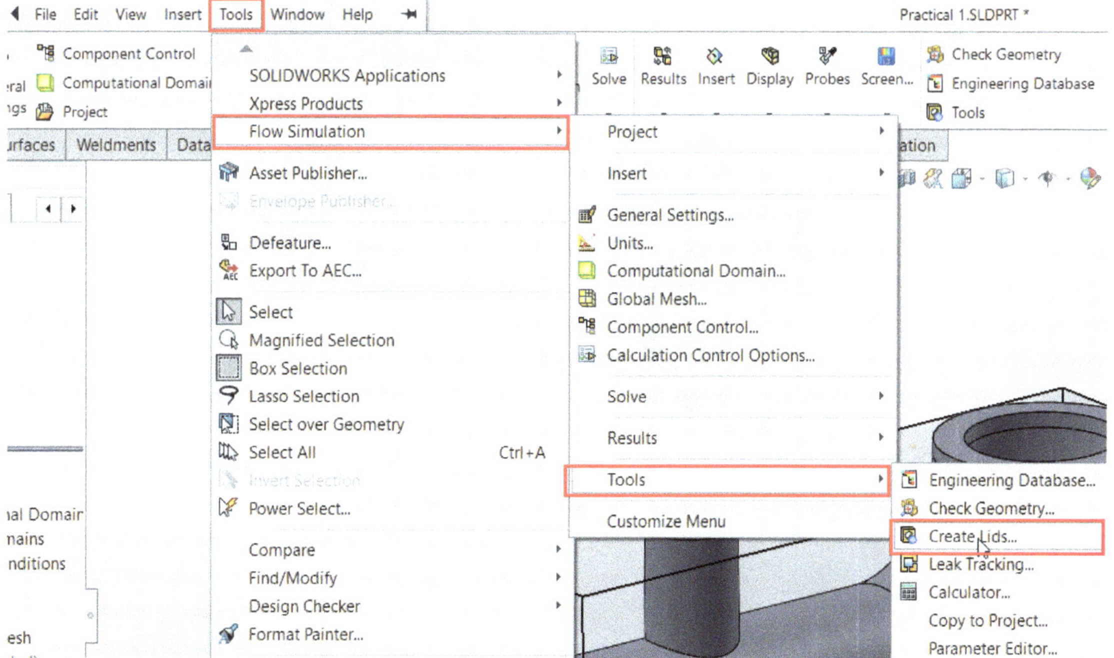

Figure-21. Create Lids button

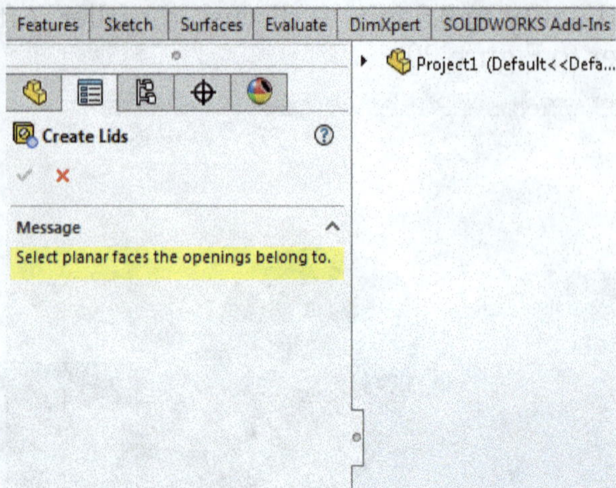

Figure-22. The Create Lids feature

Note that to perform any flow simulation analysis, you need to close all openings of components by creating lids.

- Click on the face of opening, the preview of lids will be displayed; refer to Figure-23.

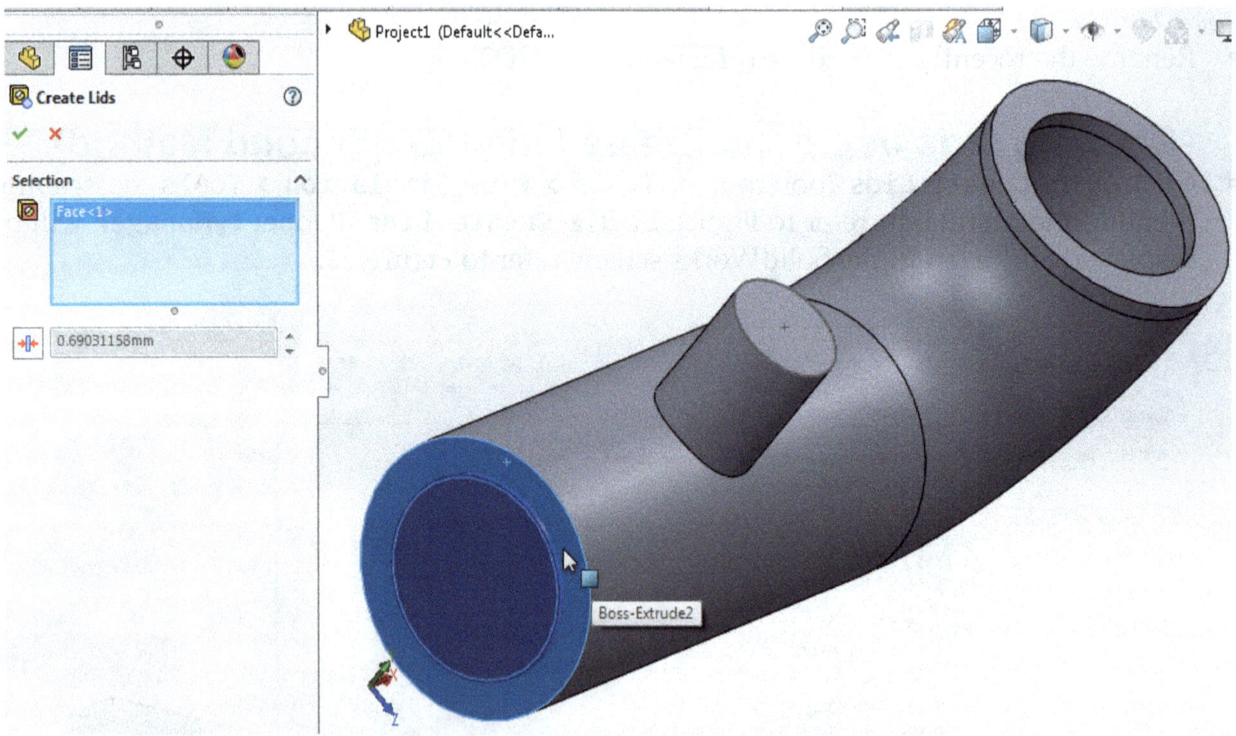

Figure-23. Selecing face for creating lid

- Similarly, select the other faces to create the lids; refer to Figure-24.

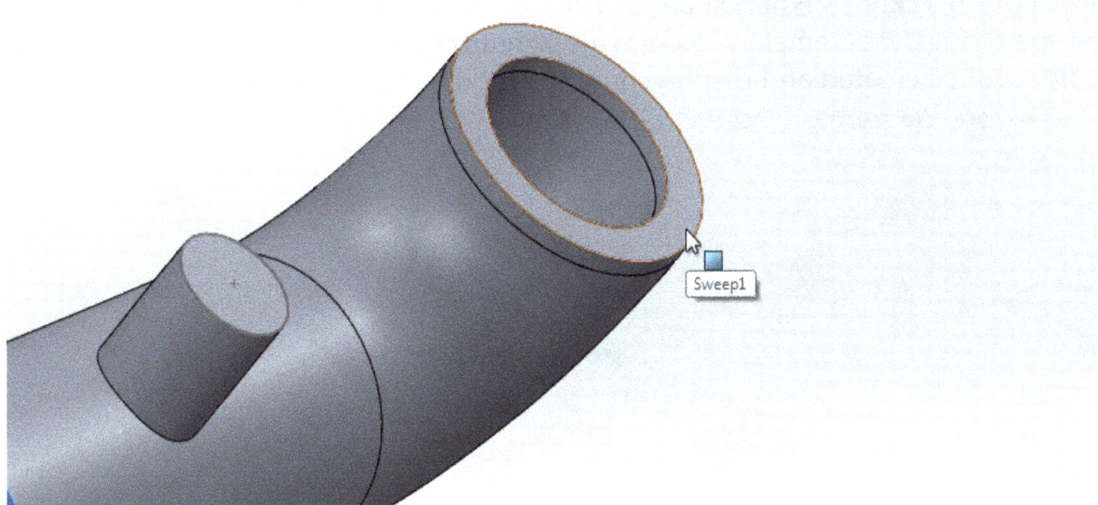

Figure-24. Selecting other face

- The selected faces will be displayed in the selection box of **Create Lids PropertyManager**; refer to Figure-25.

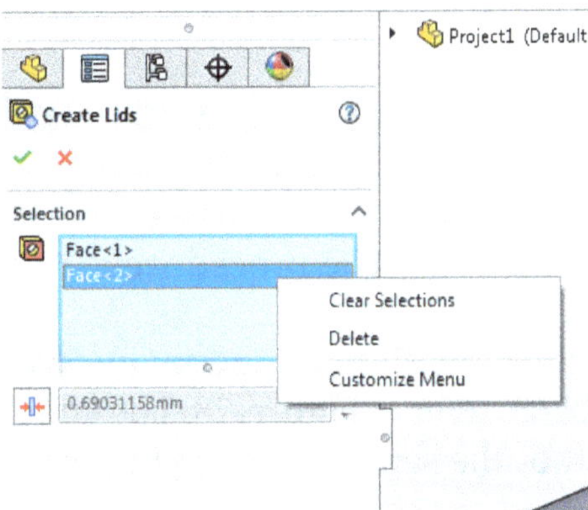

Figure-25. Selection box

- If you want to clear or delete the selected faces then right-click on the respective face. The menu will be displayed. Click on the **Delete** option to delete the selection.
- To remove all the selections, click on the **Clear Selection** option from the right-click menu. All the selected face will be cleared.
- To specify the thickness of lid, click on the **Adjust Thickness** button from **Create Lids PropertyManager**. The **Thickness** edit box will be activated; refer to Figure-26

Figure-26. Adjust Thickness button and Thickness edit box

- Click in the **Thickness** edit box and specify the value of thickness for lids as desired.
- After specifying the parameters, click on the **OK** button from **Create Lids PropertyManager**. The lids will be created and displayed on the model; refer to Figure-27.

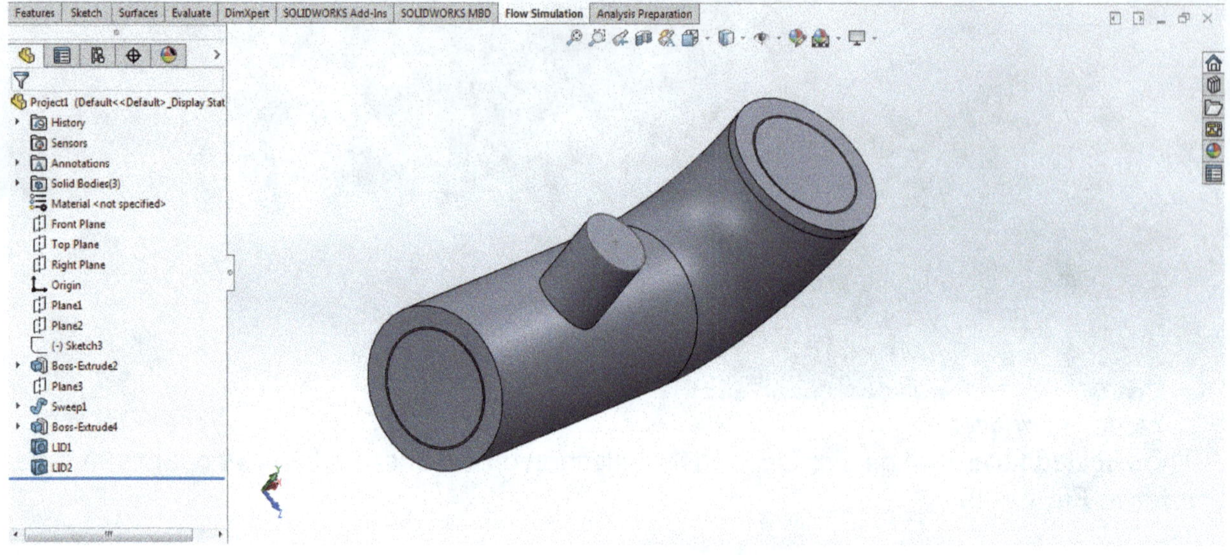

Figure-27. Created lids

Checking Geometry for leakage

The **Check Geometry** tool is generally used to check any leakage in the component. The leakage might occurs through created lids. Note that we have started a flow simulation project with internal flow before using this tool. The procedure to use this tool is discussed next.

- After creating the lids, click on the **Check Geometry** tool from the **Flow Simulation Command Manager** in the **Ribbon**; refer to Figure-28. The **Check Geometry PropertyManager** will be displayed on the left of the screen; refer to Figure-29.

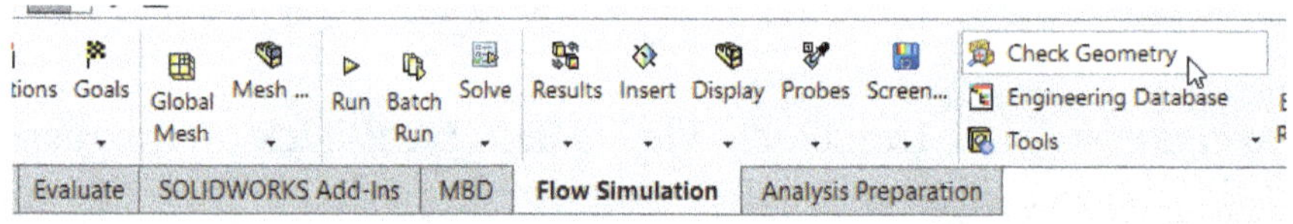

Figure-28. Check Geometry tool

- In **State** box of **Check Geometry PropertyManager**, the created lids and closed component of the part will be displayed. Select the required check box for analysis.
- If you have not started project then select the **Internal** radio button to analyze internal fluid flow in specified geometry. Select the **External** radio button to analyze external fluid flow around a geometry.
- Select the **Exclude cavities without flow conditions** radio button from **Analysis Type** rollout to exclude the internal area or volume from the part for analysis.
- Select the **Exclude internal space** radio button from **Analysis Type** rollout to exclude the internal area or volume from the part for external flow analysis.

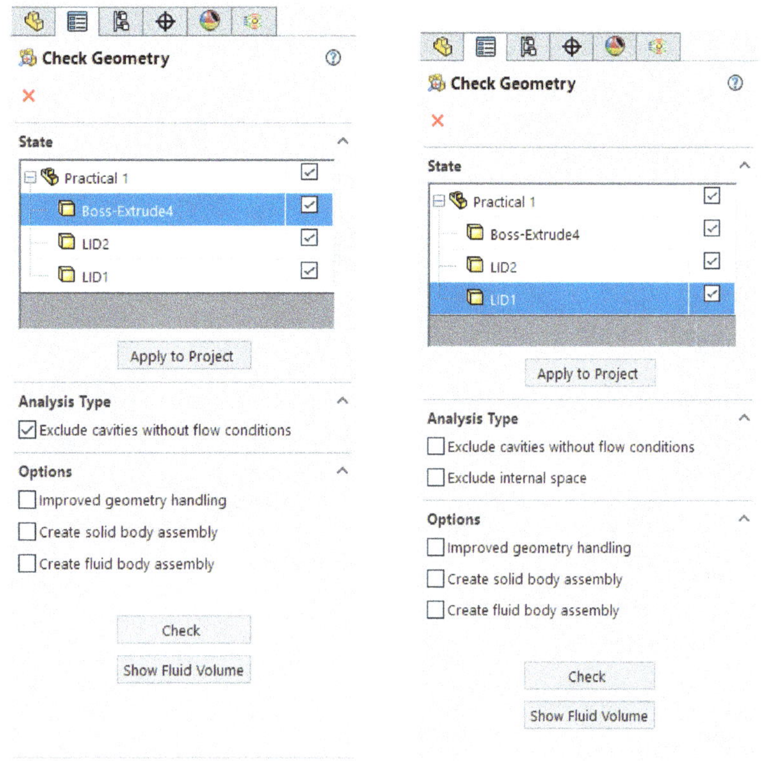

Figure-29. Check Geometry PropertyManager with internal flow

Figure-30. Check Geometry PropertyManager with external flow

- Select the **Improved geometry handling** check box so that solids and fluids use different set of boolean operations in their respective domains. Note that fluids will use their own algorithm of boolean operations like how water and oil will mix and so on. Selecting this check box will increase the processing time.
- Select the **Create Solid Body Assembly** check box from **Options** section to create an assembly of solid bodies.
- Select the **Create Fluid Body Assembly** check box from **Options** section to create an assembly of fluid bodies.
- Click on the **Show Fluid Volume button** from **Check Geometry** feature to display the volume occupied by fluid in the model body graphically. The fluid will be displayed in the model; refer to Figure-31.

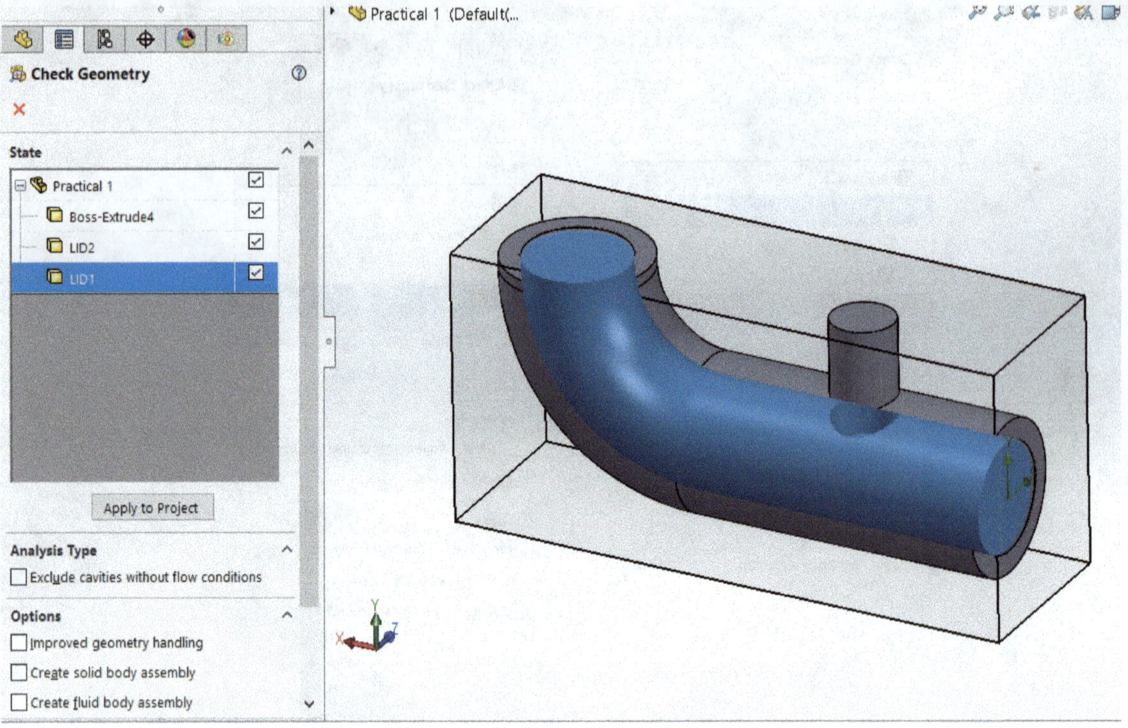

Figure-31. Fluid shown in the model

- Clicking the same button will hide the displayed fluid.
- Click on the **Check** button from **Check Geometry PropertyManager** for analysis of any leaks present in the part. The **Check Geometry** tab will be displayed in the bottom display bar; refer to Figure-32.

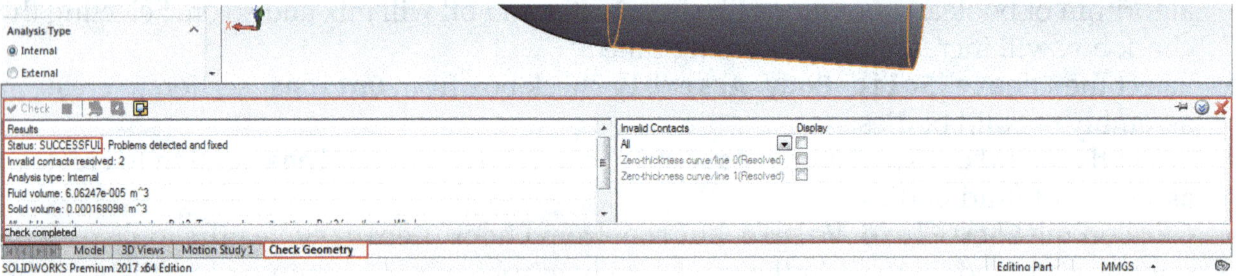

Figure-32. Check Geometry tab

Check the status of the component. If **Successful** is displayed as status then there is no leak in the whole component.

Leak Tracking

There might be chances when any leakage is left in the model. To remove the leakage, you need to check component. The procedure is discussed next.

- After creating the component, click on the **Check** button from **Check Geometry** toolbar at the bottom. The **Check Geometry** tab will be displayed in the bottom display bar; refer to Figure-33.

Figure-33. Check Geometry box of FAILED status

- If the **Failed** notification is displayed in the **Status** row of **Check Geometry** tab then it means that some kind of leakage is present in the component.
- To check the location of leakage, click on the **Open Leak Tracking** button. The **Leak Tracking** tab will be displayed; refer to Figure-34.

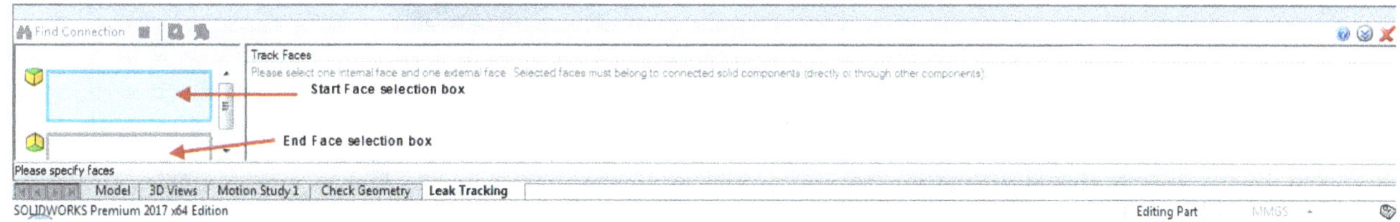

Figure-34. The Leak Tracking tab

- The **Start Face** selection box is active by default in the **Leak Tracking** tab. You need to click on the face of component from where flow starts; refer to Figure-35. The face will be selected and displayed in the selection box.

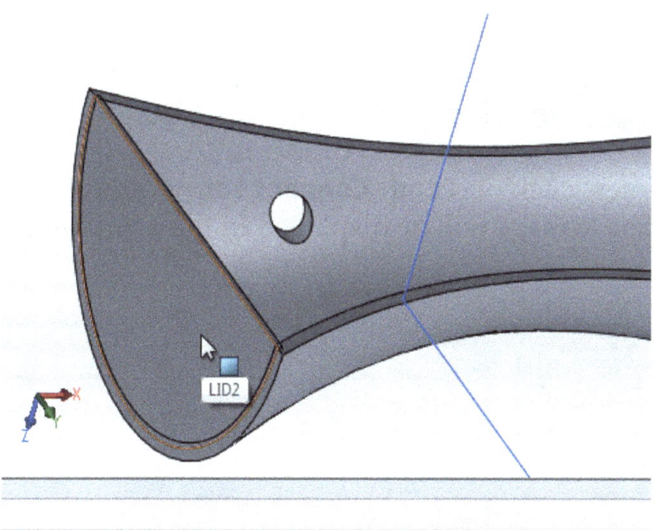

Figure-35. Selecting first face for checking leak

- Click in the **End Face** selection box. You will be asked to select the face where fluid will exit.
- Right-click on the other face and click on the **Select Other** button; refer to Figure-36. The **Select Other** dialog box will be displayed where you can select hidden faces; refer to Figure-37.

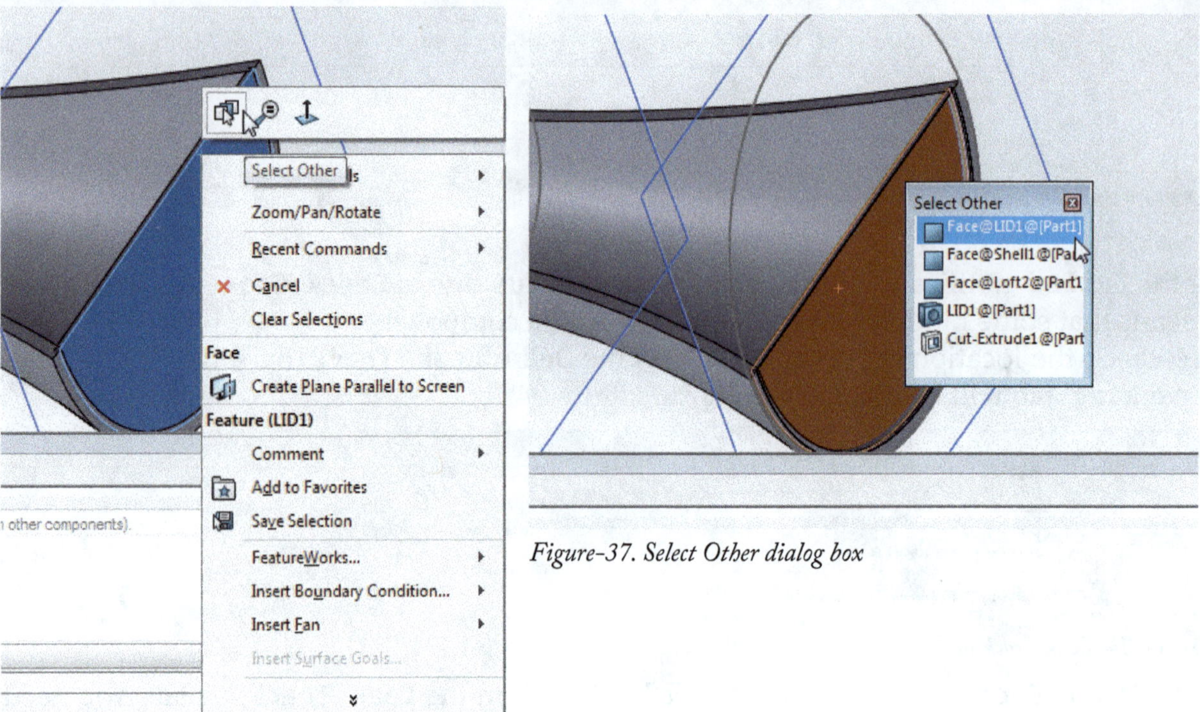

Figure-37. Select Other dialog box

Figure-36. Right click shortcut menu

- Click on the inner side face from **Select Other** dialog box. The inner face will be selected.

Note- In the leak tracking process, always remember that you need to select inner side face of one lid and outer side face of the other lid.

- After selection of faces, click on **Find Connection** button from **Leak Tracking** tab; refer to Figure-38. The leakage will be displayed in the component; refer to Figure-39 and Figure-40.

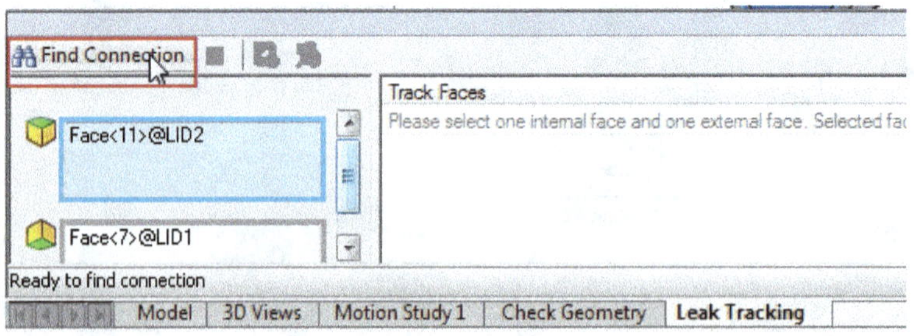

Figure-38. Find Connection button

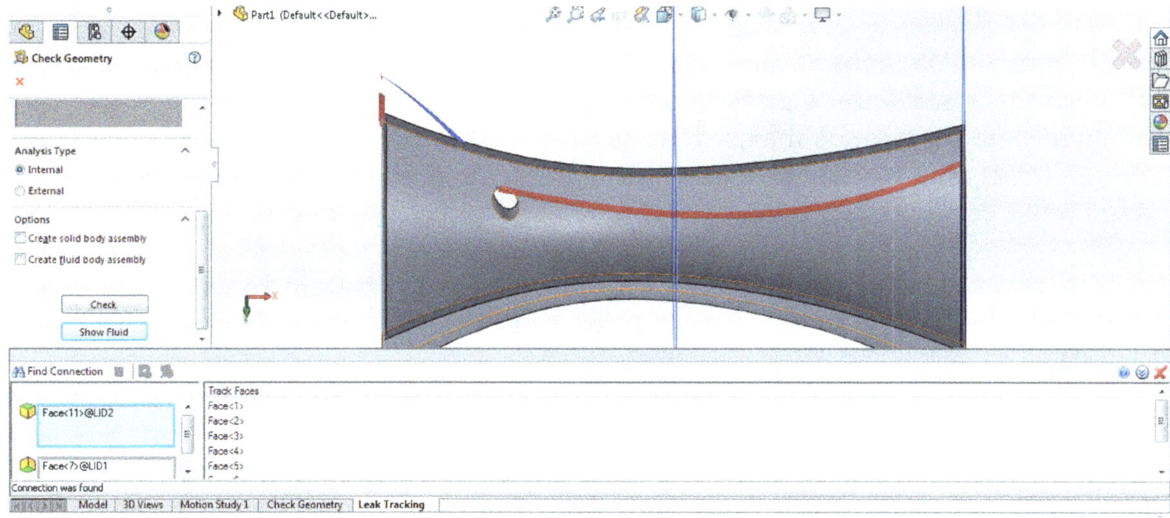

Figure-39. Leakage in component

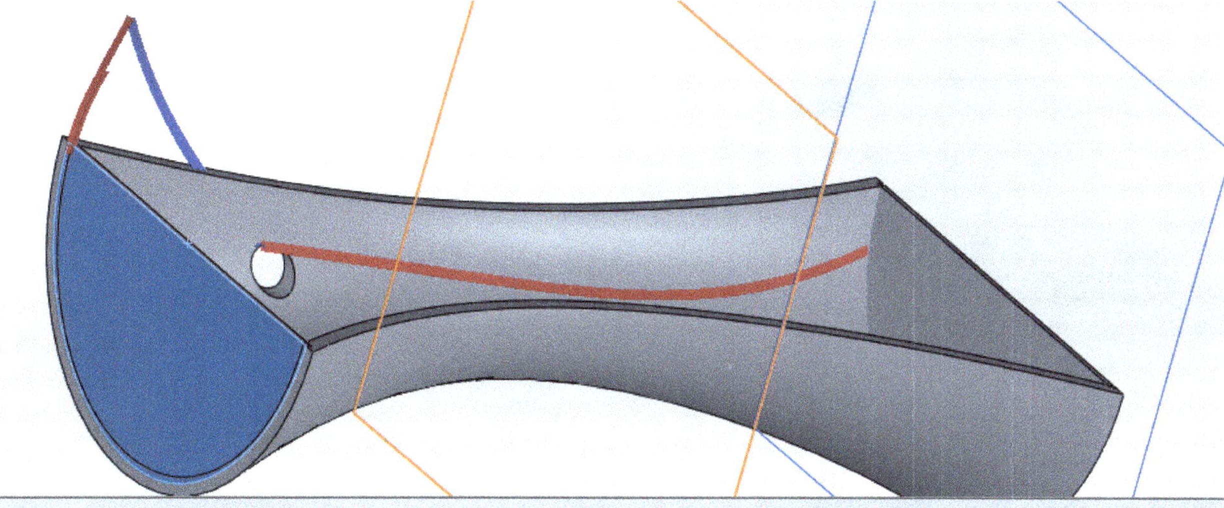
Figure-40. Leakage from component

- Close the leakage in the component for fluid simulation by creating lid or performing Solid Modeling operations. After closing the leakage in the component, the check geometry results will be displayed as successful; refer to Figure-41. Click on the **Close** button to exit the analysis.

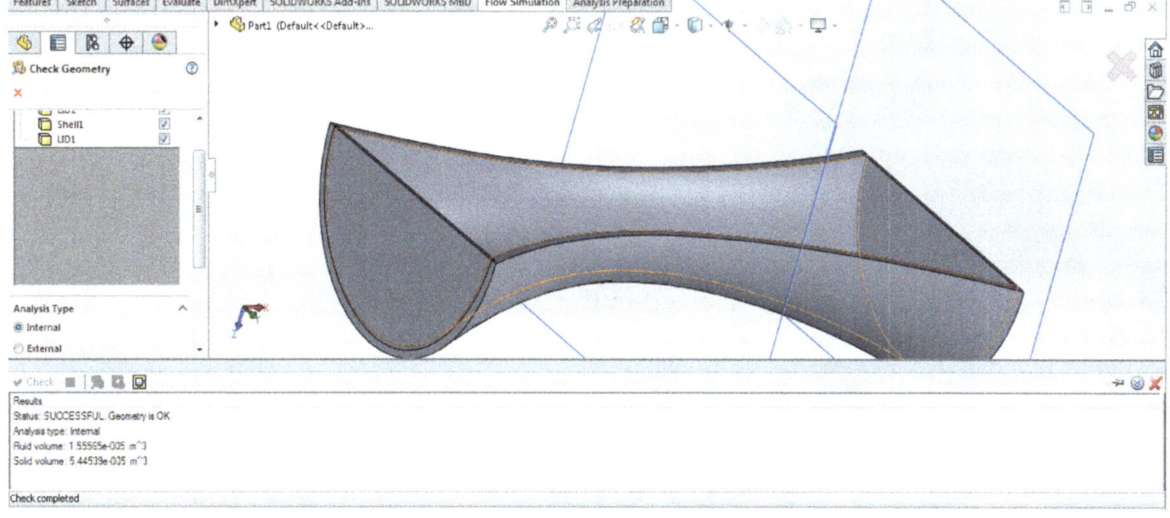

Figure-41. Leakage free model

Engineering Database

The **Engineering Database** tool is used to check and modify the engineering database of the SolidWorks software. Database consists of parameters for performing analysis like environment data for various major cities, contact thermal resistance for various materials, physical, thermal, and electric properties of material, and so on. The procedure to do so is discussed next.

- Click on the **Engineering Database** button from **Flow Simulation CommandManager**; refer to Figure-42. The **Engineering Database** dialog box will be displayed; refer to Figure-43.

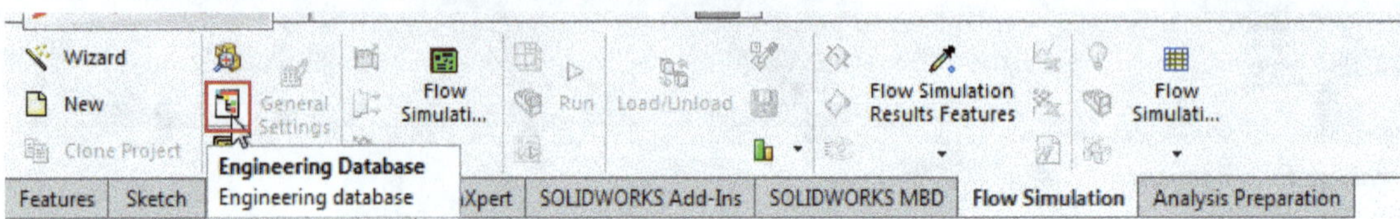

Figure-42. Engineering Database button

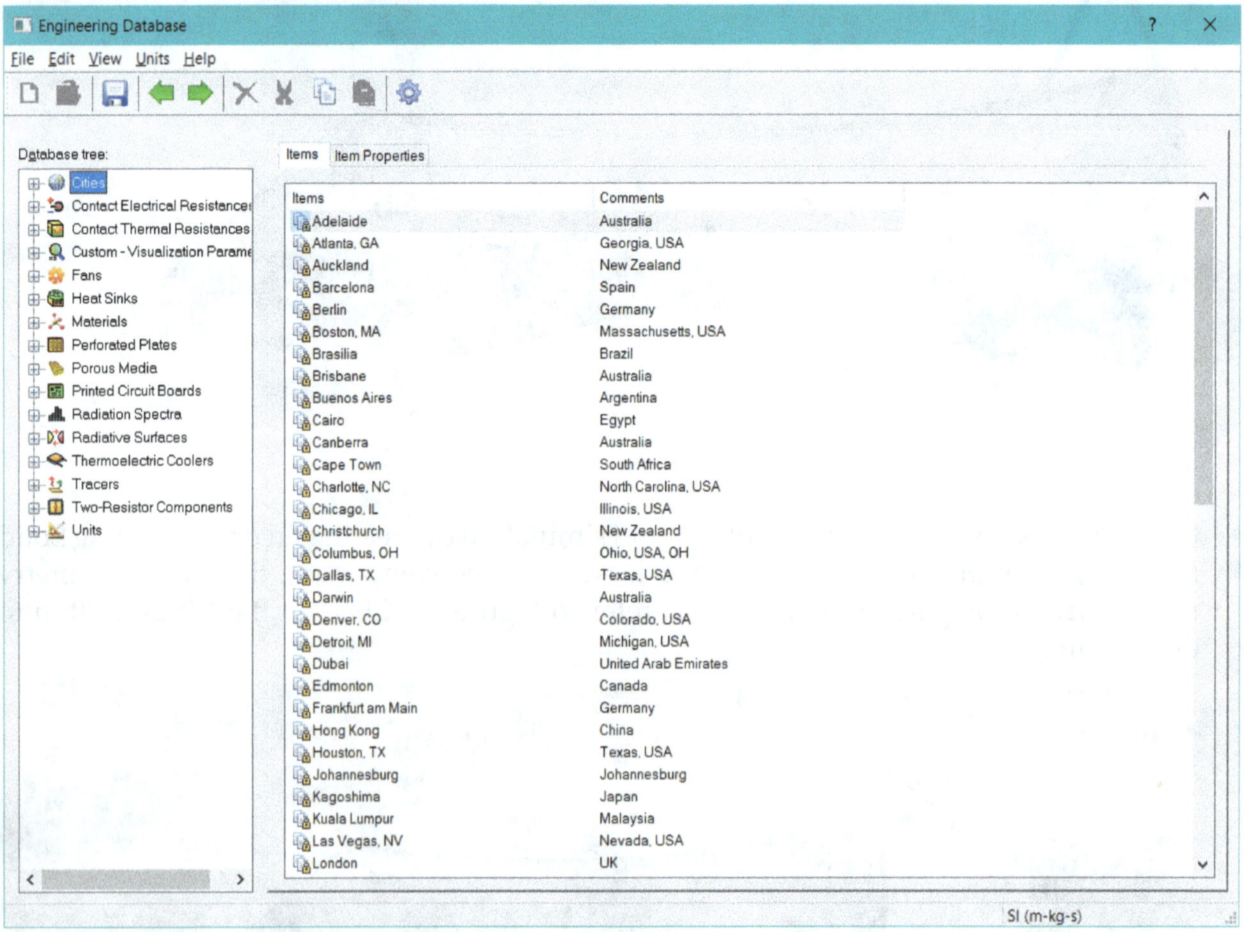

Figure-43. Engineering Database dialog box

- By default, the **Cities** node is selected in the **Database Tree** box at the left in the dialog box and hence list of cities is displayed. To check the properties of any of the city, double-click on it from the list.
- Expand the node in **Database Tree** box to modify its content. Each of the node has two categories: **Pre-Defined** and **User Defined**; refer to Figure-44. By default, the content of **Pre-Defined** category are locked so you cannot modify them although you can use them

in your project. The **User Defined** category can be used to create new properties in the database.

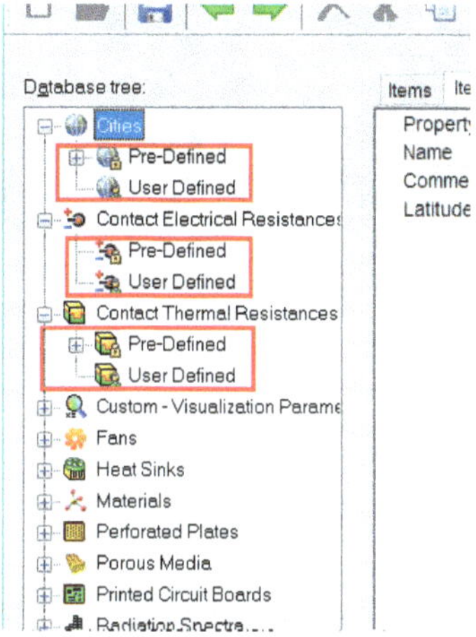

Figure-44. Categories in Engineering Database nodes

- To add a new city in the database, expand the **Cities** node and click on the **User Defined** category in the list. The **New Item** button will become active in the toolbar. Click on this button. A new item will be added and its properties will be displayed; refer to Figure-45.
- Double-click in the desired fields to edit the data. Now, click on the **Items** tab in the dialog box. The name of property will be displayed.

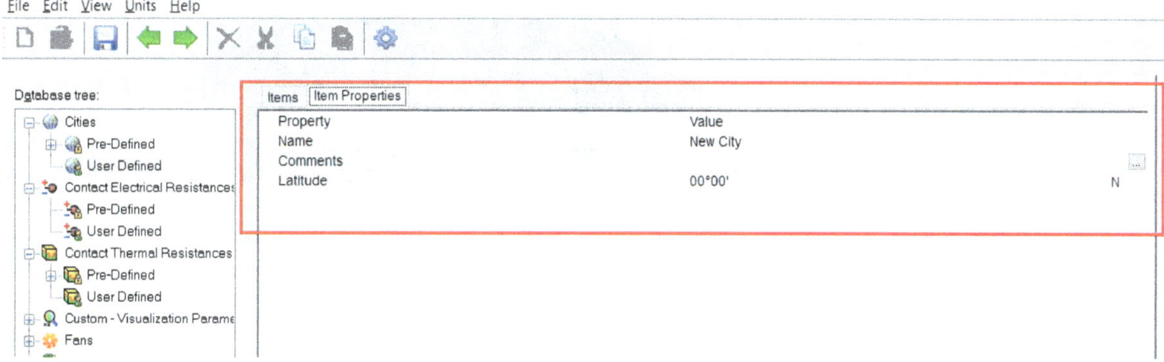

Figure-45. New property added in the database

- Similarly, check and modify the other parameters in **Engineering Database** dialog box as desired. Click on the **Close** button.

PRACTICAL 1

Find the leakage in the model displayed in Figure-46 with the use of leakage tracking method.

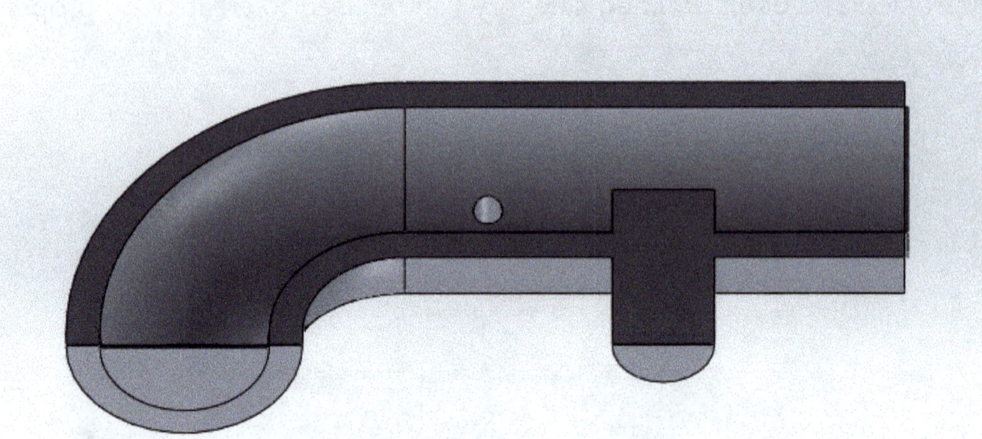

Figure-46. Practical 1

- Double-click on **Practical 1** file from the resource kit of this book. The file will open in the SolidWorks software; refer to Figure-47.

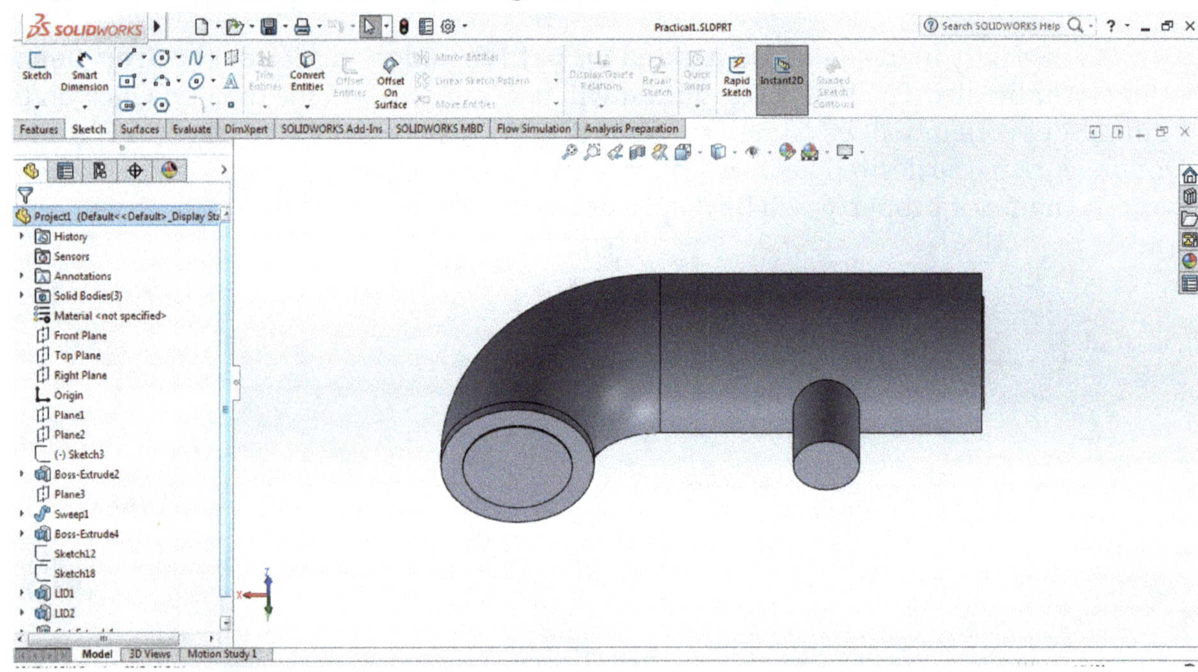

Figure-47. File opened in SolidWorks

- Start the SolidWorks Flow Simulation Add-In if not started yet.
- Click on the **Section View** button from **Heads-Up View Toolbar**; refer to Figure-48. The **Section View PropertyManager** will be displayed at the left of the screen; refer to Figure-49.

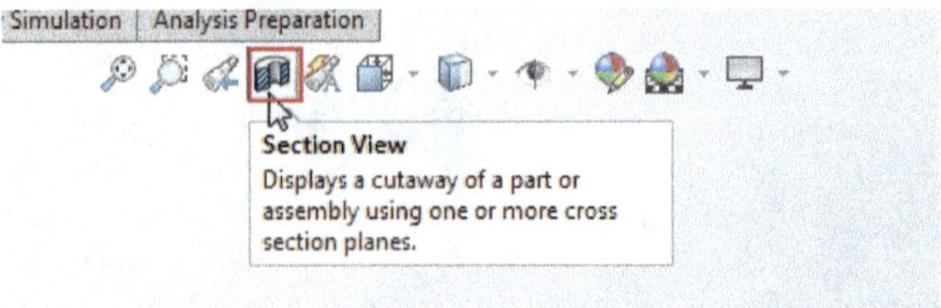

Figure-48. Section View button

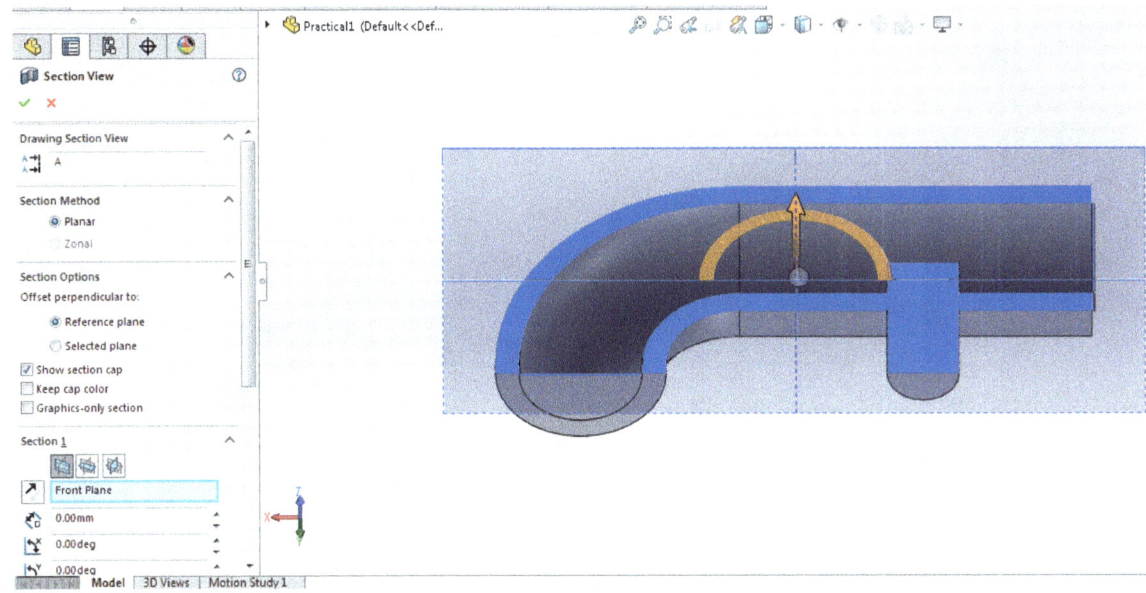

Figure-49. Section View feature

- The **Front Plane** is selected by default. Click on the **OK** button from **Section View** feature. The section view of component is displayed; refer to Figure-50. Note that the lids have already been created in the model for two sides of pipe section.

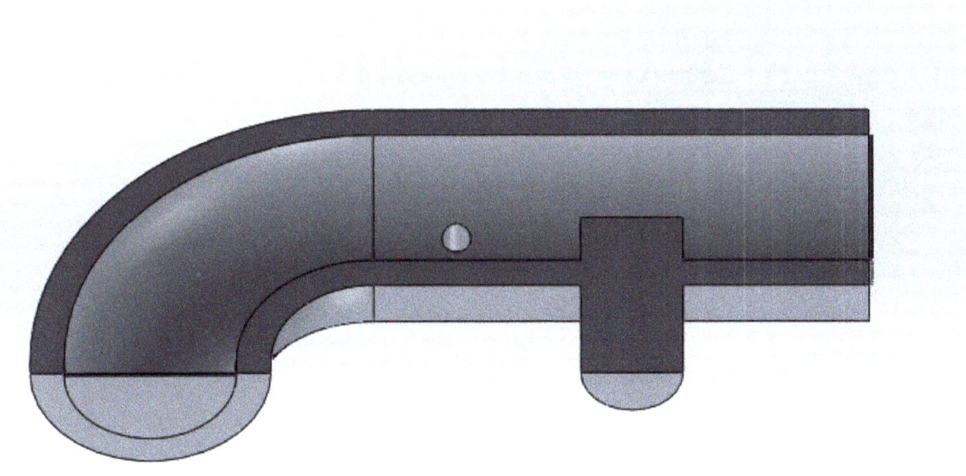

Figure-50. Section view of Practical 1

- Click on the **Check Geometry** button from the **Flow Simulation CommandManager**. The **Check Geometry PropertyManager** will be displayed; refer to Figure-51.

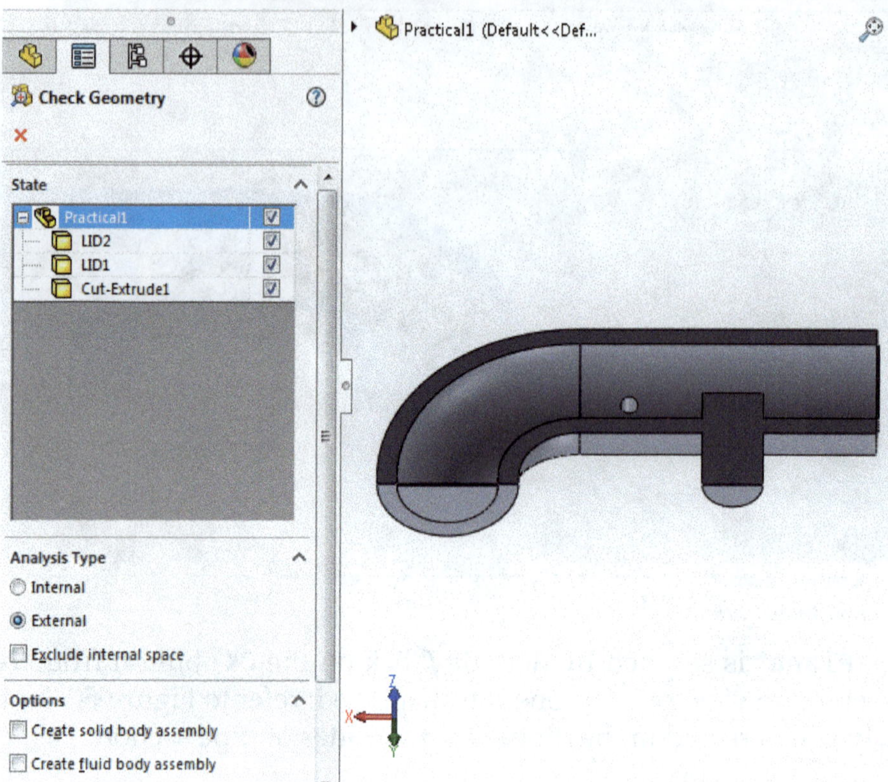

Figure-51. Check Geometry PropertyManager for practical

- Select the **Internal** radio button from the **Analysis Type** rollout and click on the **Check** button from **Check Geometry PropertyManager**. The **Check Geometry** tab will be displayed; refer to Figure-52.

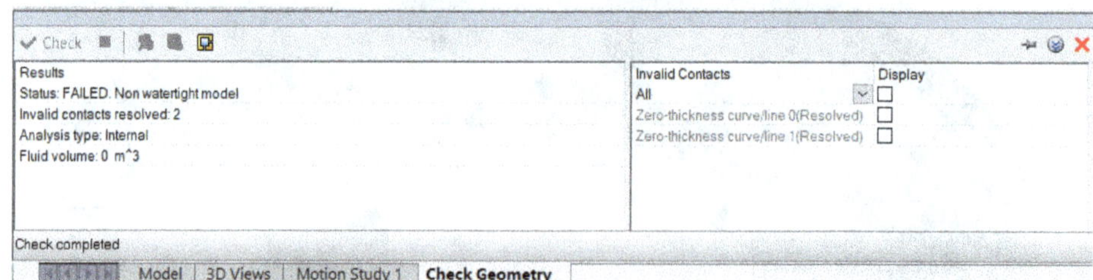

Figure-52. Check Geometry tab for Practical 1

- Click on the **Open Leak Tracking** button from **Check Geometry** tab. The **Leak Tracking** tab will be displayed; refer to Figure-53.

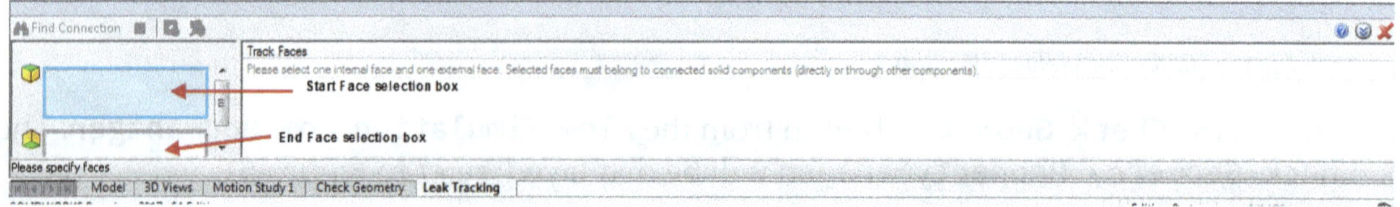

Figure-53. The Leak Tracking tab

- The **Start Face** box is active by default. Click on the face to select; refer to Figure-54.

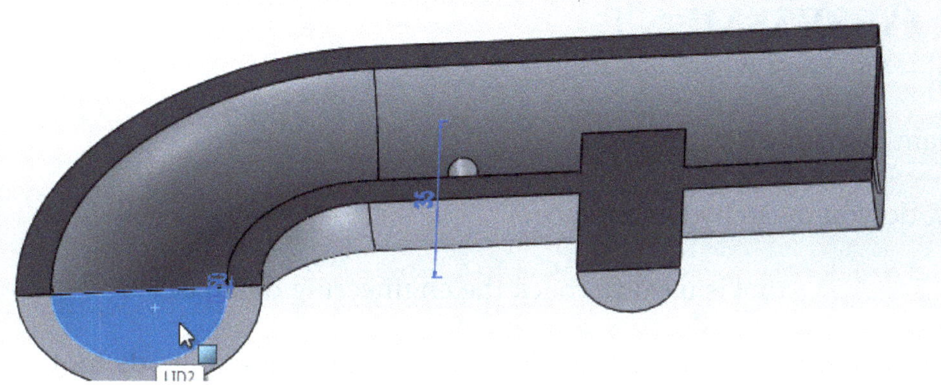

Figure-54. Selection of first face

- After selecting the first face, the **End Face** selection box is activated by default. Click on the inner face as shown in Figure-55.

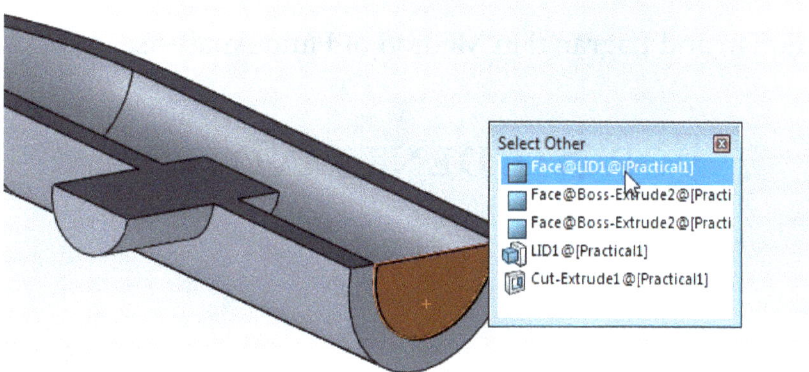

Figure-55. Selection of other face

- After selecting both the faces, click on the **Find Connection** button from **Check Geometry** tab. The analysis for leakage will be shown; refer to Figure-56.

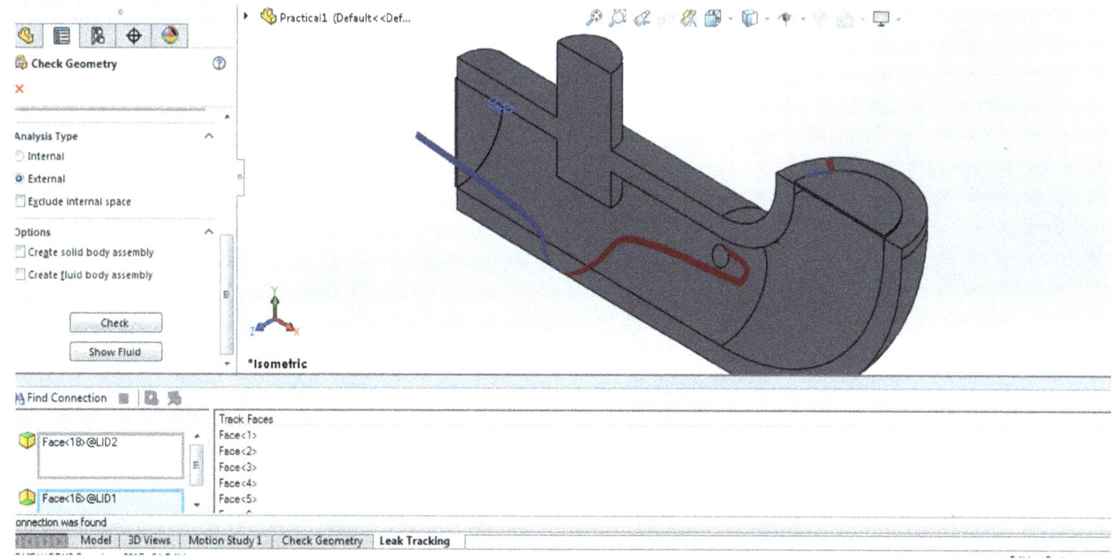

Figure-56. Analysis for fluid leakage

- Click on the **Close** button from **Leak Tracking** tab and close the **Check Geometry PropertyManager**. Remove the leakage.

SELF ASSESSMENT

Q1. What is CFD and where is it used?

Q2. Explain Dynamic Viscosity.

Q3. What do you mean by Divergence?

Q4. The .... tool is used to check the engineering database.

Q5. Explain the procedure to create lids without use of **Create Lids** tool.

Q6. What is the use of **Open Leak Tracking** tool and why it is used?

Q7. What is the importance of Navier-Stokes equation?

Q8. Discuss the Eulerian and Lagrangian Method of Fluid Analysis.

FOR STUDENTS NOTES

Chapter 2

Creating and Managing Flow Simulation Project

Topics Covered

The major topics covered in this chapter are:

- ***Creating Project***
- ***Computational Domain***
- ***Fluid Sub-domains***
- ***Boundary Conditions***
- ***Goals***
- ***Mesh***

STARTING PROJECTS

In the last chapter, we have discussed the procedure of activating SolidWorks Flow Simulation Add-In. Also, we have learned the procedure of checking geometry for leak tracking. In this chapter, we will discuss the procedure of creating projects for flow simulation.

Creating Project

The projects created in the **Flow Simulation** are used for performing setup of fluid simulation. Various parameters like domain of analysis (internal or external), boundary conditions, and so on can be specified in the project setup. The procedure to create a project is discussed next.

- Click on the **Wizard** tool from **Flow Simulation CommandManager**; refer to Figure-1. The **Wizard - Project Name** dialog box will be displayed; refer to Figure-2.

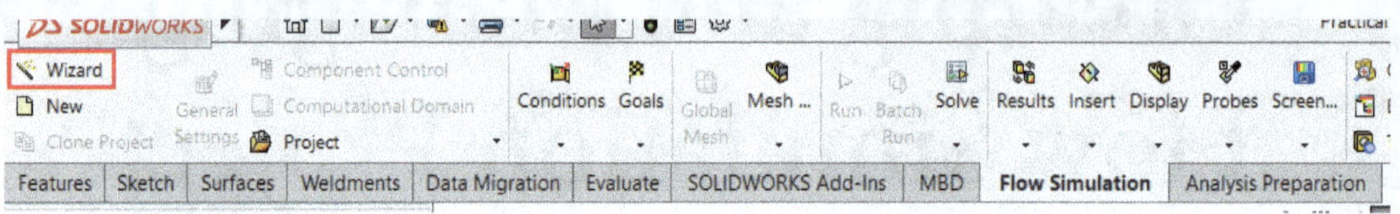

Figure-1. Wizard button

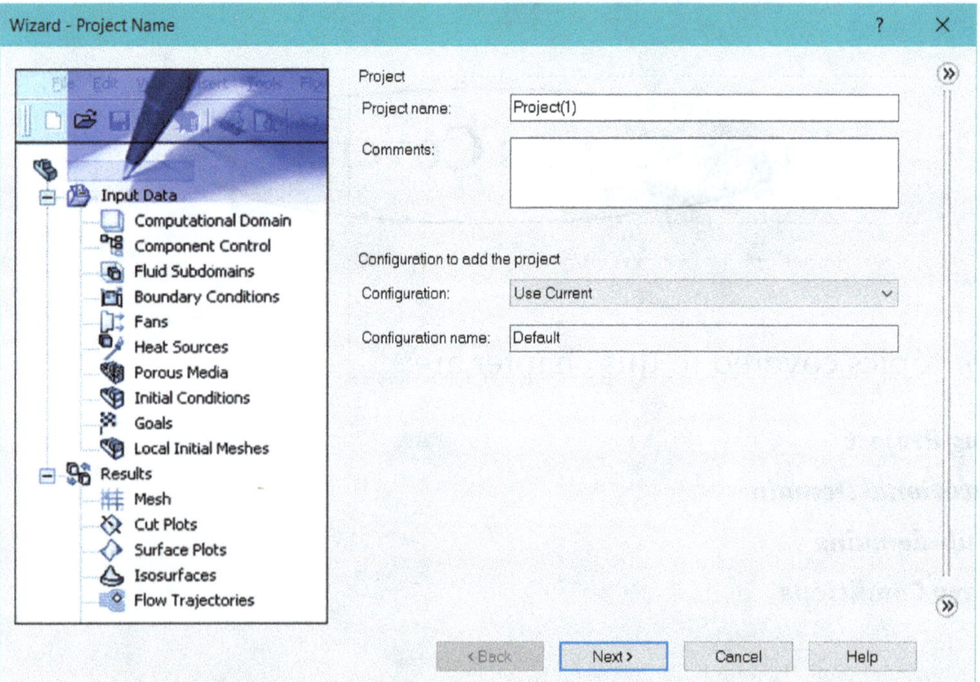

Figure-2. Wizard-Project Name dialog box

- Click in the **Project Name** edit box from **Project** section of dialog box and enter the desired name of the project.
- Click in the **Comments** edit box from **Project** section and enter the desired comments for the project which you are going to create.
- Select the **Use Current** option of **Configuration** drop-down from **Configuration to add the project** section to create a new flow simulation project using the currently active configuration of SolidWorks part/assembly.
- Select the **Select** option from **Configuration** drop-down in **Configuration to add the project** section if you want to select the desired configuration from the list in the **Configuration Name** drop-down of the dialog box.

- Select the desired configuration from **Configuration name** drop-down.
- Select the **Create New** option of **Configuration** drop-down from **Configuration to add the project** area to define a new configuration for the flow simulation project based on existing configuration. Click on the **Configuration name** edit box and enter the desired name.

Note that in a SolidWorks part/assembly configuration, you can add or remove features while keeping the other configuration unchanged.

Unit System

- After specifying the parameters, click on the **Next** button from **Wizard - Project Name** dialog box. The **Wizard - Unit System** dialog box will be displayed; refer to Figure-3.

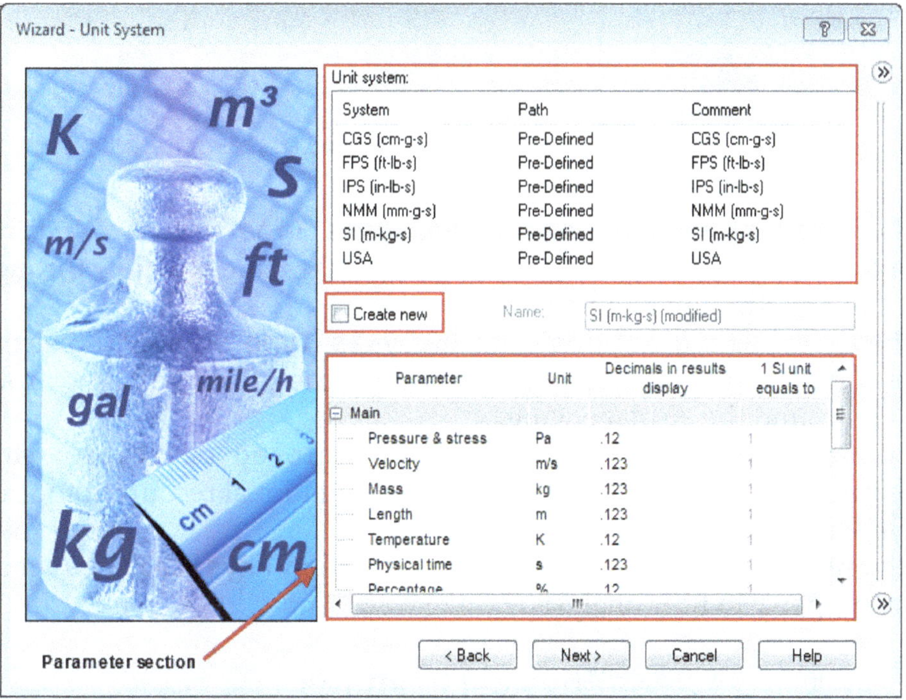

Figure-3. Wizard-Unit Name dialog box

- Select the desired unit from **Unit system** section of **Wizard - Unit System** dialog box.
- If you want to create a new unit system then select the **Create New** check box. The **Name** edit box will be displayed. Enter the parameters as desired for the newly created unit system.
- If you want to change the precision or unit of any parameter then click on the respective field for the parameter in the **Parameters** section of the dialog box.
- After specifying the parameters, click on the **Next** button from **Wizard - Unit System**. If you want to edit any previous parameter then you can click on the **Back** button to edit the submitted parameters.
- On clicking on **Next** button, The **Wizard - Analysis Type** dialog box will be displayed; refer to Figure-4.

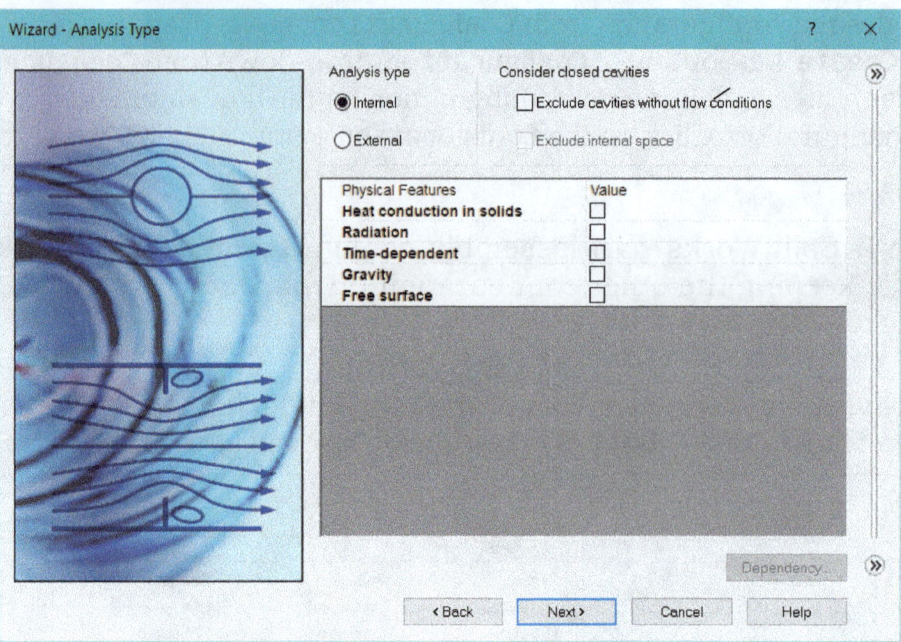

Figure-4. The Wizard-Analysis Type dialog box

- Click on the **Internal** radio button from **Analysis Type** section to specify the internal analysis of component for fluid simulation. In this case, fluid will flow through solid body like water flowing through a pipe system.
- Click on the **External** radio button from **Analysis Type** section to specify the external analysis of component for fluid simulation. In this case, fluid will flow around the solid body like air around the airplane wing.
- Select the desired check boxes from the **Consider closed cavities** area of the dialog box to select additional options for reducing computational resource requirements in the flow analysis.
- Click on the required physical feature check box from **Physical Features** section to include the property in flow analysis.

Physical Features

Heat Conduction in Solids

Select the **Heat conduction in solids** check box if you want to include the heat conduction property of solids in analysis. Note that when any viscous fluid moves through a conducting solid then temperature of solid rises. If you have selected the **Heat conduction in solids** check box then temperature plot will be generated including over solid body. After selecting the **Heat conduction in solids** check box, the **Heat conduction in solids only** check box will be displayed. Select this check box if there is no fluid in the system and you are performing heat transfer analysis through solids only.

Radiation

Select the **Radiation** check box if you want to include heat transfer by radiation in the study. Three options will become active below the selected check box; refer to Figure-5. Select the desired heat transfer method of radiation from the **Radiation model** drop-down. Select the **Discrete Transfer** option from the drop-down if you want to consider radiation leaving the surface in certain angle range as a single ray. When heat transfer occurs using this method then an exchange factor is calculated for each ray which defines the heat transferred between two radiative surfaces.

This method is recommended for cases with high-temperature gradients or with compact radiation sources (i.e. with high power heat sources such as incandescent lamps, furnaces or other heat sources of temperatures grater then 1000K) and geometric-optical effects of the shadows, focusing and etc. But it does not allow to simulate the absorption and/or spectral dependencies.

Select the **Discrete Ordinates** option if you want to consider finite number of angles in a range for solving heat transfer equation. This method allows the solution of radiation in absorptive (semi-transparent) media and to model spectral dependencies. The accuracy of the solution depends on the number of discrete directions used.

On selecting this option from the drop-down, **Absorption in solids** and **Spectral characteristics** check boxes are displayed. Select the **Absorption in solids** check box if the solid body in analysis is not transparent and you want to the solids to absorb heat in analysis. Select the **Spectral characteristics** check box if you want to consider radiation spectrum in your analysis.

On selecting the **Spectral characteristics** check box, the options to define spectral characteristics will be displayed; refer to Figure-6. Specify the desired number of bands to be used for spectral distribution of heat energy from the source in Number of bands drop-down. Based on selected number of bands, the **Band edge** edit boxes will be displayed to define the boundary edge values. If there are two bands then one edit box will be displayed to specify the wavelength value dividing two bands. If there are 5 bands then 4 edit boxes will be displayed to specify wavelength values for dividing edges. In the **Environment radiation** drop-down, select the desired type of environmental radiation source. You can select **Blackbody Spectrum**, **Daylight Spectrum**, and **Extraterrestrial Solar Spectrum** option from the drop-down. Note that these sources have been defined in the **Engineering Database** by default and you can check their properties in the database. You can define more sources under **User-Defined** category if you want. Those user defined sources will also be displayed in the drop-down here.

This model is recommended for most cases with low-temperature variations or without compact radiation sources (for example, for electronics cooling, the greenhouse effect, and cabin comfort analysis). At the same time in most cases with low-temperatures, it is not necessary to consider spectral dependencies (i.e. Number of bands can be set to 0). Also, the Discretization level can be set to 2 or 3 to get an acceptable accuracy. If the geometric-optical effects of the shadows are significant, it is recommended to increase the Discretization level to improve the accuracy of geometrical optics modelling. If that is not enough, the using of Discrete Transfer model is recommended.

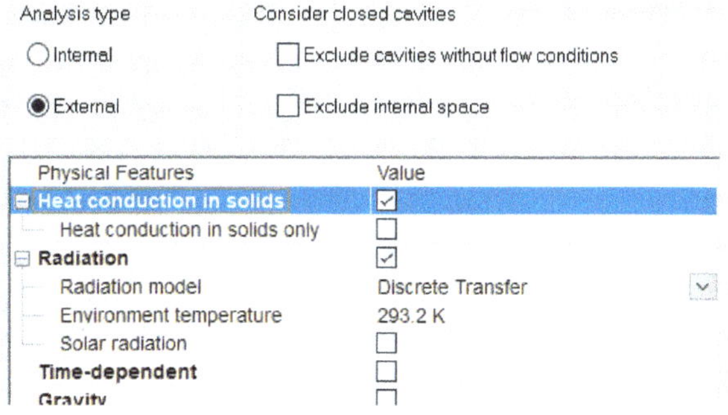

Figure-5. Radiation check box and options

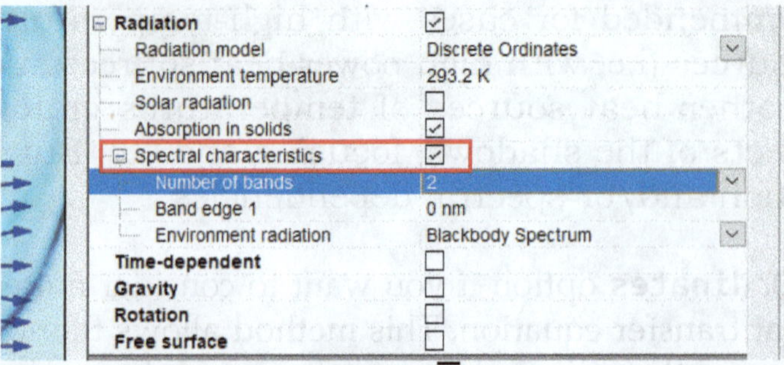

Figure-6. Spectral characteristics options

Solar Radiation: Select the **Solar radiation** check box if you want to include radiation effect of Sun on our component during analysis. This option is generally selected if the designed fluid system will be working directly under Sun light. On selecting this check box, the options related to Solar radiation will be displayed in the dialog box; refer to Figure-7.

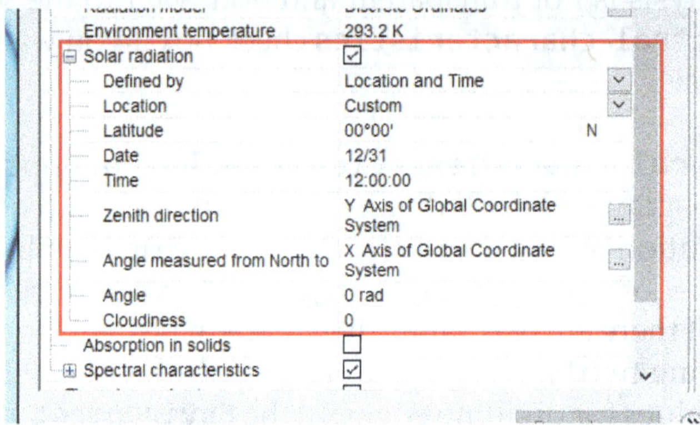

Figure-7. Solar radiation options

Select the desired option in the **Defined by** drop-down to specify how solar radiation will be defined. Select the **Location and Time** option if you want to specify location of your model on Earth. Select the **Direction and Intensity** option if you want to specify direction and intensity of solar radiation. Select the **Azimuth and Altitude** option if you want to specify direction of solar radiation in the form of azimuth and altitude coordinates. After selecting the desired option from the drop-down, specify related parameters in the fields below drop-down. Note that arrows of radiation will be displayed on the model showing direction of radiation.

Time-dependent

Select the **Time-dependent** check box if you want to set time limits for the analysis. On selecting this check box, the **Total analysis time** and **Output time step** edit boxes are displayed. Specify the desired value of total time during which the analysis will be performed in the **Total analysis time** edit box. Specify the desired value of time step at which output will be displayed in the **Output time step** edit box.

For Transient or say Time-dependent problems, Flow Simulation "time marches" the solution from initial conditions for the problem's physical time to specified duration. The initial conditions must be precise, with the exception of unsteady problems which have a steady periodic solution (e.g. in the case of periodic boundary conditions) that can be obtained from arbitrary initial conditions. But additional time will be required to eliminate the influence of specified initial conditions in such cases.

Gravity

Select the **Gravity** check box if you want to include the effect of gravity in the analysis. On selecting this check box, three edit boxes are displayed defining gravity acceleration in each of

X, Y, and Z direction; refer to Figure-8. Specify the value of gravity in desired edit box. Note that an arrow will be displayed on model showing the direction of gravity.

Rotation

Select the **Rotation** check box if you want to set rotating frame for analysis. On selecting this check box, the **Type** drop-down will be displayed.

Select the **Local region(s)(Averaging)** option from the drop-down if there is a steady flow of fluid in multiple streams due to rotating part. In this case, fluid stream will be divided into different regions and flow parameters for each region will be calculated. At the end, an average of all the regions will be taken to present result.

Select the **Local region(s)(Sliding)** option from the drop-down if there is unsteady flow of fluid or you are performing time dependent study. In this approach, the rotor and stator cells remain in contact with each other during the flow with an arbitrary sliding interface. Note that when using this option, the solid body motion is not taken into account while performing heat transfer analysis. Also, this option is not compatible with very high speed fluid flow.

Select the **Global rotating** option from the drop-down if when whole fluid system is rotating about an axis of global coordinate system. The options to specify angular velocity and rotation axis will be displayed

Free Surface

Select the **Free surface** check box if you want to model two immiscible fluids with an interface between them. Note that the **Free surface** check box is applicable for **Time-dependent** study only.

Default Fluid

- After specifying the parameters, click on the **Next** button from **Wizard - Analysis Type** dialog box; refer to Figure-8. The **Wizard - Default Fluid** dialog box will be displayed; refer to Figure-9.

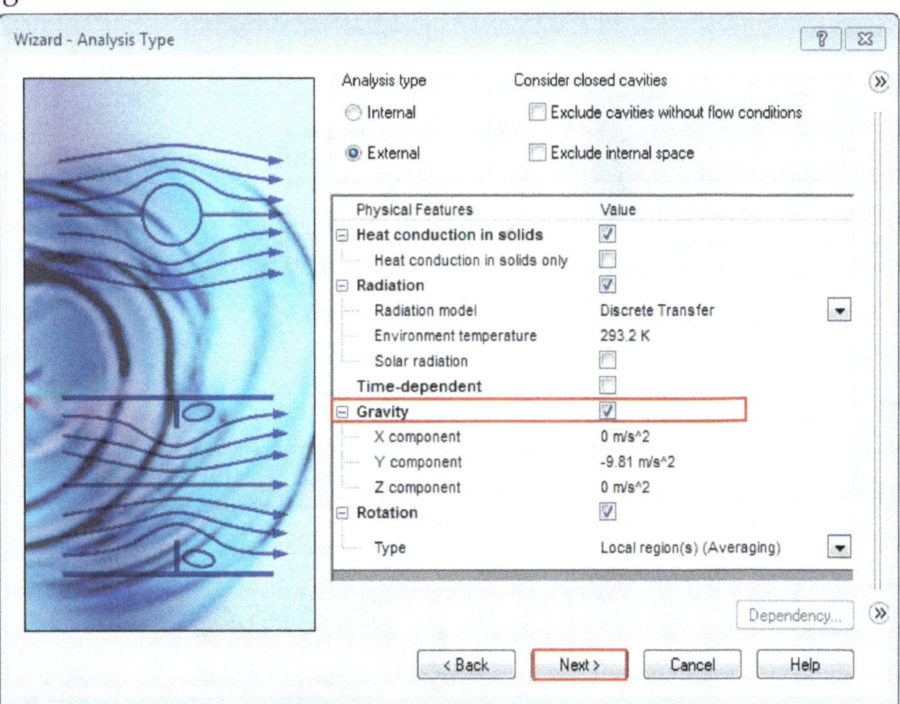

Figure-8. Updated wizard-Analysis Type dialog box

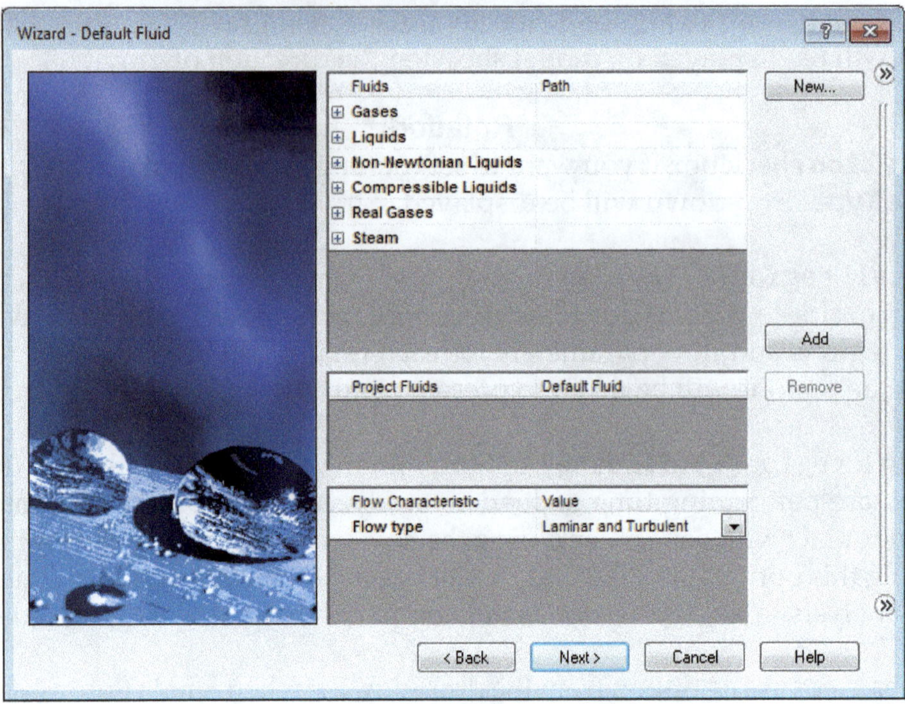

Figure-9. The Wizard-Default Fluid dialog box

- Expand the desired category of fluids and click on the fluid from **Fluids** section that you want to include in analysis. Click on the **Add** button from the dialog box. The fluid will be selected and added in **Project Fluids** section.
- To add a new fluid for the specific project, click on the **New** button. The **Engineering Database** dialog box will be displayed. You can create a new user defined fluid as discussed earlier.
- Click on the **Flow Type** drop-down and select the required flow type; refer to Figure-10. Laminar Flow is the flow of a fluid when each particle of the fluid follows a smooth path. One result of laminar flow is that the velocity of the fluid is constant at any point in the fluid. Turbulent Flow is the irregular flow that is characterized by tiny whirlpool regions. The velocity of this fluid is definitely not constant at every point.

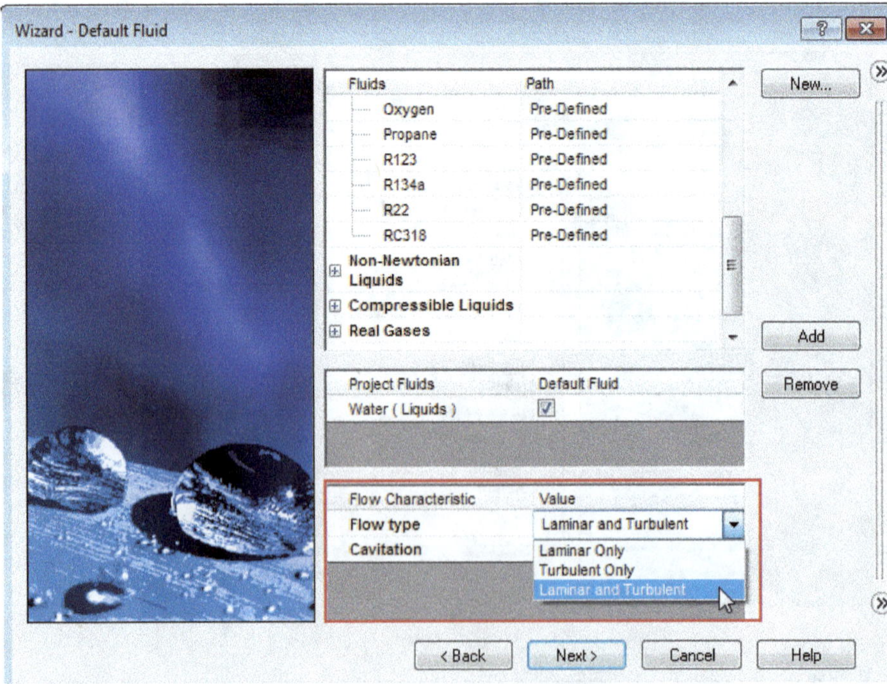

Figure-10. Selecting required flow type

- Select the **Cavitation** check box to consider the cavitation in the analysis. Cavitation is the formation of vapor cavities in a fluid that are the consequence of forces acting upon the liquid. It usually occurs when a liquid is subjected to rapid changes of pressure that cause the formation of cavities in the liquid where the pressure is relatively low. When subjected to higher pressure, the voids implode and can generate an intense shock wave.
- If you want to remove the selected fluid then click on the **Remove** button from **Wizard - Default Fluid** dialog box or just double-click on the selected fluid. The added fluid will be removed.
- If you have selected a gas (note that it is not real gas but it is gas) as fluid then the **High Mach number flow** and **Humidity** check boxes are displayed. Select the **High Mach number flow** check box if the flow of fluid is high in Mach number. Mach number is the ratio of speed of object to speed of sound. Select the **Humidity** check box if you want to consider humidity in the gas during analysis. Note that you can either select the **High Mach number flow** or the **Humidity** check box.

Default Solid

- After specifying the parameter, click on the **Next** button if Heat conduction in solids check box was selected in **Wizard - Analysis Type** page of dialog box then the **Wizard - Default Solid** dialog box will be displayed; refer to Figure-11.

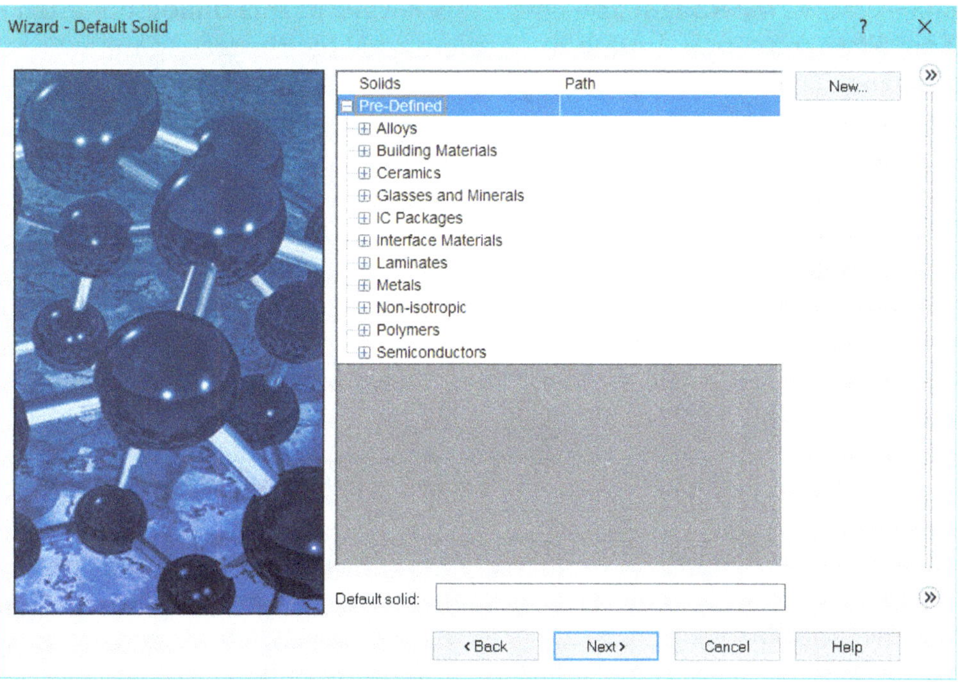

Figure-11. Wizard-Default Solid dialog box

- Click on the required solid material from **Solids** section to select as a material of component for analysis.
- On selection, the selected metal will be displayed in **Default Solid** box; refer to Figure-12.

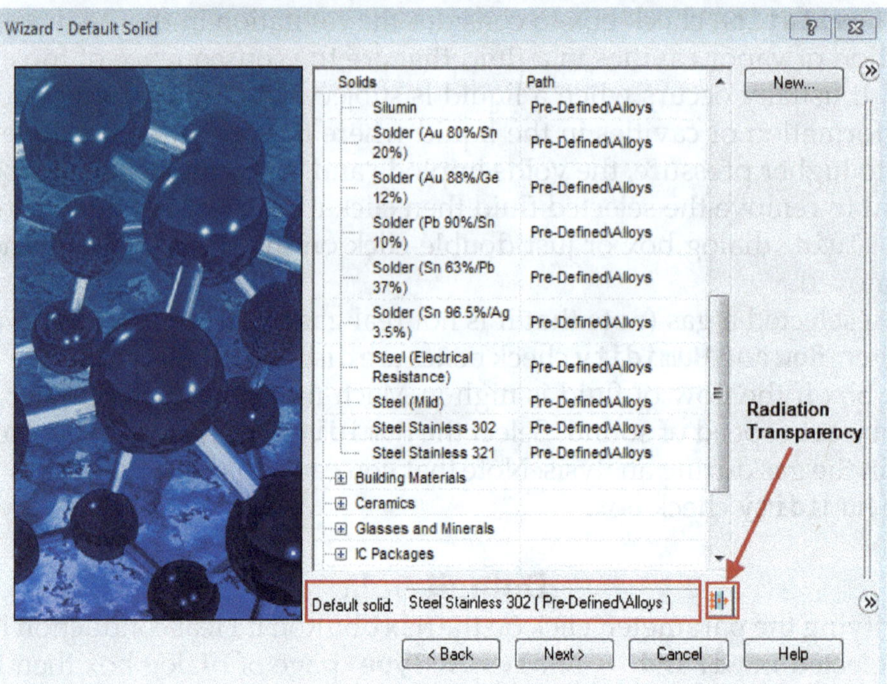

Figure-12. Selected solid

- If you have selected the **Radiation** check box earlier in this dialog box then click on the **Radiation Transparency** button to enable or disable the transparency of component. The **Radiation Transparency** dialog box will be displayed; refer to Figure-13.

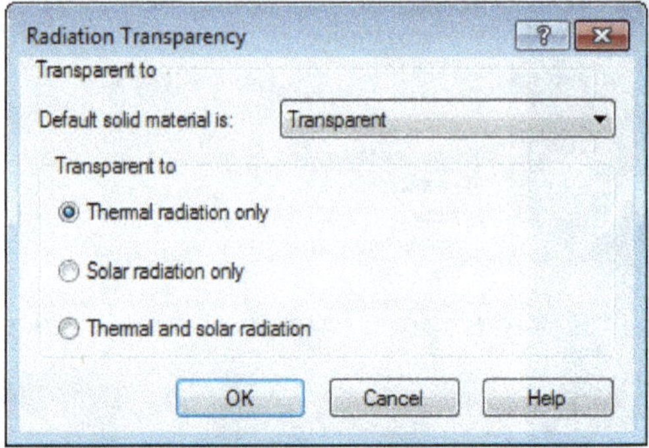

Figure-13. Radiation Transparency dialog box

- Click on the **Default solid material is** drop-down and select the required type of transparency according to the material used for simulation.
- After specifying the parameters, click on the **OK** button from **Radiation Transparency** dialog box. The option will be selected.

Wall Conditions

- After specifying the parameters in **Wizard - Default Solid** dialog box, click on the **Next** button. The **Wizard - Wall Conditions** dialog box will be displayed based on your previous selections; refer to Figure-14.

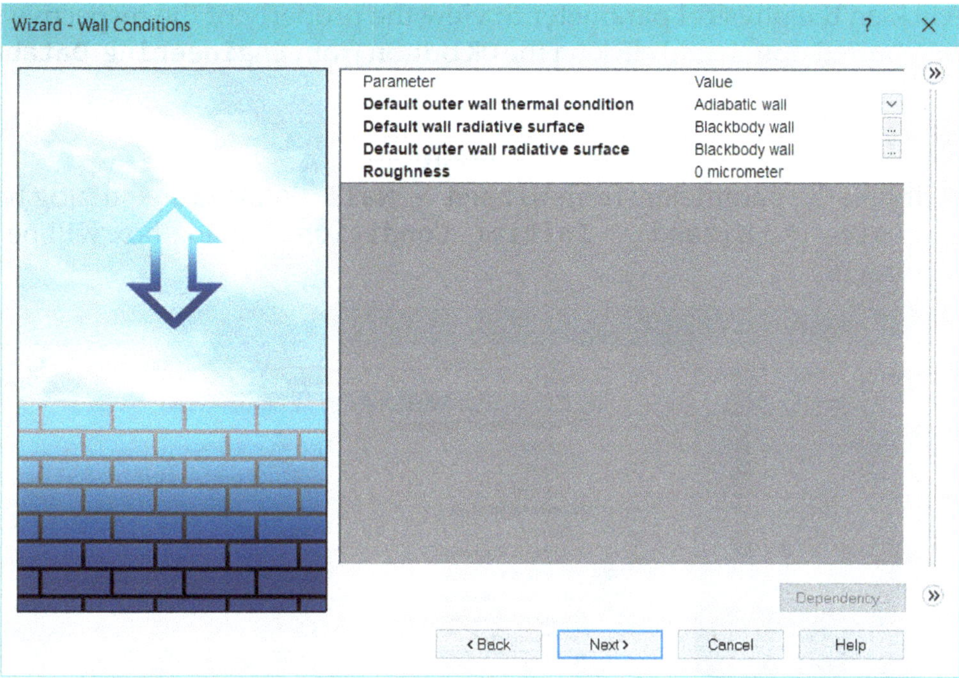

Figure-14. Wizard-Wall Conditions dialog box

- Select the desired option from the **Default outer wall thermal condition** drop-down to define how heat transfer will occur through the outer wall. Select the **Adiabatic wall** option if no heat is allowed to escape through the wall. Select the **Heat transfer coefficient** option from the drop-down to specify general temperature outside the wall and heat transfer rate for wall. Select the **Heat generation rate** option from the drop-down to specify desired heat transfer capacity in the **Heat generation rate** edit box. Select the **Surface heat generation rate** option from the drop-down to specify heat transfer rate per area. Select the **Wall temperature** option from the drop-down to define temperature of outer wall.
- Click on the **Browse** button for **Default wall radiative surface** and **Default outer wall radiative surface fields** from **Parameter** section to select the radiative surface types. The **Engineering Database** dialog box will be displayed; refer to Figure-15.

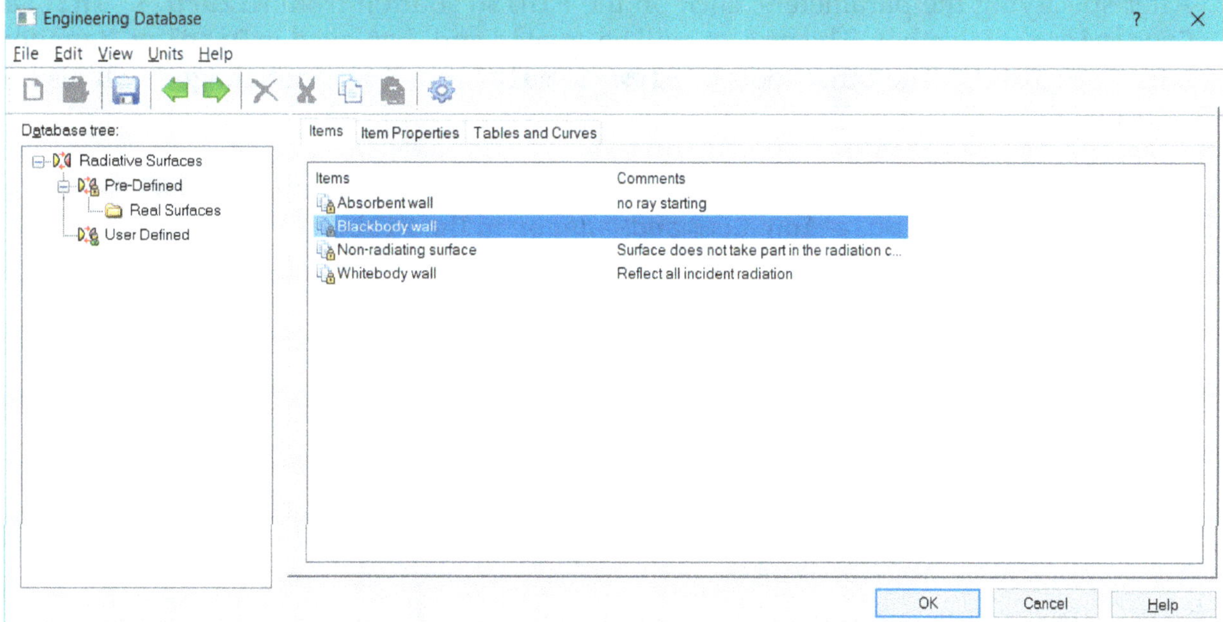

Figure-15. Engineering Database dialog box

- Double-click on the physical parameter to view the property of the respective parameter.
- Select the required option and click on the **OK** button from **Engineering Database** dialog box.

Initial Conditions

- After specifying the parameters from **Wizard - Wall Conditions** dialog box, click on the **Next** button. The **Wizard - Initial Conditions** dialog box will be displayed; refer to Figure-16.

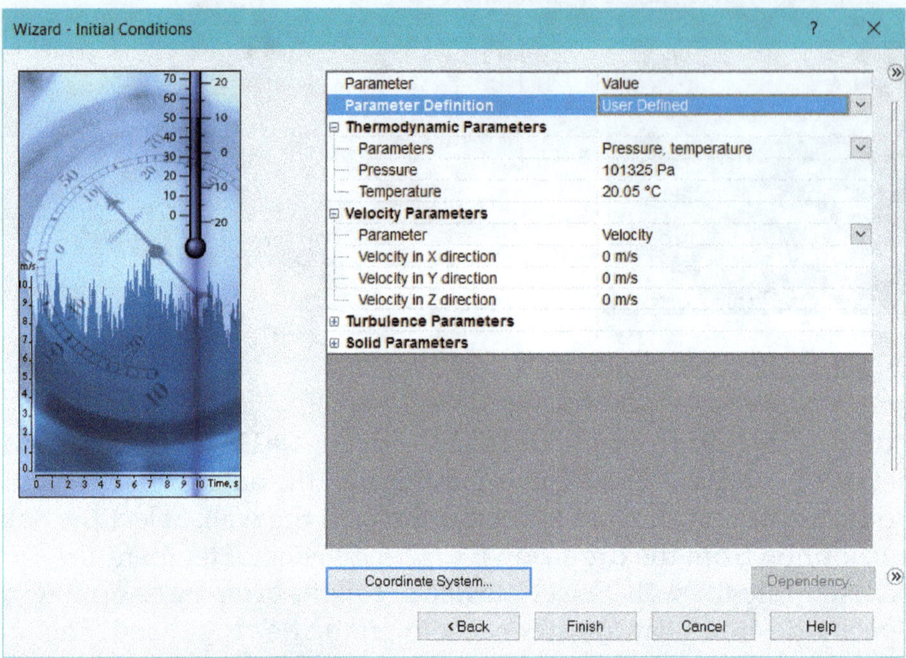

Figure-16. Wizard-Initial Conditions dialog box

- Specify the desired values for various parameters like thermodynamic parameters, velocity parameters, turbulence, solids, and so on.
- Click on the **Coordinate System** button from the dialog box to set coordinate system and reference axis for analysis. The reference axis specified here defines the parameters in the **Dependency** dialog box. After setting the parameters, click on the **OK** button.
- After specifying the parameters, click on the **Finish** button from **Wizard - Initial Conditions** dialog box. The project will be created and displayed in **Display Manager**; refer to Figure-17. The other tools of **Flow Simulation CommandManager** will also be activated.

Note that you can edit the project properties at any time by clicking on the **General Settings** button from the **Flow Simulation CommandManager** in the **Ribbon**.

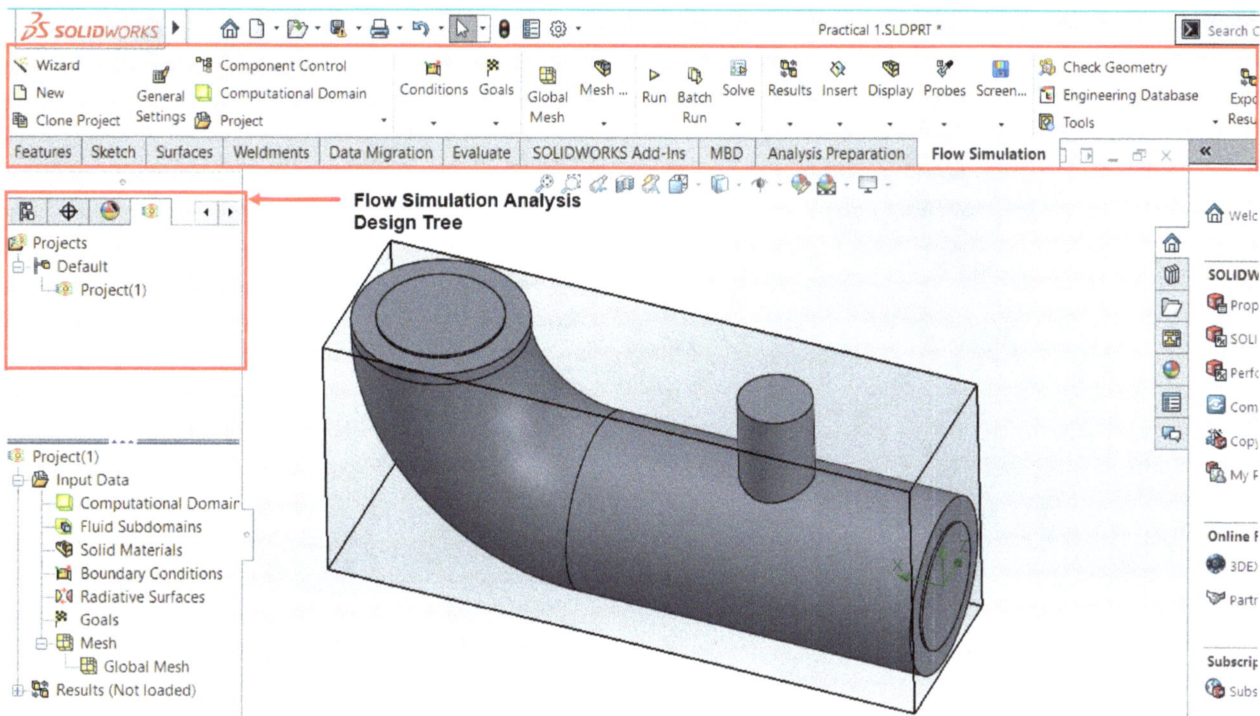

Figure-17. Project displayed in Display Manager

COMPUTATIONAL DOMAIN

The computational domain is the region where the flow and heat transfer calculations are performed. When you create a new project with the **Wizard** tool, Flow Simulation automatically creates the **Computational Domain** enclosing the model.

For External flows, the computational domain's boundary planes are automatically distanced from the model.

For Internal flows, the computational domain's boundary planes automatically envelop either the entire model if Heat Conductions in Solids is considered. If heat Conductions in Solids is not considered then the model's flow passage only is considered in domain.

The procedure to set computational domain is discussed next.

- Right-click on the **Computational Domain** option from the **Flow Simulation Analysis Tree** or click on the **Computational Domain** tool from the **CommandManager** in the **Ribbon**. The right-click shortcut menu will be displayed; refer to Figure-18.
- Click on the **Edit Definition** button from the displayed menu. The **Computational Domain PropertyManager** will be displayed; refer to Figure-19.

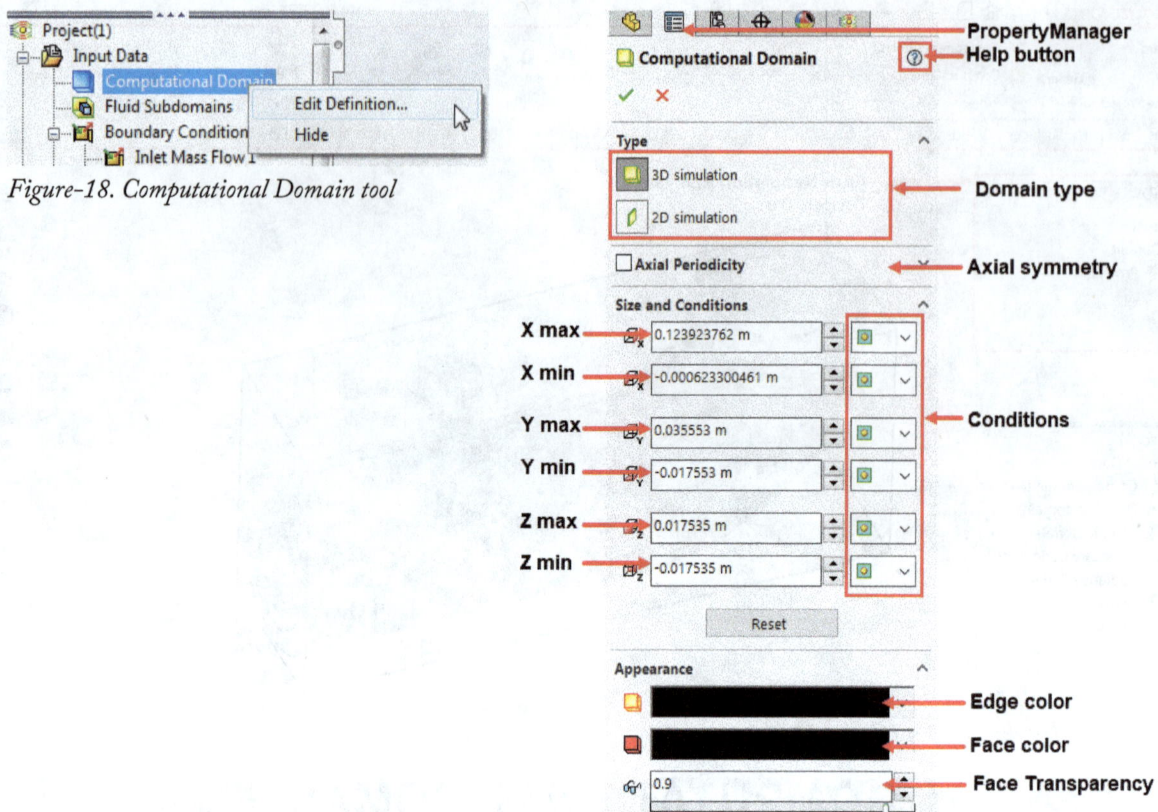

Figure-18. Computational Domain tool

Figure-19. Computational Domain PropertyManager

- Click on the **3D simulation** button from **Type** rollout if you want to create a 3 dimensional domain for analysis.
- Click on the **2D simulation** button from **Type** rollout if you want to create a 2 dimensional computational domain for analysis.
- If you have clicked on **2D simulation** button then select desired radio button for the desired plane below **2D simulation** button. The computational domain of specified size and conditions will be displayed; refer to Figure-20.

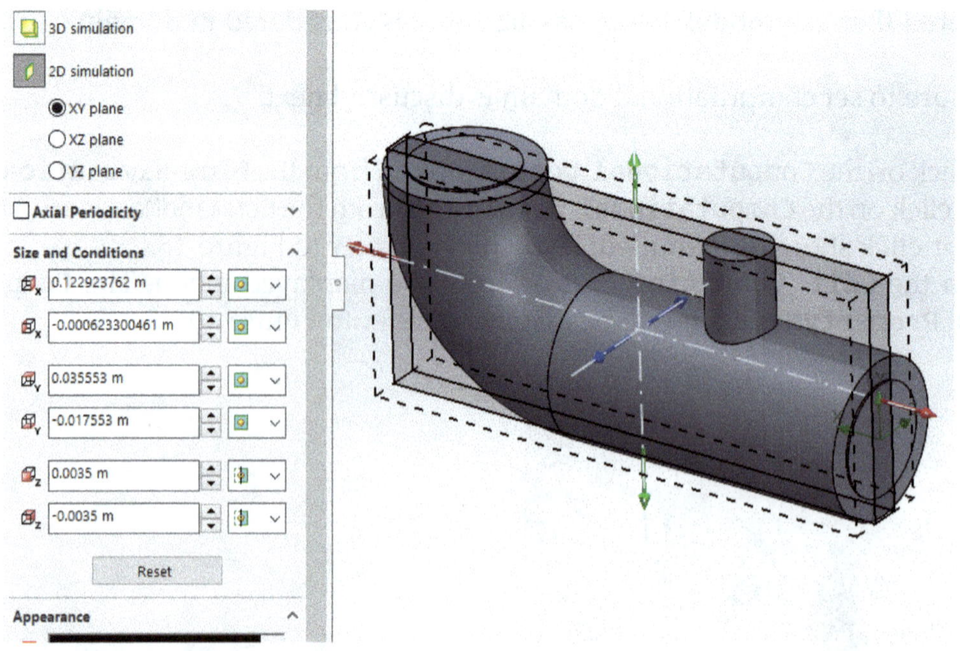

Figure-20. 2D computational domain

- Select the **Axial Periodicity** check box if your model is in the form of a tube symmetric about an axis and flow in one sector of the tube can represent the flow in full round of the tube. On selecting this check box, the options for axial periodicity will become active. Select a planar face or plane which cuts through the part. You will be asked to select an axis. Select the axis about which the tube is symmetric. Specify the other parameters of periodicity in **Start Angle**, **Number of Sectors**, **Max Radius**, and **Min Radius** edit boxes, as required; refer to Figure-21.

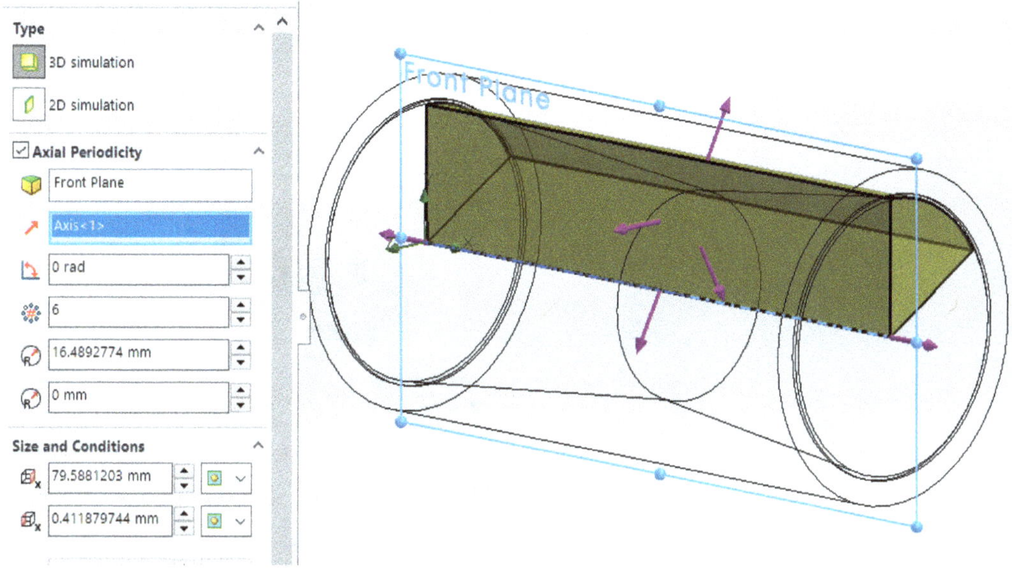

Figure-21. Setting axial periodicity

- Click in the desired edit boxes from **Size and Conditions** section and enter the desired value to specify limits of computational domain along X,Y,and Z-axes.
- If you want to reset the entered value of X,Y, and Z then click on the **Reset** button.
- Click on the **Edge Color** edit box from **Appearance** section and click on the specific color to select the color of the edge.
- Click on the **Face Color** edit box from **Appearance** section and click on the specific color to select the color of front or face color.
- Click on the **Face transparency** edit box and enter the desired value for the transparency of face. You can also select the transparency of face by moving the slider.
- After specifying the parameters, click on the **OK** button from **Computational Domain** option. The computational domain will be updated.

FLUID SUBDOMAINS

The **Fluid Subdomains** option is used to define a fluid sub domain with same fluid type or fluids different from the specified in the **Wizard** dialog box. A fluid sub domain is closed region of fluid different from the main fluid. The procedure to use this tool is discussed next.

- Right-click on the **Fluid Subdomains** option from the **Flow Simulation Analysis Tree**. The right-click shortcut menu will be displayed; refer to Figure-22.
- Click on the **Insert Fluid Subdomain** button from the displayed menu. The **Fluid Subdomain PropertyManager** will be displayed; refer to Figure-23.
- Click in the **Selection** box from **Fluid Subdomain** option and click on the faces to apply the fluid subdomain; refer to Figure-24.

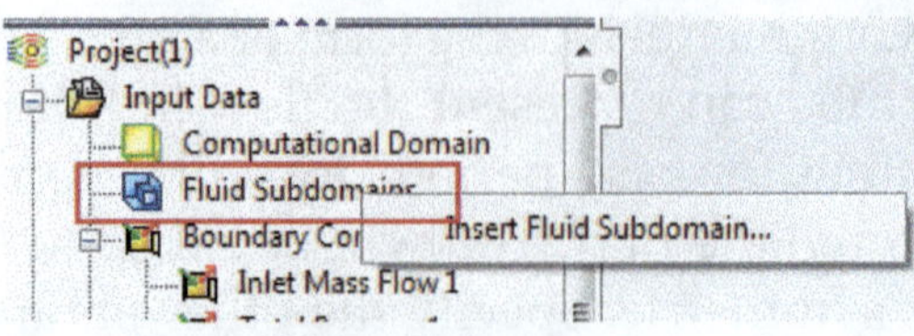

Figure-22. Fluid Subdomains tool

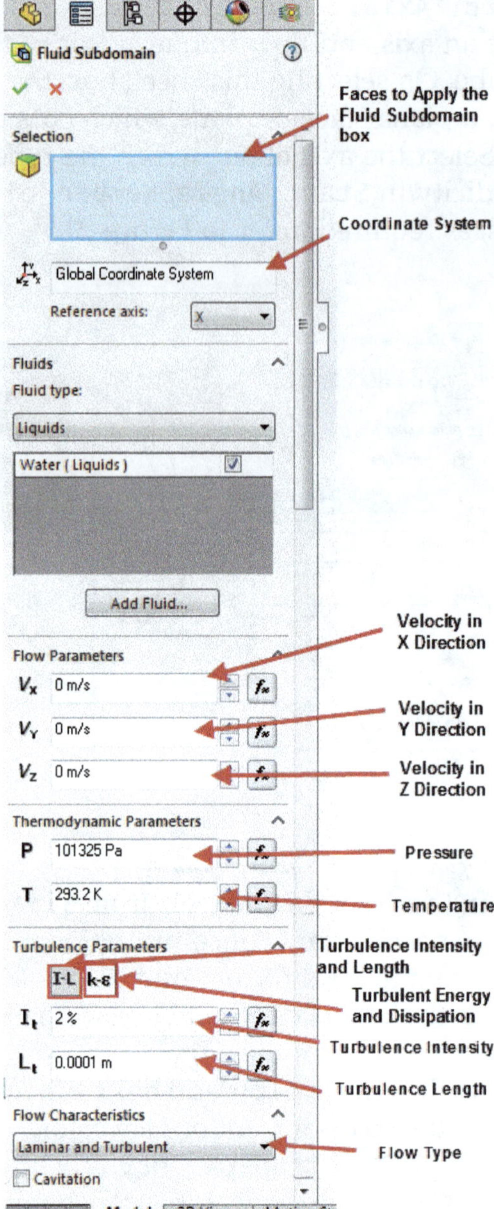

Figure-23. Fluid Subdomain Feature

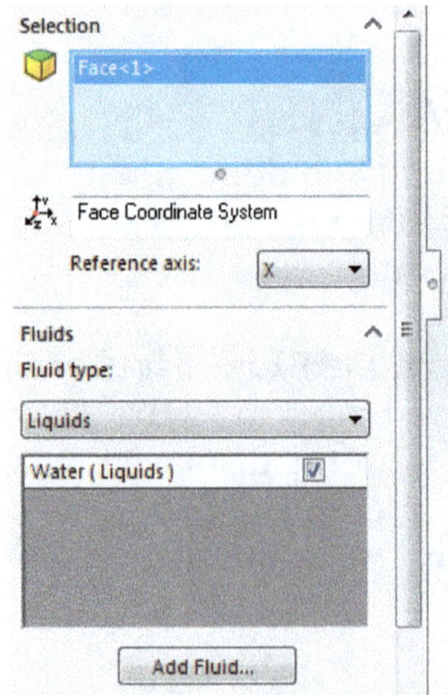

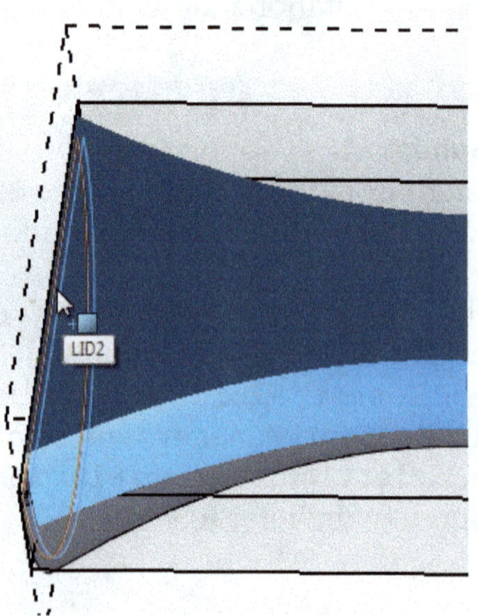

Figure-24. Selection of face

- Click on the **Reference axis** drop-down and select the desired axis as a reference for fluid flow.
- Click on the **Fluid type** drop-down from **Fluids** section and select the desired type of fluid.
- If you want to add another fluid for analysis then click on the **Add Fluid** button from **Fluids** section. The **General Settings** dialog box will be displayed; refer to Figure-25.

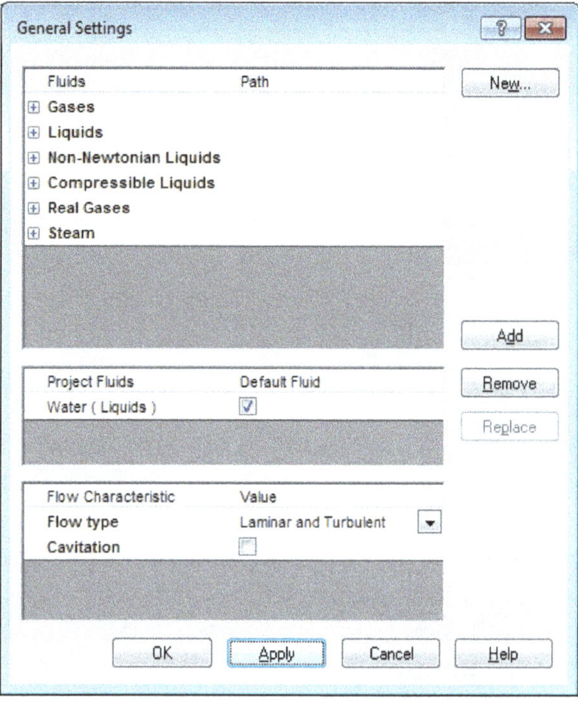

Figure-25. General Settings dialog box

- You can also open the **General Setting** dialog box by clicking on the **General Settings** tool from **CommandManager**; refer to Figure-26.

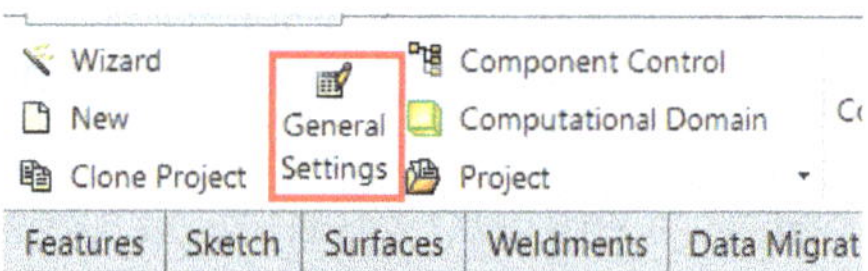

Figure-26. General Settings tool

- Select the required fluid type from the **General Settings** dialog box and click on the **Apply** button. The selected fluid will be added.
- Click in the respective edit box of **Flow Parameters** section and enter the value to specify the velocity in X,Y, and Z direction.
- Click on the **Dependency** button from **Flow Parameters** section to specify data in a suitable manner; as a constant, as a tabular, or formula dependent on x, y, z, radius (r), phi Φ, theta (θ) coordinates and time **t.** The **Dependency** dialog box will be displayed; refer to Figure-27.

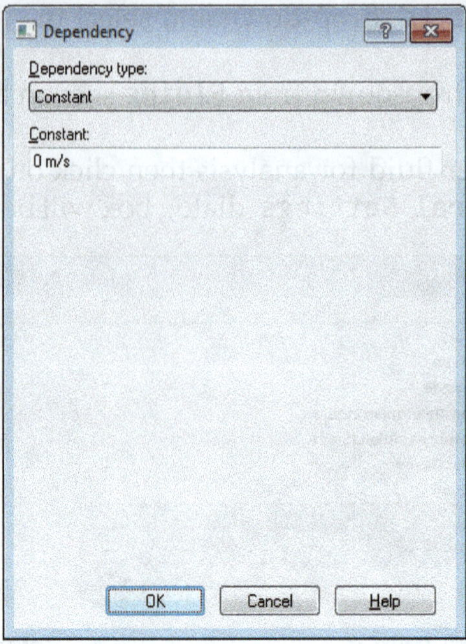

Figure-27. Dependency dialog box

- Select the required option from **Dependency Type** drop-down and specify the parameter.
- After specifying the parameters, click on the **OK** button from **Dependency** dialog box. The data will be added.
- Click in the **Pressure** edit box from **Thermodynamic Parameters** section and enter the desired value of pressure in sub-domain.
- Click in the **Temperature** edit box from **Thermodynamic Parameters** section and enter the value of temperature in sub-domain.
- If you have selected two or more fluids for subdomain then the **Substance Concentrations** section is also displayed in the **PropertyManager**; refer to Figure-28. Expand the **Substance Concentrations** rollout to specify ratio between two or more subdomain fluids. You can the set the ratio in mass or volume by selecting the respective radio button.
- Click on the **Turbulence Intensity and Length** button from **Turbulence Parameters** rollout and specify the value of **Turbulence Intensity** and **Turbulence Length** in their respective edit box for analysis if you have opted to include turbulence in the analysis.
- If you want to specify turbulence energy then click on the **Turbulent Energy and Dissipation** button from **Turbulence Parameters** section and specify the value of **Turbulent Energy and Turbulent Dissipation** in their respective edit box.
- Expand the **Flow Characteristic** section and click in the **Flow Type** drop-down from to select the required type of flow.
- Click on the **OK** button from the **PropertyManager**. The **Fluid Subdomain 1** will be created in the project tree.

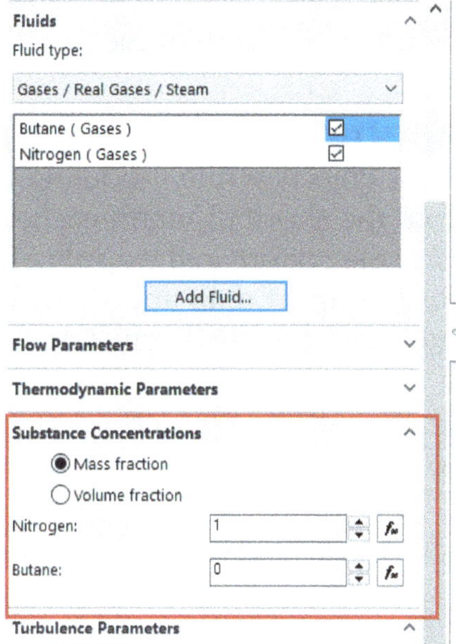

Figure-28. Substance Concentrations section

ROTATING REGION

If you have enabled rotation in analysis then you can specify rotating regions and their rotational speeds. The procedure to do so is given next.

- Right-click on the **Rotating Regions** option from the **Flow Simulation Analysis Tree** and select the **Insert Rotating Region** option from the right-click shortcut menu. The **Rotating Region PropertyManager** will be displayed; refer to Figure-29.

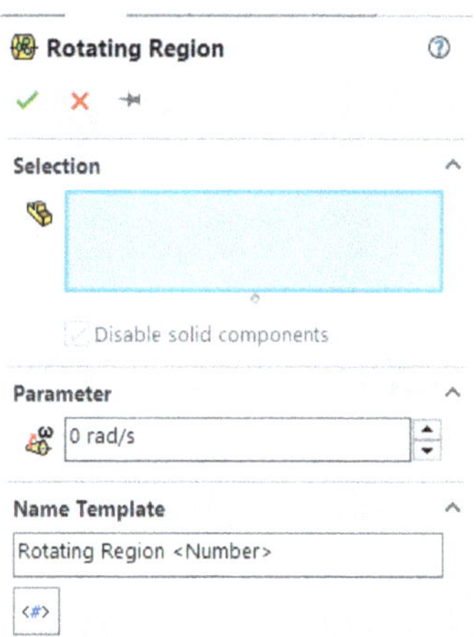

Figure-29. Rotating Region PropertyManager

- Select the part which you want to be rotating during analysis.
- Click in the **Angular Velocity** edit box and specify desired value of velocity.
- Click on the **OK** button from the **Rotating Region PropertyManager** to apply settings.

SETTING MATERIAL FOR MODEL

Generally, material for solid objects in the model are defined while setting the project details but if you have not defined it at that time or you have an assembly of components to define material then follow the steps given next. Note that the options to apply material to solids will only be displayed when you have included heat conduction in solids.

- Right-click on the **Solid Materials** option from the **Flow Simulation Analysis Tree** at the left in the application window. A shortcut menu will be displayed.

Importing Material Data

- If you have defined material for objects while modeling them then click on the **Import Data From Model** option from the shortcut menu earlier displayed. The **Import Data From Model** tab will be added at the bottom in the application window; refer to Figure-30.
- Select the desired check boxes to import data and then click on the **Import** button. The material data will be imported. Close the tab by clicking on the **Close** button at top-right corner of the box.

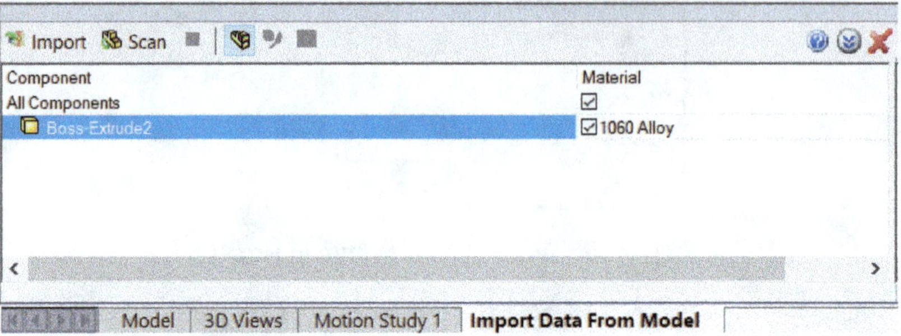

Figure-30. Import Data From Model tab

- If you have not specified any material while modeling then select the **Insert Solid Material** option from the shortcut menu after right-clicking on **Solid Materials** option in the **Flow Simulation Analysis Tree**. The **Solid Material PropertyManager** will be displayed; refer to Figure-31.

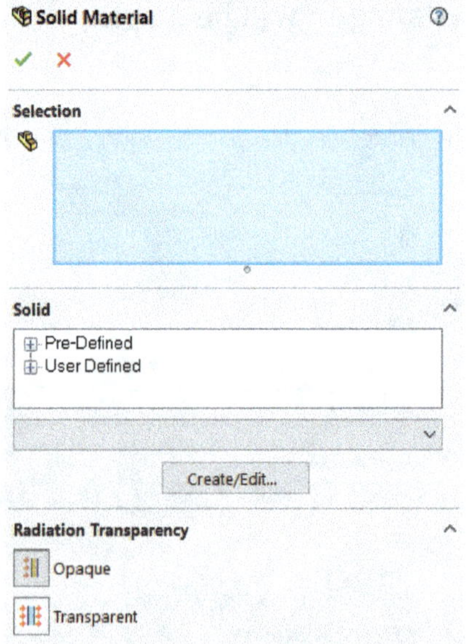

Figure-31. Solid Material PropertyManager

- Select the part from the model to which you want to apply material.
- Select the desired material from the **Solid** section of the **PropertyManager**.
- Set the desired radiation transparency by clicking on the button from the **Radiation Transparency** section of the **PropertyManager** if radiation is applicable in the analysis.
- Click on the **OK** button from the **PropertyManager** to apply solid material.

BOUNDARY CONDITIONS

The **Boundary Conditions** option is used to create flow inlet and outlet boundary conditions as well as wall conditions on selected fluid-contacting faces for both internal and external flow analyses. Also, thermal wall conditions can be created on selected external walls for internal flow analyses with enabled heat conductions in solid. The procedure to use this tool is discussed next.

- Right-click on the **Boundary Conditions** option from **Flow Simulation Analysis Tree**. The right-click shortcut menu will be displayed; refer to Figure-32.

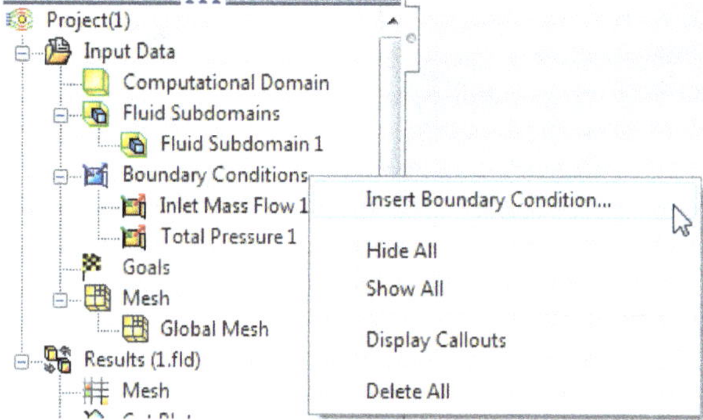

Figure-32. Right click shortcut menu for Boundary Conditions

- Click on the **Insert Boundary Condition** option from the displayed menu. The **Boundary Condition PropertyManager** will be displayed on the left of the screen; refer to Figure-34.
- You can also open the **Boundary Condition PropertyManager** by clicking on the **Boundary Condition** tool from **CommandManager**; refer to Figure-33.

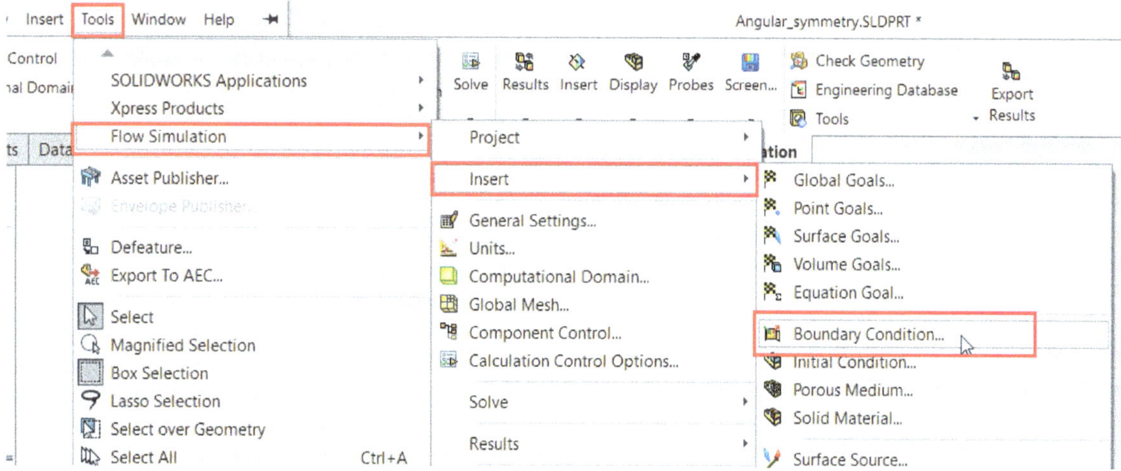

Figure-33. Boundary Conditions tool

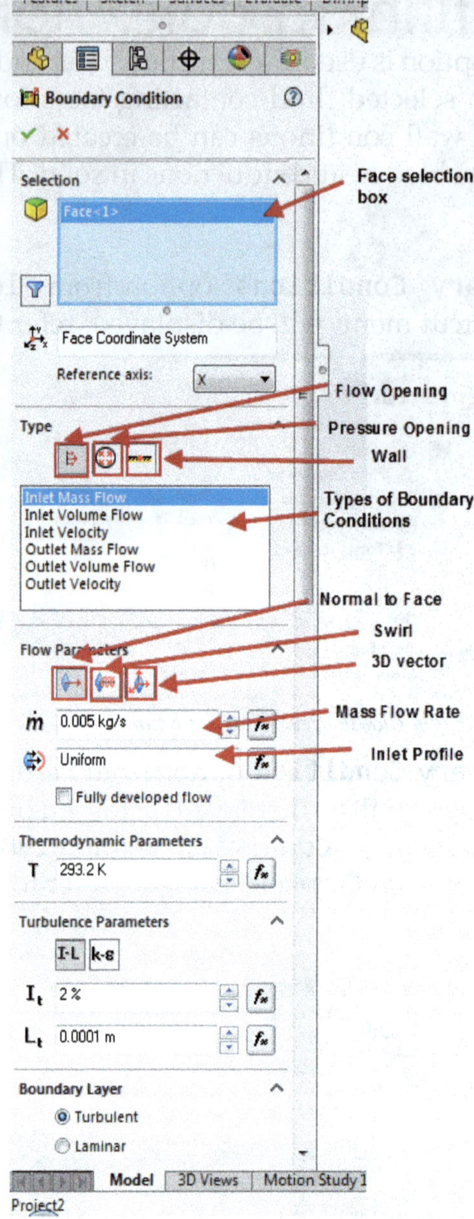

Figure-34. Boundary Conditions feature

- The selection box is active by default. You need to click on the inner face of the component to apply boundary conditions; refer to Figure-35.

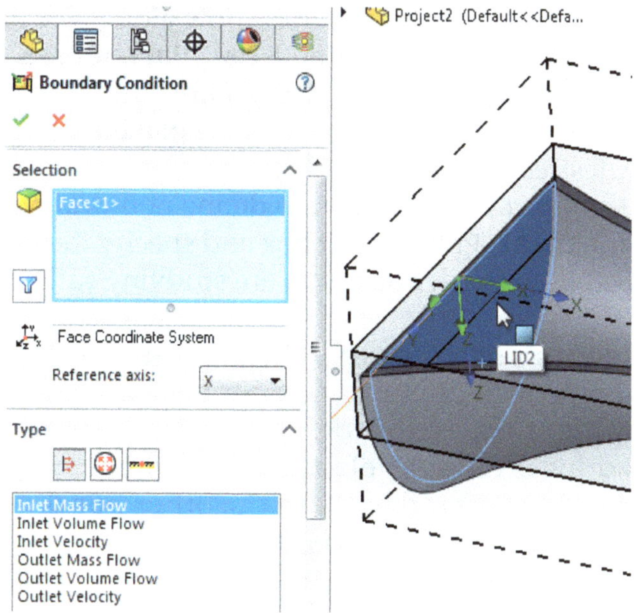

Figure-35. Selection of face for boundary condition

- Click in the **Coordinate System** selection box and select the desired coordinate system.
- Click on the **References axis** drop-down and select the required axis along which you want the fluid to flow.
- Click on the desired button from **Type** section of the **PropertyManager**. If you want to define the inlet or outlet flow of fluid from a lid then select the **Flow Opening** button. If you want to define pressure at an opening of the part then select the **Pressure Opening** button. If you want to define condition of fluid contacting wall in the model then select the **Wall** button from the section.

Specifying Flow Opening

- Click on the **Flow Opening** button from **Type** section to specify inlet or outlet flow of fluid.
- If you have activated Mach flow while defining the study then 9 options will be displayed in the **Type** list box viz. **Inlet Mass Flow**, **Inlet Mass Flux, Inlet Volume Flow**, **Inlet Velocity**, **Inlet Mach Number**, **Outlet Mass Flow**, **Outlet Volume Flow**, **Outlet Velocity**, and **Outlet Mach Number**. If you have not activated Mach flow then the options related to Mach number will not be displayed.

Specifying Inlet Mass Flow

- Select the **Inlet Mass flow** option if you want to specify the inlet fluid flow in mass per second.
- Select the desired button from **Normal to Face**, **Swirl** or **3D Vector** buttons. If you want the flow to be perpendicular to selected face then select the **Normal to Face** option. You will be asked to specify the value of **Mass Flow Rate** and **Inlet Profile**. Enter the desired value in **Mass Flow Rate** edit box and set the desired flow profile by clicking on the **Dependency** button. On selecting the **Dependency** button, the **Dependency** dialog box is displayed. The options of this dialog box have been discussed earlier.
- If you want the fluid at inlet to swirl while entering the analysis volume then select the **Swirl** button from **Flow Parameters** section of the **PropertyManager**. The **Angular Velocity** and **Radial Velocity** edit boxes will be displayed along with **Mass Flow Rate** edit box and **Inlet Profile** option. Set the desired parameters in the edit boxes.

- If you want to define three components of mass flow along X, Y and Z axes individually, then select the **3D Vector** button from the **Flow Parameters** section. The **Relative Component in X Direction**, **Relative Component in Y Direction**, and **Relative Component in Z Direction** edit boxes will be displayed along with **Mass Flow Rate** edit box. Specify the desired values to define flow.
- By default the flow is uniform. If you want to define non-uniform flow then click on the **Dependency** button for **Inlet Profile** edit box and specify the desired parameters. Click on the Fully developed flow check box if you are studying fully developed flow.

Fully developed flow occurs when the viscous effects due to the shear stress between the fluid particles and pipe wall create a fully developed velocity profile. In order for this to occur the fluid must travel through a length of a straight pipe. In addition, the velocity of the fluid for a fully developed flow will be at its fastest at the center line of the pipe (equation 1 laminar flow). On the other hand, the velocity of the fluid at the walls of the pipe will theoretically be zero. As a result, fluid velocity should be expressed as an average velocity.

- Expand the **Thermodynamic Parameters** rollout if you want to specify the temperature and pressure parameters for inlet flow (if Pressure Opening button is selected). Click in the **Temperature** edit box of **Thermodynamic Parameters** section and specify the value of temperature of fluid at inlet. Similarly, specify the pressure value in **Pressure** edit box as desired. If **Rotation** is activated for the analysis then **Pressure potential** check box will be displayed in the rollout. Select the **Pressure potential** check box to calculate the pressure relative to rotating reference frame using the formula:

$$P_r = P_{abs} - \frac{1}{2}\rho\omega^2 \cdot r^2$$

Here, P_r is pressure relative to rotating frame, P_{abs} is the absolute pressure, ρ is the density, ω is angular speed, and r is radius of rotating frame.

- Expand the **Substance Concentration** section of the **PropertyManager** to define the ratio of different fluids at the inlet if you have included more than one fluid in your analysis. The edit boxes for each of the fluid will be displayed. Set the desired ratio between different fluids in their respective edit boxes; refer to Figure-36. Note that you can define the concentration both Mass-wise or Volume-wise by selecting the respective radio button from the section.

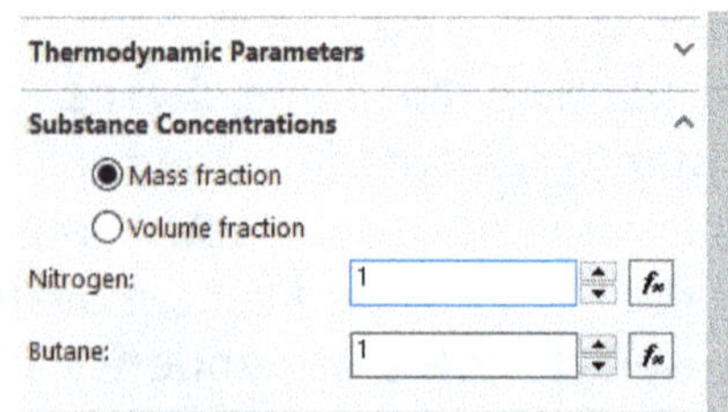

Figure-36. Substance Concentrations rollout

- Set the other parameters as discussed earlier.
- After specifying the parameters, click on the **OK** button from **Boundary Conditions** option. The boundary condition will be added in **Flow Simulation Analysis Tree**.
- Similarly, you can set other Inlet fluid flow boundary conditions.

Specifying Inlet Mass Flux

The inlet mass flux is applied when you have value of rate of mass flow per unit area. It is product of density and volume, given by the unit kg/(s*mm^2). On selecting the **Inlet Mass Flux** option from the **Type** rollout, the **Mass Flux** edit box will be displayed in the **Flow Parameters** rollout; refer to Figure-37. Specify the desired value in edit box and set the other parameters as desired.

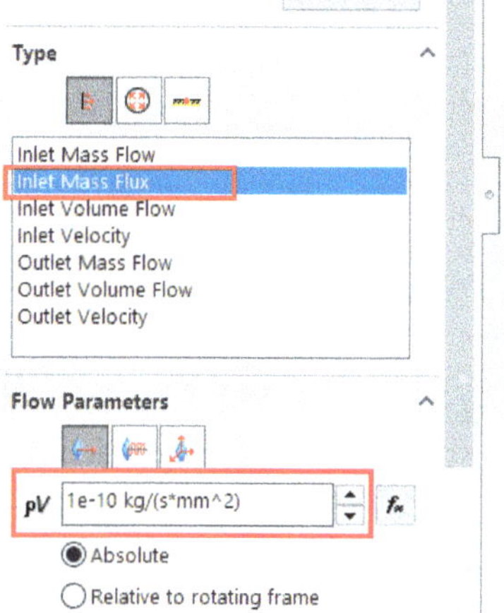

Figure-37. Specifying mass flux

Specifying Volume Flow

Select the **Inlet Volume Flow** option from the list in **Type** rollout if volume per second or liter per second is known for the inlet. On selecting this option, the **Volume Flow Rate** and **Inlet Profile** options will be displayed in the **Flow Parameters** rollout; refer to Figure-38. Specify the desired value of volumetric flow rate in the edit box and set the other parameters as discussed earlier.

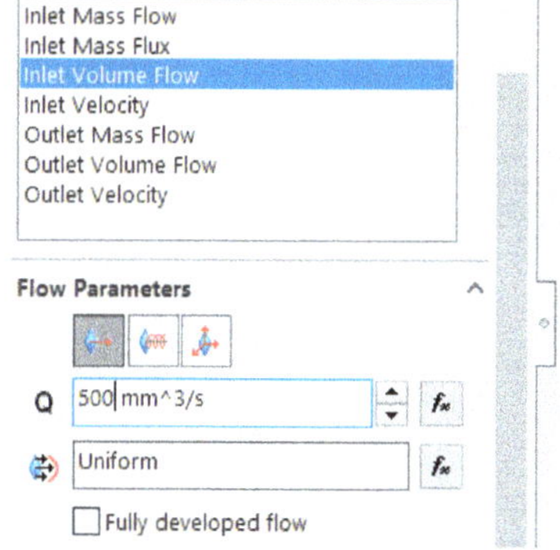

Figure-38. Inlet Volume Flow options

Specifying Inlet Velocity

The **Inlet Velocity** option from the **Type** rollout if you want to specify only the velocity of fluid at the inlet. The options will be displayed as shown in Figure-39. Specify the desired values as discussed earlier.

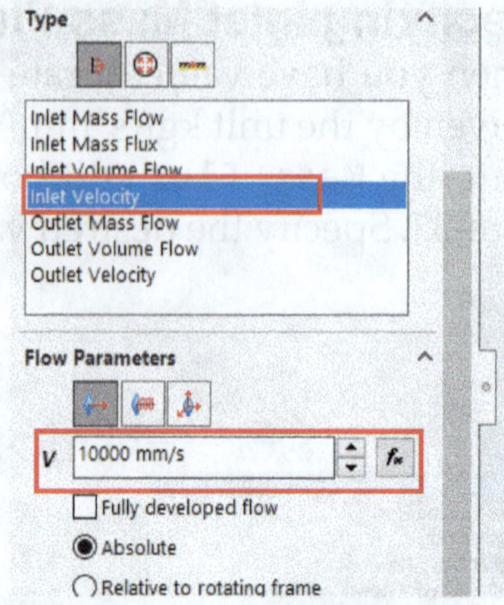

Figure-39. Inlet Velocity option

Specifying Outlet Mass Flow

- Select the **Outlet Mass Flow** option from the **Type** list box after selecting the face. The **PropertyManager** will be displayed as shown in Figure-40.

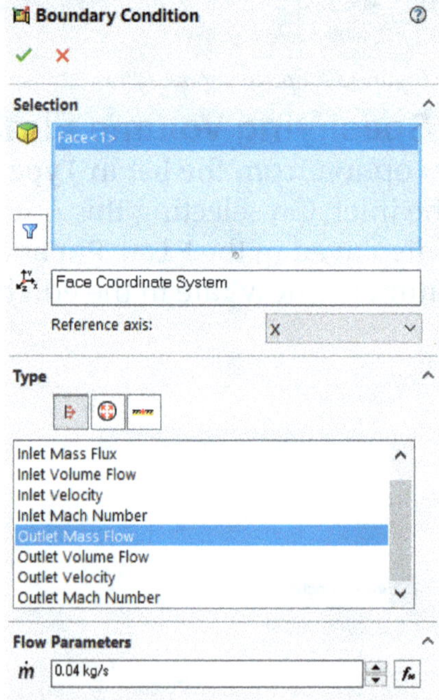

Figure-40. Boundary Condition Property Manager with Outlet Mass Flow options

- Specify the desired value of mass flow rate in the **Mass Flow Rate** edit box.
- In the **Goals** rollout, select check boxes for parameters to be associated as goal with outlet. These parameters will be used to find out result of analysis. For example, if you have selected **Static Pressure Av** check box then the analysis will terminate when value of static pressure average becomes almost same between two iterations.
- Click on the **OK** button from the **PropertyManager**. You can set the other outlet flow parameters in the same way.

Specifying Pressure Opening

- Click on the **Pressure Openings** button from the **Type** section of the **PropertyManager** if you want to specify pressure parameter.
- Select the face for which you want to specify the pressure.

- Select the desired option from the **Type** list box. Select the **Environment Pressure** option if you want to set environmental pressure at select the selected face. Note that setting the environmental pressure defines total pressure at inlet and static pressure at outlet of fluid. Select the **Static Pressure** option if you want to specify a fixed pressure at the selected face during the whole course of calculations. Select the **Total Pressure** option if you want to define a combination of static pressure and dynamic pressure at the selected face.
- After selecting the desired option, specify the related values in the **Thermodynamic Parameters** section of the **PropertyManager**; refer to Figure-41.
- Set the other parameters as discussed earlier.

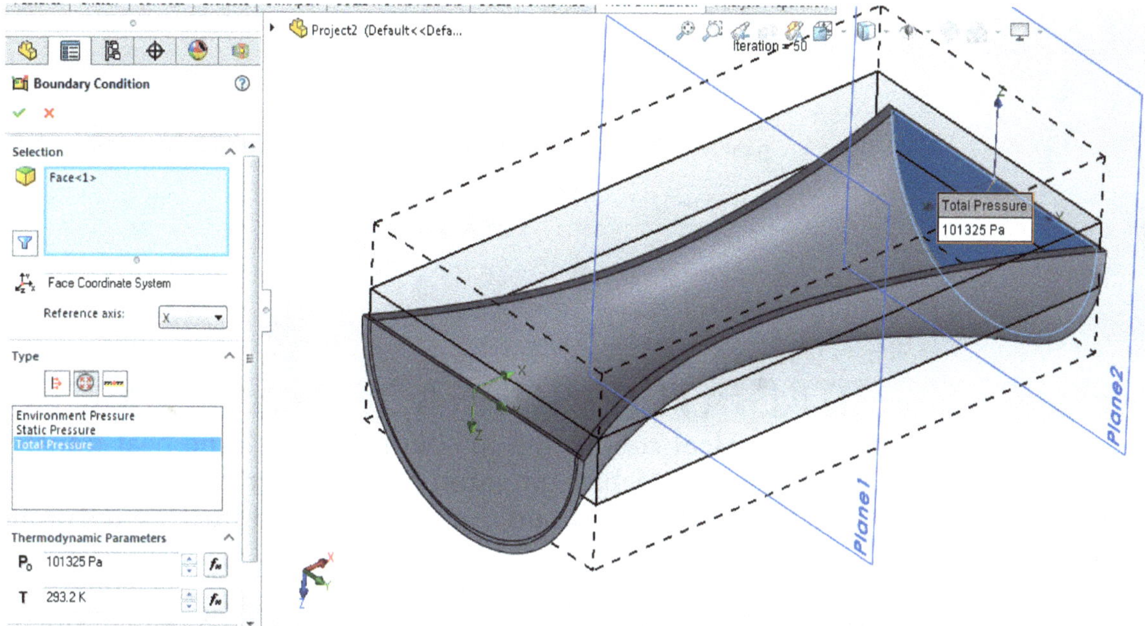

Figure-41. Creating boundary condition for other end

Specifying Wall Boundary Condition

- Click on the **Wall** button from the **Type** section. The options in the **PropertyManager** will be displayed as shown in Figure-42.
- Select the desired option from the **Type** list box. Select the **Real Wall** option if you want to specify temperature, heat transfer coefficient, and roughness of the selected face. Select the **Ideal Wall** option if you want to create an adiabatic frictionless wall using the selected face. Select the **Outer Wall** option if you want to outer wall temperature or heat transfer coefficient together with the external fluid temperature for the selected external walls in an internal flow analysis with Heat conduction in solids. Note that if you have enabled rotation for analysis then **Stator** check box and **Wall Motion** rollout will be displayed in the **PropertyManager**. Select the **Stator** check box if you want to make the wall stationary for analysis or specify the parameters related to wall movement during analysis in the **Wall Motion** rollout.
- After selecting the desired option, specify the respective parameters in the edit boxes of this section.
- Click on the **OK** button from the **PropertyManager** to apply boundary condition. Note that you can click on the **Pin** button to keep **PropertyManager** visible and apply multiple boundary conditions to different faces.

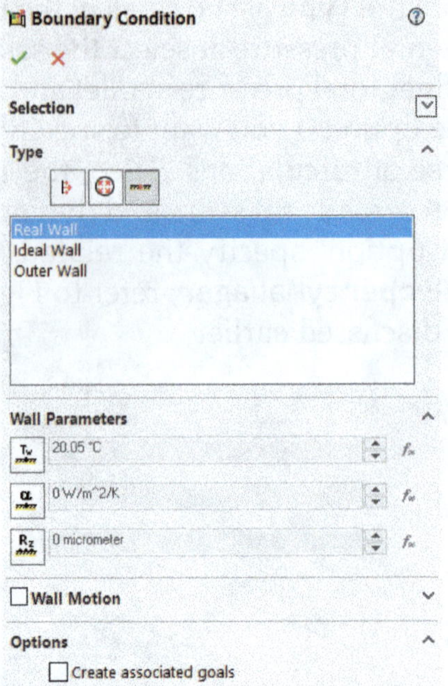

Figure-42. Boundary Condition Property Manager with Wall options

RADIATIVE SURFACES

If you solve a problem involving Heat conduction in solids, in which the heat transfer by radiation from or between solids is noticeable compared to the heat transfer by convection, you can use the radiative surface option to mark such faces. The procedure to define radiative surface is given next.

- Right-click on the **Radiative Surfaces** option from the **Flow Simulation Analysis Tree** and select the **Insert Radiative Surface** option; refer to Figure-43. The **Radiative Surface PropertyManager** will be displayed; refer to Figure-44. You can also select the tool from Menu bar; refer to Figure-45.

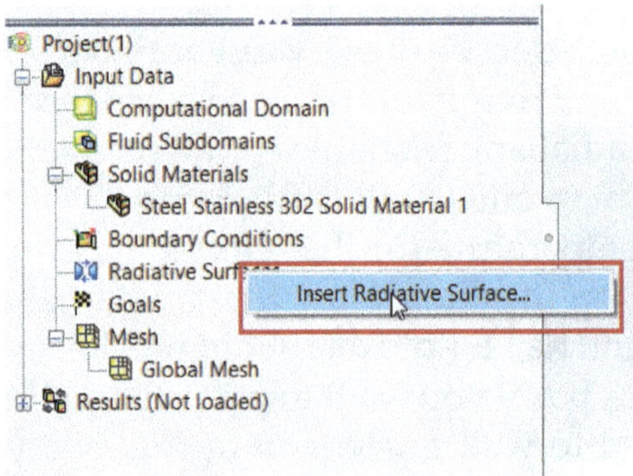

Figure-43. Right-click shortcut menu

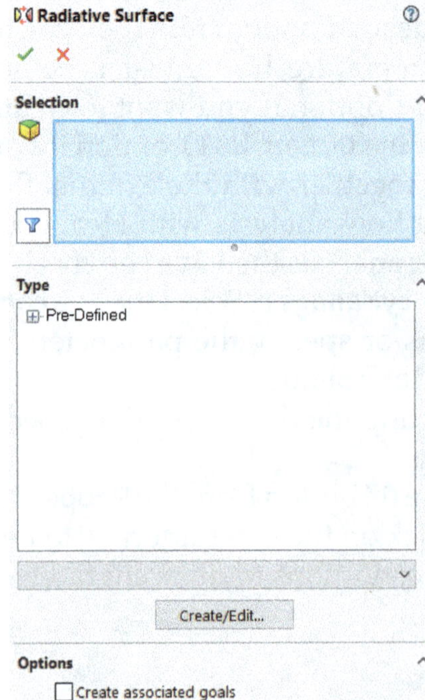

Figure-44. Radiative Surface Property Manager

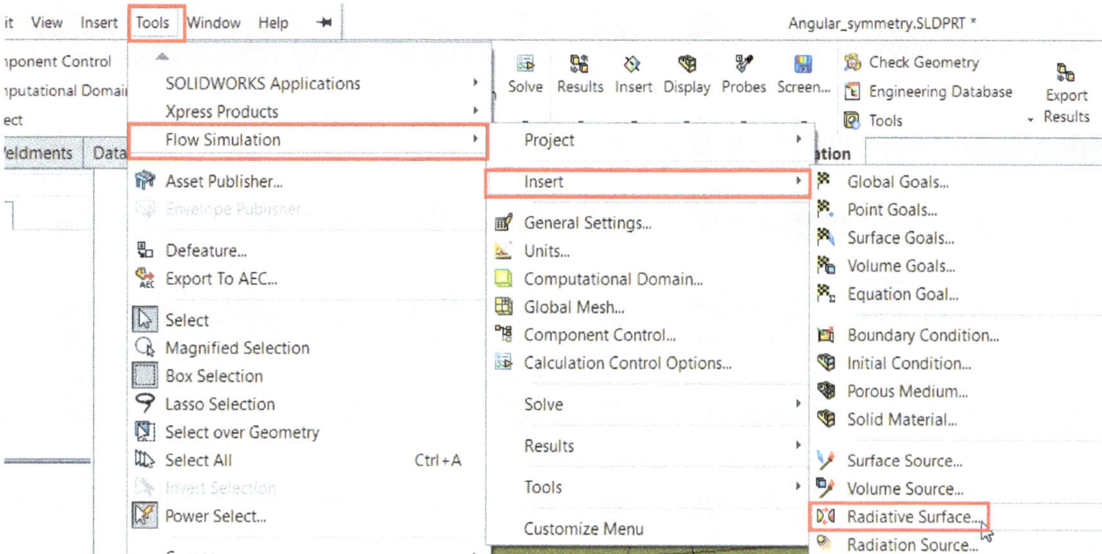

Figure-45. Radiative Surface tool

- Select the face(s) to which you want to assign radiative properties.
- Expand the **Pre-Defined** category from the **Type** selection box and select the desired option. Major types of radiative surfaces are given next.

Absorbent Wall: Absorbent wall is a type of surface which absorbs all the coming radiation and does not radiate it back.

Blackbody Wall: Blackbody wall is a type of surface which has surface emissivity of 1. This means all the coming radiation is fully absorbed and heat is emitted by Stefan-Boltzmann law.

Non-radiating surface: Non-radiating surfaces are those surfaces that do not participate in the radiation heat transfer, i.e. neither emits nor absorbs heat radiation. So rays arriving at the non-radiating surface will have no influence on the wall temperature (also the wall is not transparent so rays stopped there) and the absence of starting rays means that for all other walls the non-radiating surface is treated as a wall with zero temperature and all incoming energy is disappeared.

Real Surfaces: The Real Surfaces are those of real materials which have different emissivities based on their material properties.

Symmetry: Select the **Symmetry** option if radiative surface represents a surface of ideal reflection (a mirror) with reflectivity r = 1 and Secularity coefficient f_s = 1. If you use the **Ideal Wall** condition on a wall to specify the problem symmetry plane and the radiative heat transfer is considered in the project then you must specify the Symmetry radiative surface at this wall.

Whitebody Wall: Select the **Whitebody wall** option if the surface emissivity is equal to 0 (i.e. to that of whitebody), so that the surface fully reflects all the incident radiation (in accordance with the Lambert law) and the surface temperature does not influence the radiative heat transfer.

- After setting the desired parameters, click on the **OK** button to create radiative surface.

GOALS

The Flow Simulation contains built-in criteria to stop the solution process, but it is best to use your own criteria, which are named Goals. You need to specify the **Goals** as physical parameters of interest in your project, so that their convergence can be considered as obtaining a steady state solution from the engineering viewpoint. Note that Goal Convergence is one of the conditions for finishing the calculation.

Insert Global Goals

The **Insert Global Goals** button is used to specify global goals for your project. A global goal is a physical parameter calculated in the entire computational domain. The procedure to use this tool is discussed next.

- Right-click on the **Goals** option from the **Flow Simulation Analysis Tree**. The right-click shortcut menu will be displayed; refer to Figure-46.
- Click on the **Insert Global Goals** button from the displayed menu. The **Global Goals PropertyManager** will be displayed; refer to Figure-47.

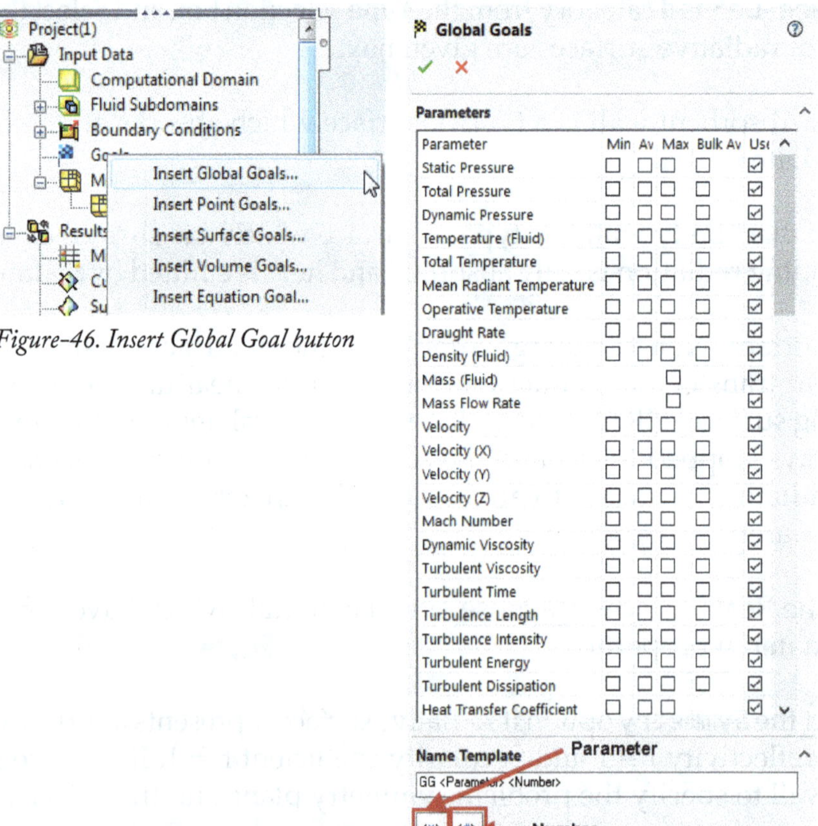

Figure-46. Insert Global Goal button

Figure-47. Global Goal feature

- Select the desired check box in **Parameters** section to define convergence condition for respective parameter in the list. You can use minimum, maximum, average, or bulk average of selected parameter as convergence condition.
- Click in the **Name Template** edit box and enter the desired parameter and number.
- Click on the **Parameter** button at the bottom of the **Global Goals PropertyManager**. The parameter will be added in **Name Template** edit box. Edit the parameter as required by using the edit box.

- Click on the **Number** button from **Global Goals** option. The number will be added in **Name Template** edit box. Edit the number as required.
- After specifying the parameters from **Global Goals** option, click on the **OK** button. The global goal will be added in the **Flow Simulation Analysis Tree**; refer to Figure-48.

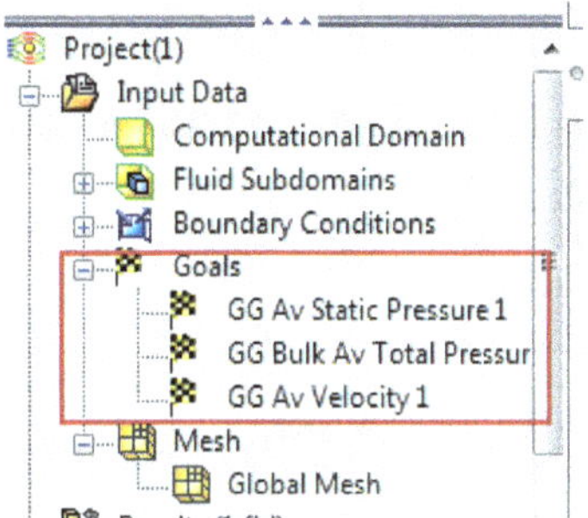

Figure-48. Added goals

- If you want to edit one of the created feature, right-click on the required feature. The right-click shortcut menu will be displayed; refer to Figure-49.

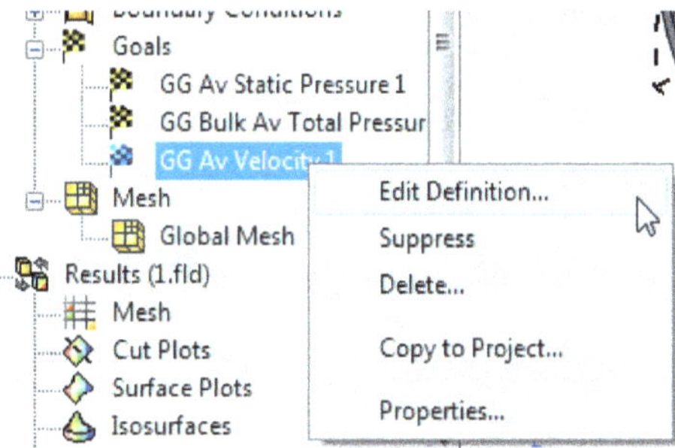

Figure-49. Editing parameter

- Click on the **Edit Definition** button. The **Global Goal PropertyManager** will be displayed; refer to Figure-50.

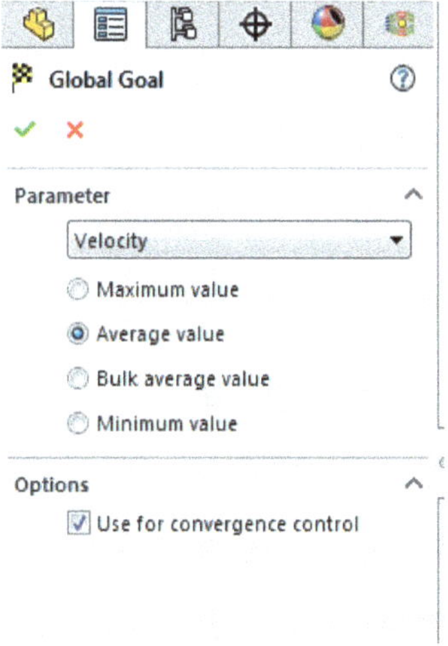

Figure-50. Global Goal edit feature

- Click on the **Parameter** drop-down and select the parameter.
- Click on the desired radio button to select the value.
- Select the **Use for convergence control** check box from **Options** section to include this goal as a condition of convergence for finishing the calculation.
- After specifying the parameters, click on the **OK** button from the **Global Goal** edit feature. The parameters will be added.

Insert Point Goals

The **Insert Point Goals** option is used to specify point goals for your project. A point goal is a physical parameter value calculated at selected point. The point can be specified either by reference or by explicit definition of its coordinates in the **Global Coordinate System**. The procedure to use this is discussed next.

- Right-click on the **Goals** option from **Flow Simulation Analysis Tree**. The right-click shortcut menu will be displayed; refer to Figure-51.

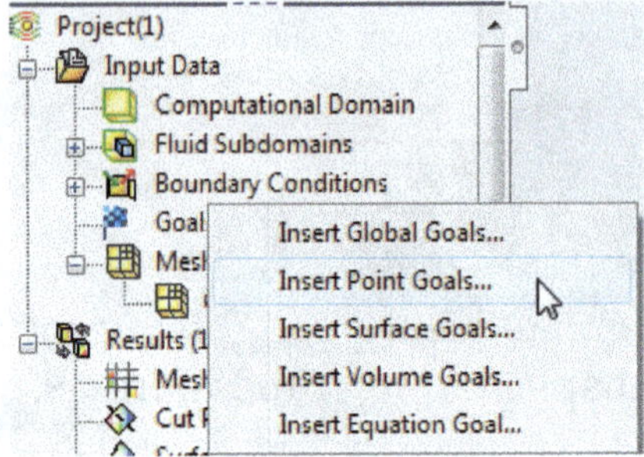

Figure-51. Insert Point Goals button

- Click on the **Insert Point Goals** button. The **Point Goals PropertyManager** will be displayed; refer to Figure-52.

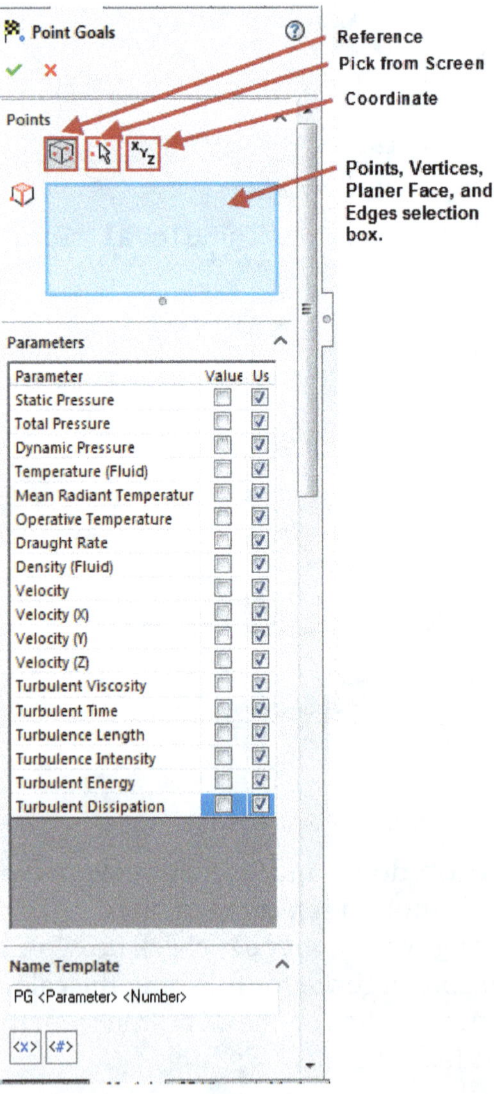

Figure-52. Point Goals feature

- Click on the **Reference** button from **Points** section to define a point by selecting geometrical features.
- Click on the **Pick from Screen** button from **Points** section to select points on a planer face in the graphics area.
- Click on the **Coordinates** button from **Points** section to specify a set of points by creating a table with points coordinates.
- After selecting the required button, specify the parameters listed under selected button.
- Click on the required parameter check boxes from **Parameters** section to select.
- After specifying the parameters, click on the **OK** button from the **Point Goals PropertyManager**. The goals will be added in **Flow Simulation Analysis Tree**.

Insert Surface Goals

The **Insert Surface Goals** option is used to specify surface goals for a project or component. A surface goal is a physical parameter calculated on the selected surfaces. The procedure to use this tool is discussed next.

- Right-click on the **Goals** option from **Flow Simulation Analysis Tree**. The right-click shortcut menu will be displayed; refer to Figure-53.
- Click on the **Insert Surface Goals** button from the displayed menu. The **Surface Goals PropertyManager** will be displayed; refer to Figure-54.
- You can also open the **Surface Goals PropertyManager** by clicking on the **Surface Goals** tool from the **CommandManager**.

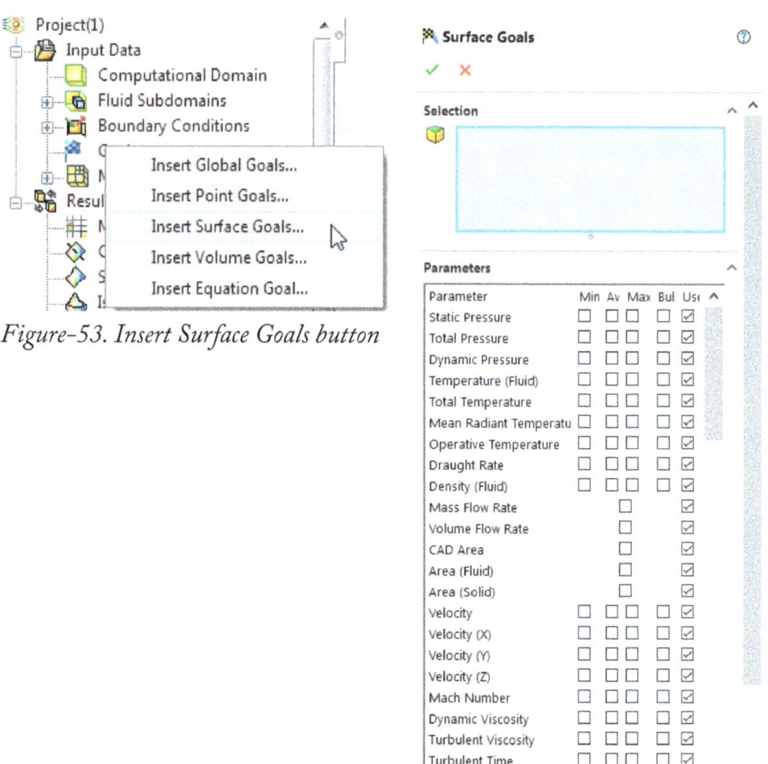

Figure-53. Insert Surface Goals button

Figure-54. Surface Goals feature

- The **Selection** box is active by default. You need to click on the faces from component which you want to check for goals.
- Click on the required parameter check box from **Parameters** section to select.
- Other tools of this feature were discussed earlier.

- After specifying the parameters, click on the **OK** button from **Surface Goals PropertyManager**. The selected parameter will be added in the **Flow Simulation Analysis Tree**.

Insert Volume Goals

The **Insert Volume Goals** button is used to specify volume goals of your project. A volume goal is a physical parameter calculated within specified volumes (part or subassembly components in assemblies, as well as bodies in multi-body parts) inside the Computational Domain. The procedure to use this button is discussed next.

- Right-click on the **Goals** option from **Flow Simulation Analysis Tree**. The right-click shortcut menu will be displayed; refer to Figure-55.
- Click on the **Insert Volume Goals** option from the displayed menu. The **Volume Goals PropertyManager** will be displayed; refer to Figure-56.

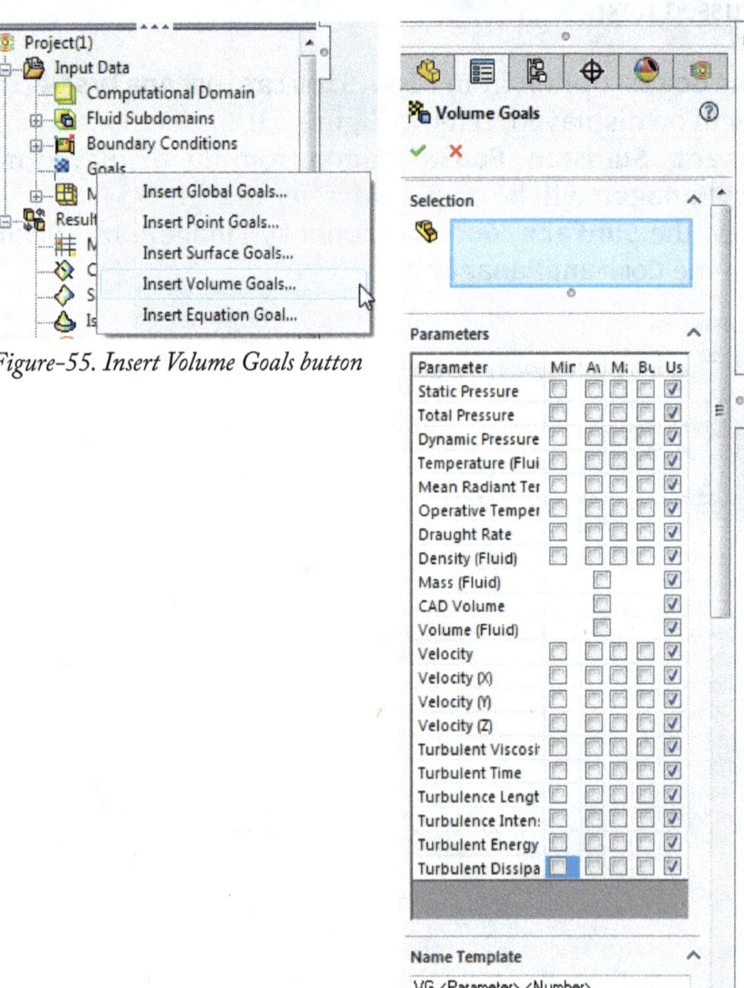

Figure-55. Insert Volume Goals button

Figure-56. Volume Goals feature

- The **Selection** box is active by default. You need to click on the face of component to select bounded volume.

- Click on the required parameter check box from **Parameters** section to select the volume goals.
- Other options of this **PropertyManager** are same as discussed earlier.
- After specifying the parameters, click on the **OK** button from **Volume Goals PropertyManager**. The selected parameter will be added in the **Flow Simulation Analysis Tree**.

Insert Equation Goals

The **Insert Equation Goals** option is used to specify a goal based on mathematical equations with the parameters like boundary conditions, fans, initial conditions, and so on. This goal can be viewed as an equation goal during the calculation and while displaying results in the same way as the other goals. The procedure to use this tool is discussed next.

- Right-click on the **Goals** option from **Flow Simulation Analysis Tree**. The right-click shortcut menu will be displayed; refer to Figure-57.

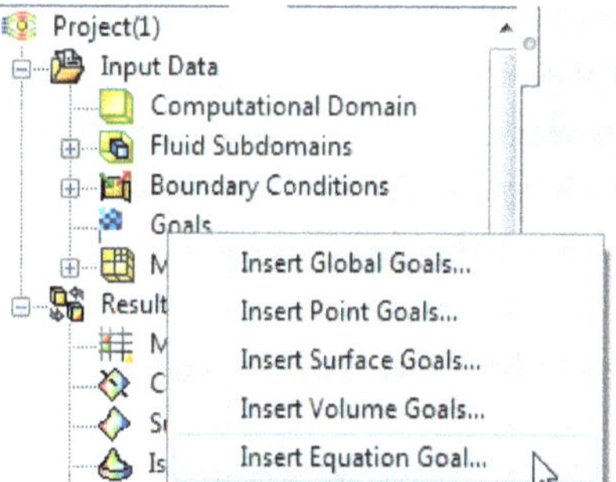

Figure-57. Insert Equation Goals button

- Click on the **Insert Equation Goal** option from the displayed menu. The **Equation Goal** tab will be displayed at the bottom of the screen; refer to Figure-58.

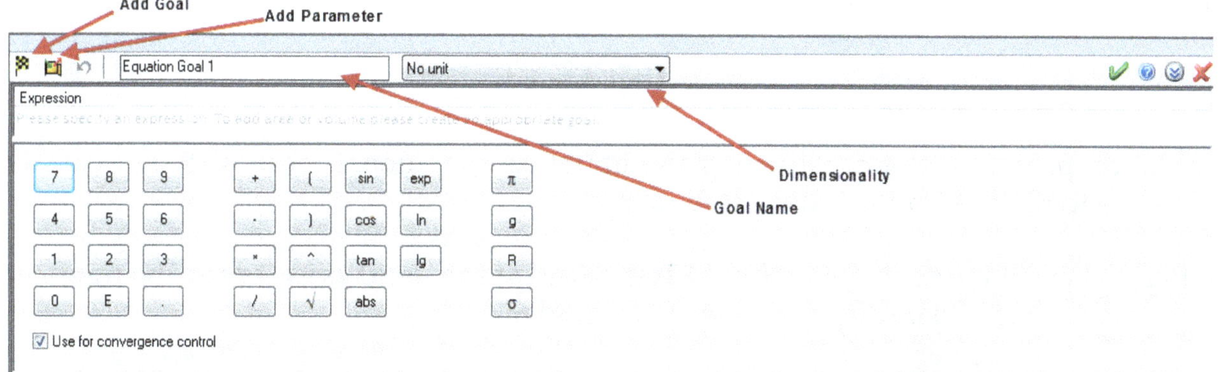

Figure-58. Equation Goal tab

- Click on the **Add Goal** button to add a previous specified goal in the expression. The **Add Goal** box will be displayed in the right of the tab; refer to Figure-59. Double-click on the desired goal value to add it in the expression.

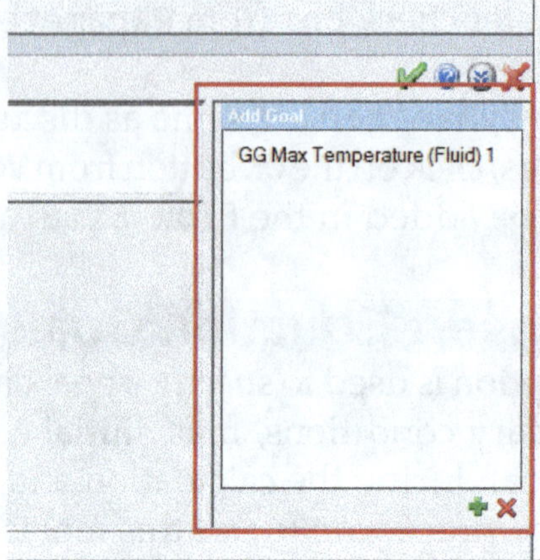

Figure-59. Add Goal box

- Click on the **Add Parameter** button to add a parameter. The **Add Parameter** box will be displayed in the right of the tab. Double-click on the desired parameter to add it in expression.
- Click in the **Goal Name** edit box and enter the name of the goal as required.
- Click on the **Dimensionality** drop-down and select the desired unit.
- Click in the **Expression** edit box from **Equation Goal** tab and enter the desired expression of equation. You can also use the calculator keys from **Equation Goals** tab to write the expression.
- After specifying the parameter, click on the **OK** button from **Equation Goal** tab.

MESH

Meshing is the process of breaking down the model into finite number of small elements. The **Mesh** feature in SolidWorks Flow Simulation is used to specify parameters governing the automatic procedure of constructing the initial computational mesh. There are various buttons or option under Mesh feature which are discussed next.

Global Mesh

The **Global Mesh** button is used to change the parameters governing the automatic Flow Simulation procedures of constructing the initial computational mesh of the model. The procedure to use this option is discussed next.

- Right-click on the **Mesh** option from **Flow Simulation Analysis Tree**. The right-click shortcut menu will be displayed; refer to Figure-60.

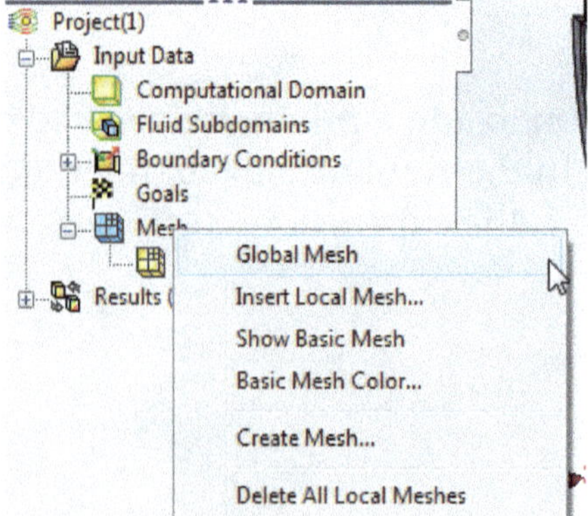

Figure-60. Global Mesh button

- Click on the **Global Mesh** button from the displayed menu. The **Global Mesh Settings PropertyManager** will be displayed; refer to Figure-61.

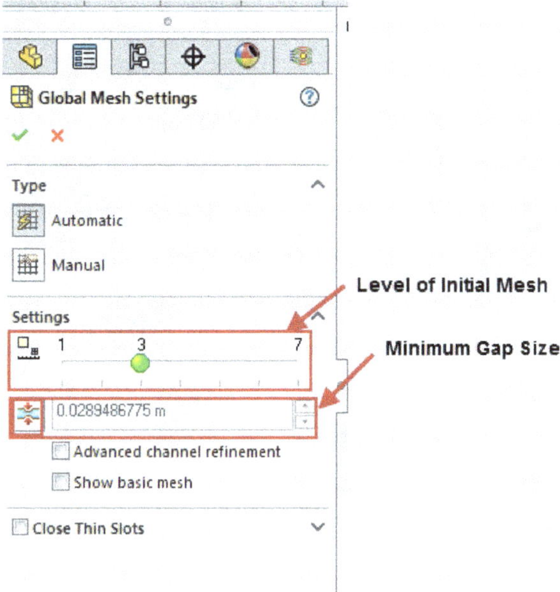

Figure-61. Global Mesh Settings Feature box

- You can also open the **Global Mesh Settings PropertyManager** by clicking on the **Global Mesh** tool from **Flow Simulation** cascading menu in the **Tools** menu; refer to Figure-62

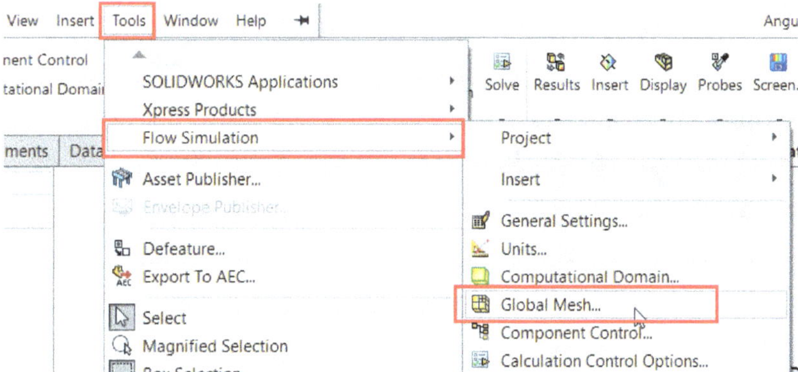

Figure-62. Global Mesh tool

Automatic

- Click on the **Automatic** button from **Type** section to automatically adjust the values of global mesh.
- Move the **Level of initial mesh** slider to adjust the number of basic mesh cells and the default procedure of mesh refining in the model's narrow channels. A higher value produces more fine cells but it will consume more CPU and computer memory.
- Click on the **Minimum Gap Size** button and enter the desired value in the edit box to specify minimum gap up to which the mesh elements can be created automatically.
- Select the **Advanced channel refinement** check box from **Setting** rollout for refinement of narrow channels during the flow.
- Select the **Show basic mesh** check box of **Settings** rollout to view the basic mesh on the component; refer to Figure-63.

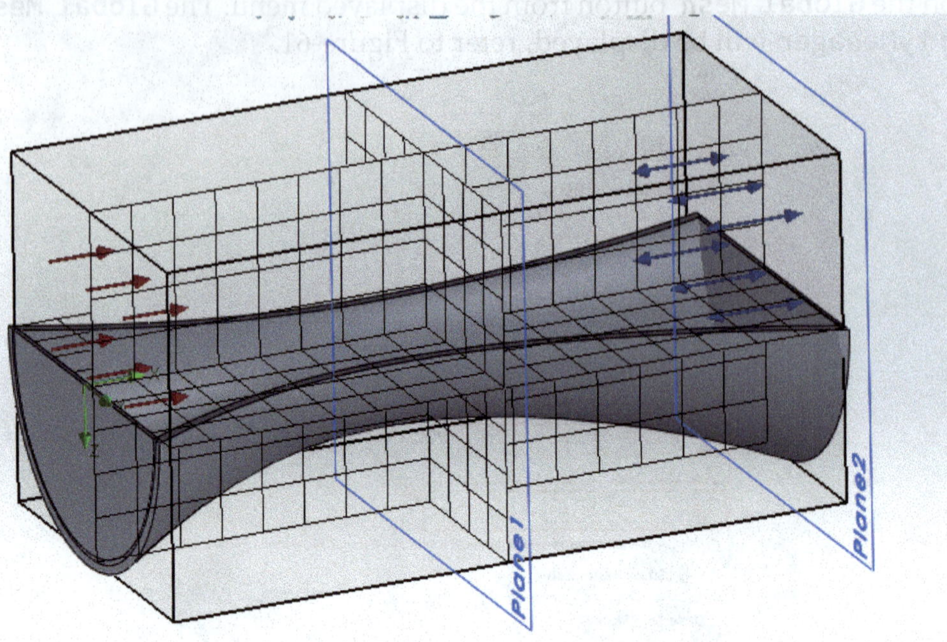

Figure-63. Basic Mesh

- Select the **Close Thin Slots** check box to close the thin slots of component which are of the size equal to or below the value specified in the **Maximum Height of Slots to Close** edit box.

Manual

- Click on the **Manual** button from **Type** section to manually set the value of mesh. The options of **Manual** button will be displayed; refer to Figure-64.

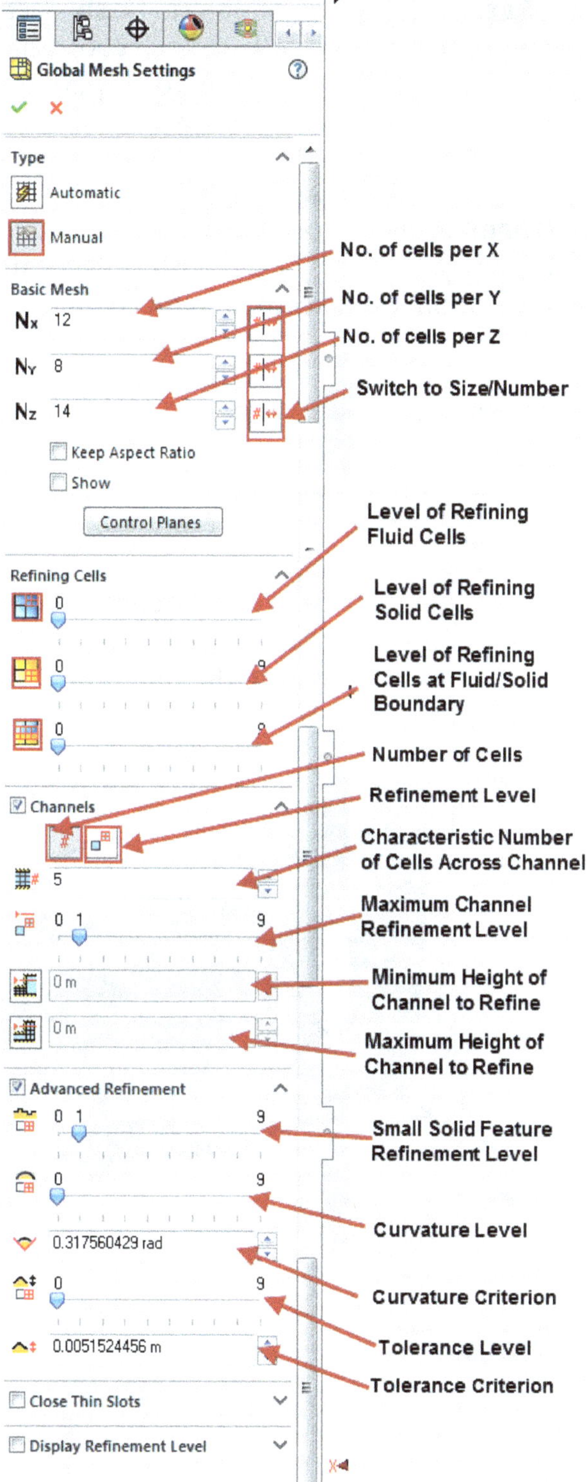

Figure-64. Options under Manual button

- Click on the **Number of Cells Per X** edit box from **Basic Mesh** section and enter the value of number of cells along X axis.
- Click on the **Number of Cells Per Y** edit box from **Basic Mesh** section and enter the value of number of cells along Y axis.
- Click on the **Number of Cells Per Z** edit box from **Basic Mesh** section and enter the value of number of cells along Z axis.
- Click on the **Switch to Size/Number of Cells Per X**,**Y**,or **Z** button from **Basic Mesh** section to switch between size of cells and number of cells.

- Select the **Keep Aspect Ratio** check box of **Basic Mesh** section to maintain the ratio between the numbers(sizes) of basic mesh cells in each coordinate direction.
- Select the **Show** check box of **Basic Mesh** section to display the obtained basic mesh of the model.

Control Planes

- Click on the **Control Planes** button from **Basic Mesh** section to rearrange the basic mesh planes and to stretch or contract the basic mesh cells locally. The **Control Planes** tab will be displayed at the bottom of the screen; refer to Figure-65.

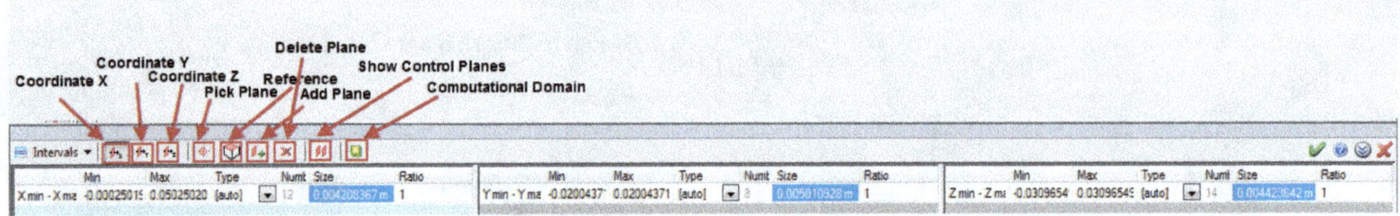

Figure-65. Control Planes tab

- Click on the **Coordinate X**, **Coordinate Y** or **Coordinate Z** button from **Control Planes** tab and specify the required parameter in the respective field.
- You can also create control planes by dragging the red, blue and green arrows displayed in the graphic area; refer to Figure-66.

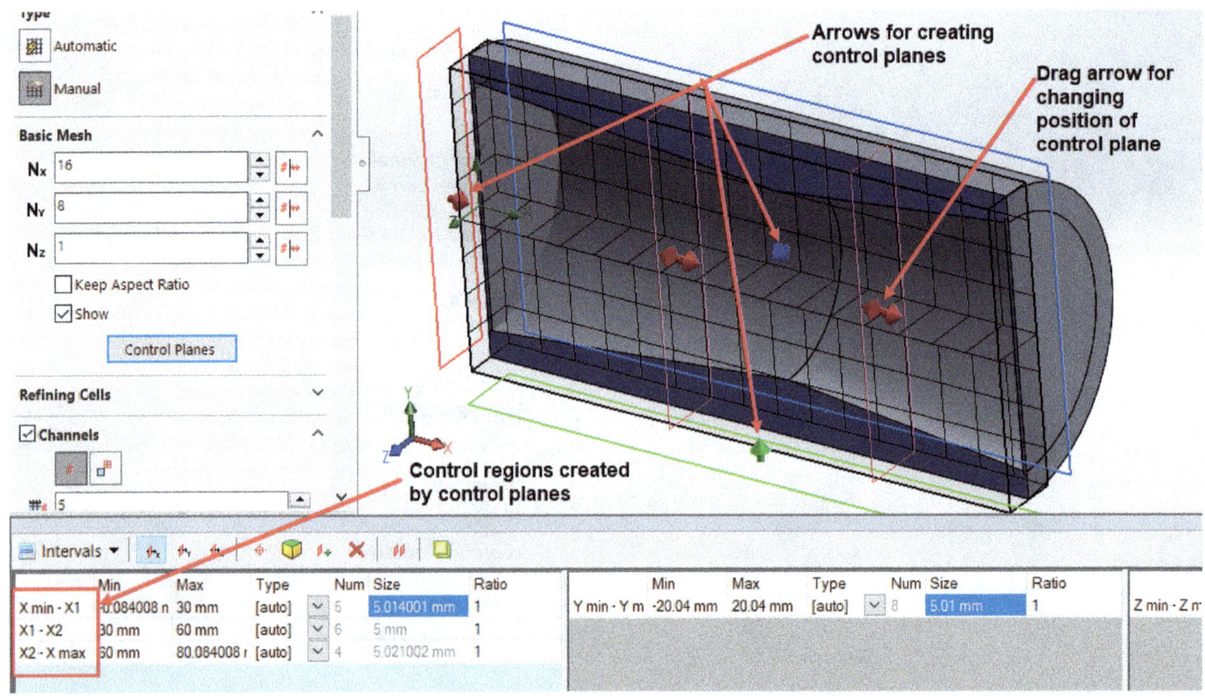

Figure-66. Creating control planes

- Click in the **Type** drop-down of each control section from the **Control Planes** tab and specify desired number of cells or size of cells to modify mesh; refer to Figure-67.
- After specifying the parameter, click on the **OK** button from **Control Planes** tab. You will return to **Global Mesh Settings** feature box.

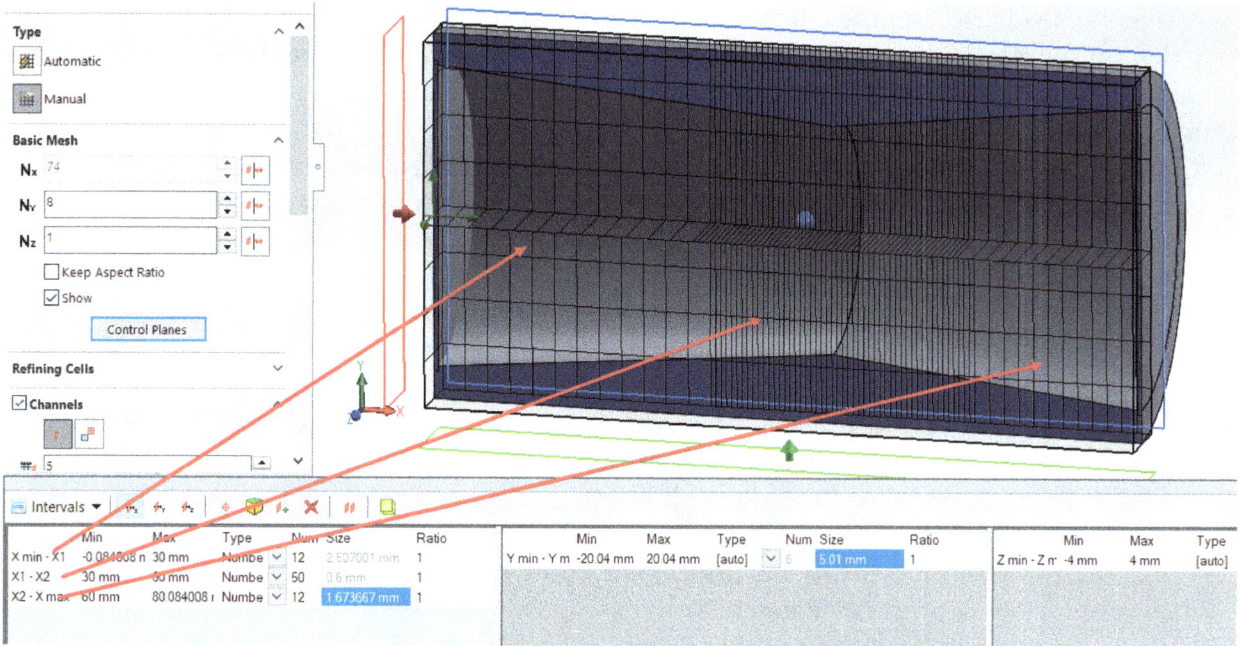

Figure-67. Mesh refinement by control planes

Refining Cells rollout options

- Move the **Level of Refining Fluid Cells** slider from **Refining Cells** to refine the mesh cells of fluid.
- Move the **Level of Refining Solid Cells** slider from **Refining Cells** to refine the mesh cells of solid.
- Move the **Level of Refining Cells at Fluid/Solid Boundary** slider from **Refining Cells** to refine the boundary mesh of solid or fluid.

Note that parameters specified in this rollout will change the cell sizes universally.

Channels Refinement options

The options in **Channels** rollout are used to refine mesh at channel like grooves or protrusions in the model through which fluid can pass; refer to Figure-68.

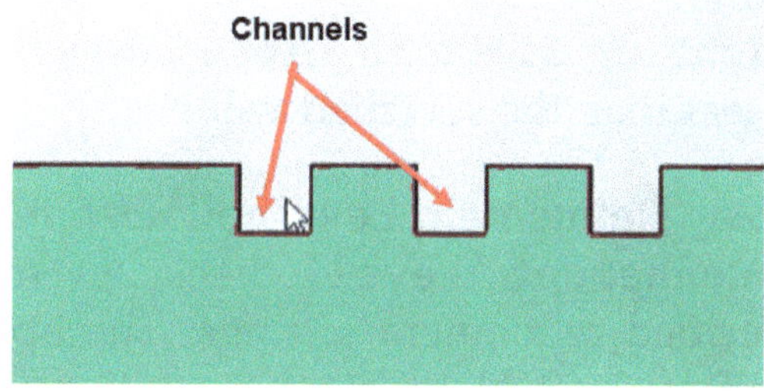

Figure-68. Channel shapes

- Click on the **Number of Cells** button from **Channels** section to specify the parameter of cells at channels. Channels are small slot type sections where fluid and solid come into contact.
- Click in the **Characteristic Number of Cells Across Channel** edit box and specify the value of number of cells across the channels up to which refinement will be applicable.

- Move the **Maximum Channel Refinement Level** slider for refinement of cells at channels.
- Click on the **Minimum Height of Channel to Refine** button from **Channels** section and enter the value to specify the minimum height of channel which qualify for refinement in the respective edit box.
- Click on the **Maximum Height of Channel to Refine** button from **Channels** section and enter the value to specify the maximum height of channel that qualify for refinement in the respective edit box.

Advanced Refinement rollout options

- Move the **Small Solid Feature Refinement Level** slider in **Advanced Refinement** section to capture relatively small features at the boundary between substances (fluid/solid, fluid/porous, solid/porous interfaces or boundary between different solids) with a denser mesh by specifying the refinement level. Example of small features can be small protrusions of the solid or grooves with lower thickness.
- Move the **Curvature Level** slider from **Advanced Refinement** section to resolve the substance interface curvature like fluid/solid, fluid/porous, and porous/solid interfaces. If the boundary of contact is curved the increasing this value will generate more mesh cells.
- Click in the **Curvature Criterion** edit box from **Advanced Refinement** section to define a critical angle above which the mesh level will be as specified in the **Curvature Level** slider. The curvature angle value is angle between normal vectors of two consecutive segments of base mesh cells at the curve.
- Move the **Tolerance Level** slider from **Advanced Level** section to restrict the smallest size of the cells, where the **Tolerance Criterion** is satisfied, by specifying the refinement value and level.
- Click in the **Tolerance Criterion** edit box from **Advanced Level** section to specify the value of tolerance criterion. It is the angular difference between normals of two cells on the surface. If angular different between two cells is higher than specified value then cell refinement level specified in **Tolerance Level** edit box will be used.

Close Thin Slot rollout options

- Click on the **Close Thin Slots** check box to specify the size of the model's flow passages that you want to fill with a solid material. The **Maximum Height of Slots to Close** edit box will be displayed.
- Click in the **Maximum Height of Slots to Close** edit box and specify the value to fill the slots whose height is less than the specified value with the Solid closest to the slot.

Display Refinement Level rollout options

- Click on the **Display Refinement Level** check box to display the refined mesh corresponding to the specified refinement settings. The options of display refinement will be displayed; refer to Figure-69. Note that these options will not create any mesh but they are just to show you what a specified level of refinement will create during meshing.

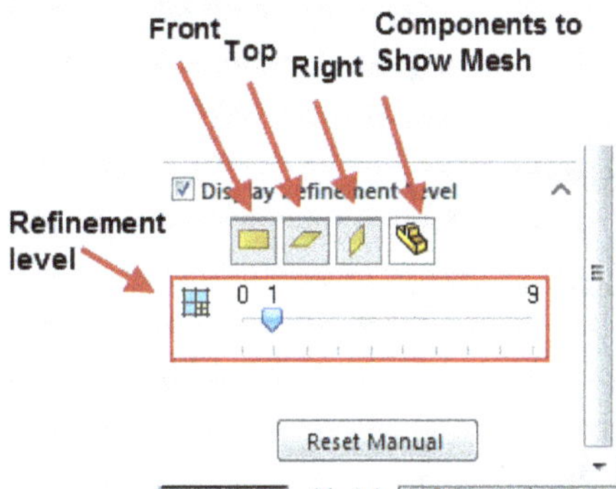

Figure-69. Display Refinement Level

- Click on the **Front** or **Top** or **Right** button to display mesh in one of the Computational Domain symmetry planes.
- Click on the **Components to Show Mesh** button to display the refined mesh on the selected components only. The **Component to Show Mesh** selection box will be displayed. Click on the components to select.
- Click on the **Use all components** check box to select all the components.
- Move the **Refinement Level** slider to specify the refinement level which you want to display.
- If you want to reset the selected or specified parameter from **Global Mesh Settings** feature then click on the **Reset Manual** button. All the parameters will be reset.
- After specifying the parameters, click on the **OK** button. The **Global Mesh** will be added in the **Flow Simulation Analysis Tree**.

Insert local Mesh

The **Insert Local Mesh** button is used to specify an initial mesh in a local region of the computational domain to better resolve the model specific geometry and/or flow (and/or heat transfer in solids) peculiarities in this region, which cannot be resolved well with the global initial mesh settings. The procedure to create local mesh is discussed next.

- Right-click on the **Mesh** option from **Flow Simulation Analysis Tree**. Click on the **Insert Local Mesh ProperyManager** from the displayed right-click shortcut menu; refer to Figure-70. The **Local Mesh Settings PropertyManager** will be displayed at the left of screen; refer to Figure-71.

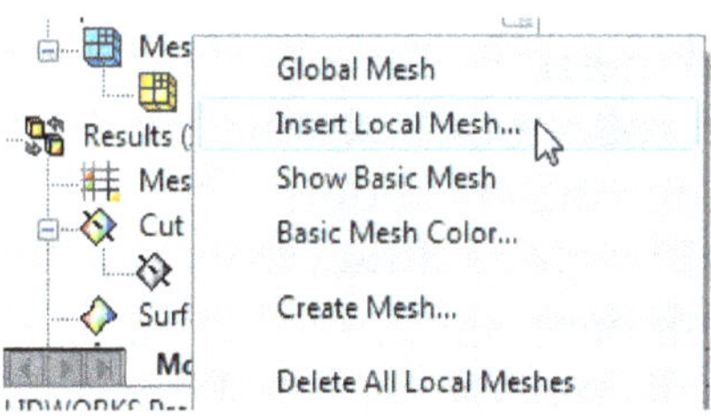

Figure-70. Insert Local Mesh button

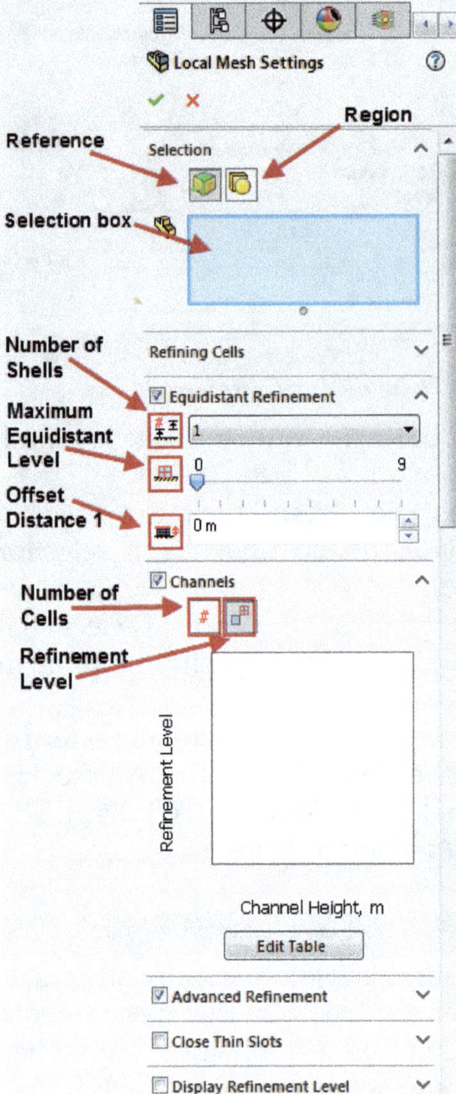

Figure-71. Local Mesh Settings feature box

Reference

- Click on the **Reference** button from **Selection** section to select a component(s), face(s), edge(s), or vertex(s) presenting the local region in which the initial mesh will be constructed.
- The **Selection** box is activated by default. You need to click on the component, face, vertex, or edge to create local mesh.
- Expand the **Refining Cells** rollout and set the level of refinement for fluid cells, solid cells, and boundary cells using various sliders.
- Click on the **Number of Shells** drop-down from the **Equidistant Refinement** section and select the required number of shells. Based on the specified number, the original mesh cells will be by respective shells planes; refer to Figure-72.

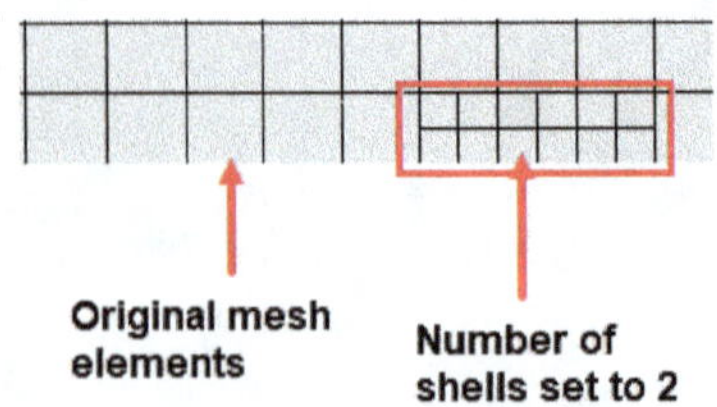

Figure-72. Shell division of mesh elements

- Move the **Maximum Equidistant Level** slider from **Equidistant Refinement** section to specify the desired equidistant level of refinement within elements earlier refined by shells.
- Click in the **Offset Distance 1** edit box from **Equidistant Refinement** section and enter the desired offset value. If you have set more than one shell then specify the offset distance in their respective edit boxes.
- Expand the **Channels** rollout if not displayed already. The options to perform channel refinement will be displayed. The options displayed on clicking **Number of Cells** button have been discussed earlier.

Refinement Level Options

- Click on the **Refinement Level** button from **Channels** section to define a uniform mesh across each model's flow passage by specifying the refinement level depending on the channel height, i.e. you can specify a particular refinement level for each range of the channel height.
- Click on the **Edit Table** button from **Channels** section to specify the refinement as a tabular dependency on the channel height. The **Dependency** dialog box will be displayed; refer to Figure-73.

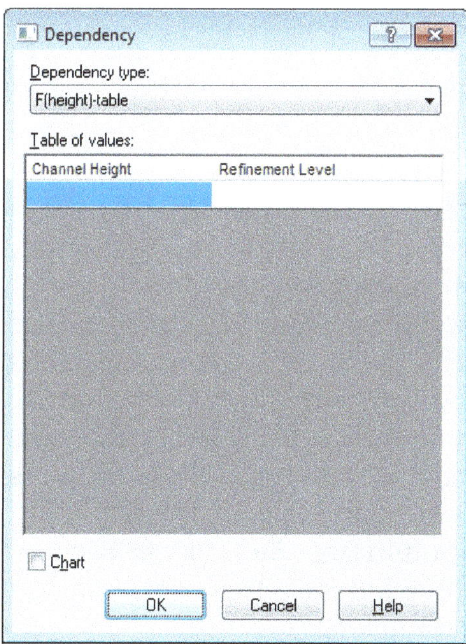

Figure-73. Dependency dialog box

- Click on the **Channel Height** edit box and specify the height of channel.
- Click in the **Refinement Level** edit box and specify the value.
- Click on the **Chart** check box from **Dependency** dialog box to select. The chart will be displayed; refer to Figure-74.

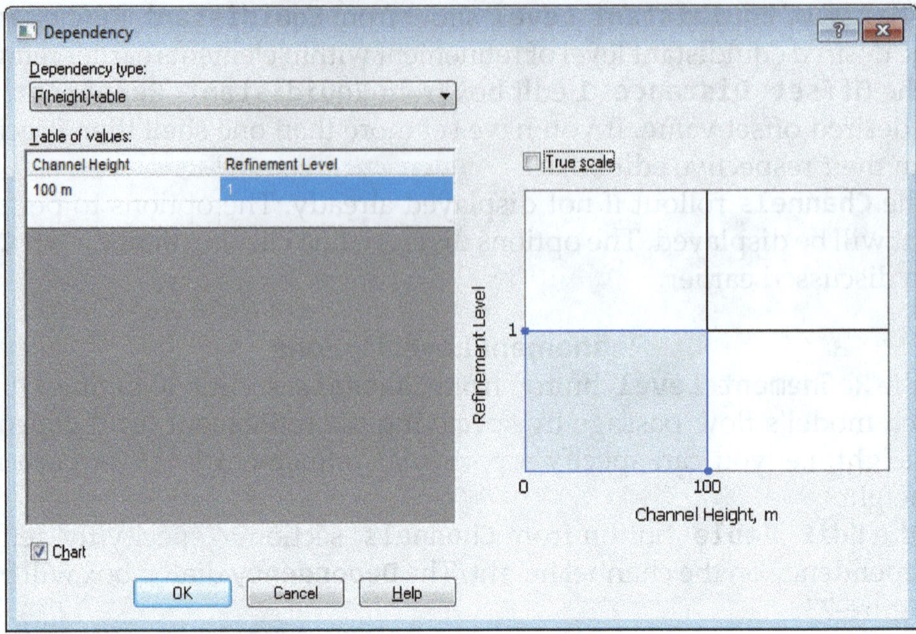

Figure-74. Chart for given values

- Select the **True Scale** check box to view actual scale of the input value.
- After specifying the parameters, click on the **OK** button from **Dependency** dialog box. The scale will be displayed in the channels section.

Region

- Click on the **Region** button of **Selection** section from **Local Mesh Settings PropertyManager** to define the local region.
- Click on the **Cuboid** radio button from **Selection** section to define a 3D box which bounds a local region, specify the distances from the Global Coordinate System origin to the region box sides and specify the required parameters.
- Click on the **Cylinder** radio button from **Selection** section to define a cylinder that bounds a local region, specify its radius and the coordinates of the cylinder bases centers in the Global Coordinate System and specify the required parameters.
- Click on the **Sphere** radio button from **Selection** section to define a sphere that bounds a local region, specify its radius and the center coordinates in the Global Coordinate System and specify the required parameters.
- After specifying the parameters, click on the **OK** button from **Local Mesh Settings PropertyManager** box. The **Local Mesh 1** will be added in **Flow Simulation Analysis Tree**.

Show Basic Mesh

The **Show Basic Mesh** button is used to view or hide the basic mesh of component. The procedure to use this button is discussed next.

- Right-click on the **Mesh** option from **Flow Simulation Analysis Tree**. The right-click shortcut menu will be displayed; refer to Figure-75.
- Click on the **Show Basic Mesh** button from the displayed menu. The basic mesh on the component will be displayed; refer to Figure-76.

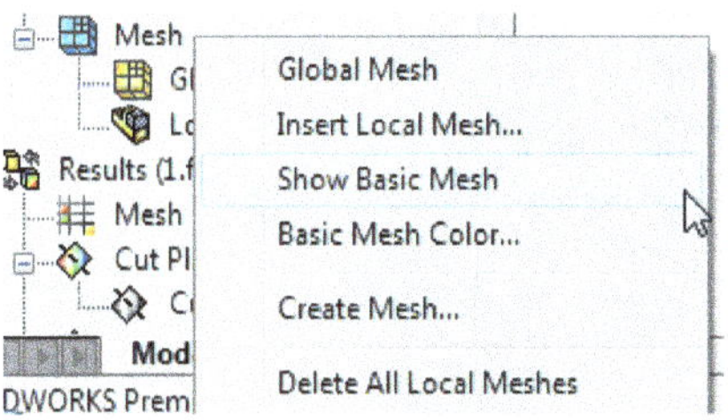

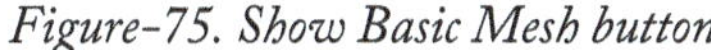

Figure-75. Show Basic Mesh button

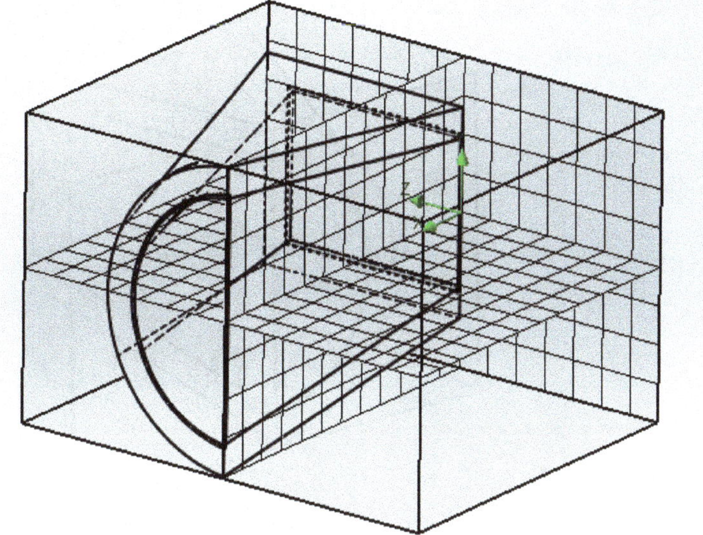

Figure-76. Basic mesh on component

- To hide the mesh, click on the **Hide Basic Mesh** button from the same right-click shortcut menu of **Mesh** option.

Basic Mesh Color

The **Basic Mesh Color** button is used to apply the desired color to the basic mesh. The procedure to use this tool is discussed next.

- Right-click on the **Mesh** option from **Flow Simulation Analysis Tree**. The right-click shortcut menu will be displayed; refer to Figure-77.
- Click on the **Basic Mesh Color** button from the displayed menu. The **Color** dialog box will be displayed; refer to Figure-78

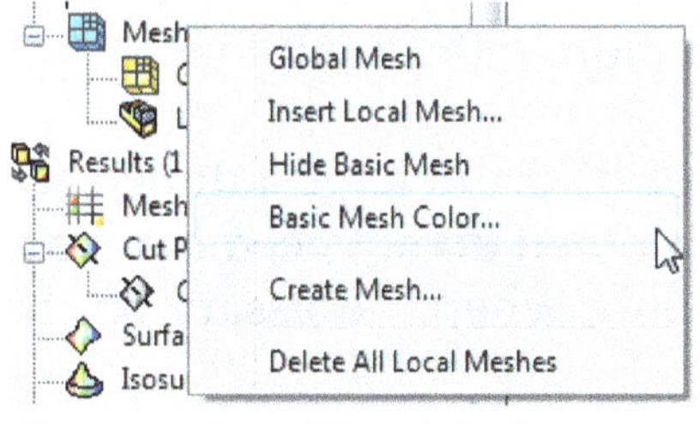

Figure-77. Basic Mesh Color button

Figure-78. Color dialog box

- Click on the desired color from **Basic Colors** section to select.
- If you want to add custom color then click on the **Define Custom Colors** and select the desired color.
- After selecting the color, click on the **OK** button from **Colors** dialog box. The color of basic mesh will be changed; refer to Figure-79.

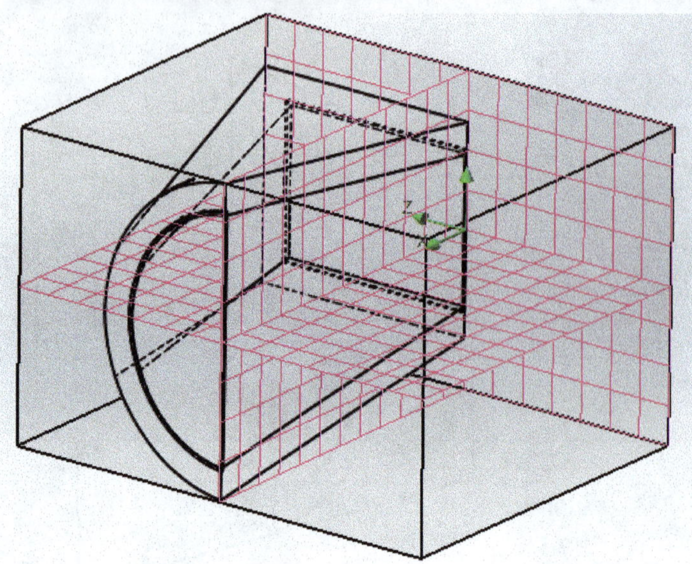

Figure-79. Changed color of mesh

Important Tips about Mesh

- In mesh creation, the cell meshing rule states that the level of two neighboring can differ by one level or need to be the same. For example, if you have cells of level 4 in the model mesh then only level 3 or level 5 mesh can be neighbor cells.
- The maximum level of refinement can be 9 whether it is for channel, fluids, or solids.
- The maximum ratio of cell size based on refinement can be 2^9:1 which is 512:1. It means if model has basic mesh size of 512 mm then size of cell will be 1 mm in mesh for maximum refinement.
- The maximum number of cells that can be created from 1 basic cell is $(2^3)^9$ which is 2^{27} = 134217728 about 134 million. So a refinement level of 9 can create millions of refined mesh cells creating a very large problem to solve. Sometimes when there is very high level of refinement then you will run out of RAM in the system and an error message of ABNORMALLY TERMINATED will be displayed. A RAM of 16 GB will permit 5 millions of cells.
- When you run out of mesh cells and need more refinement then you should decrease the size of basic mesh.
- Automatic refinement can give you ratio of 1000:1 in largest dimension and smallest size cells. So, if there is ratio in permissible limit then do not need manual refinement.
- Mesh refinement should be performed only where necessary so that the total size of problem can be minimized.
- The cell level of basic mesh is 0.

Create Mesh

The **Create Mesh** button is used to run the calculation of current project. The procedure to use this is discussed next.

- Right-click on the **Mesh** option from **Flow Simulation Analysis Tree**. The right-click shortcut menu will be displayed; refer to Figure-80.

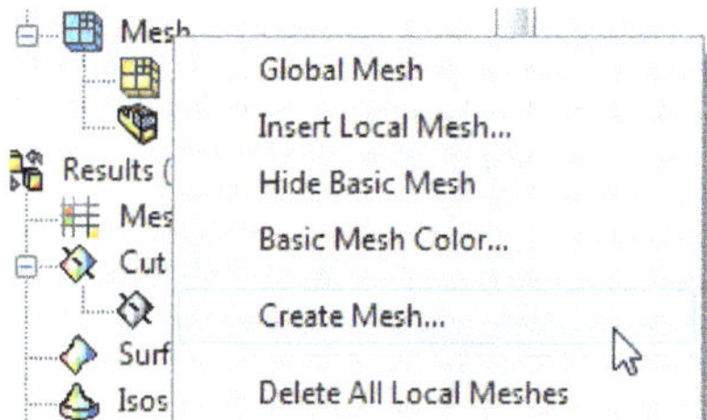

Figure-80. Create Mesh

- Click on the **Create Mesh** button from the displayed menu. The **Run** dialog box will be displayed; refer to Figure-81. You can click on the **Run** button from the **Flow Simulation CommandManager** to activate the same dialog box.

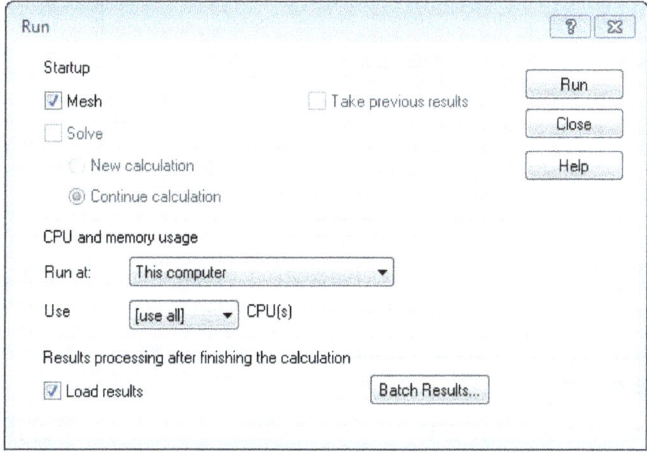

Figure-81. Run dialog box

- Click on the **Mesh** check box to select from **Startup** section to create the new computational mesh for a project that is already meshed or calculated. If the project was never meshed or calculated before, the **Mesh** check box is selected automatically.
- Click in the **Run at:** drop-down from **CPU and Memory Usage** section and select **This Computer** button to run the solver on the current computer as a separate process.
- Click on the **Run at:** drop-down from **CPU and Memory Usage** section and select **Add Computer** button to select a network computer to add it in list. Once the computer is added, you can select it to calculate the project. This option allows you to use the other computer's CPU and memory resources at the expense of additional calculation time due to data transfer through the network. The **Add Computer** dialog box will be displayed; refer to Figure-82.

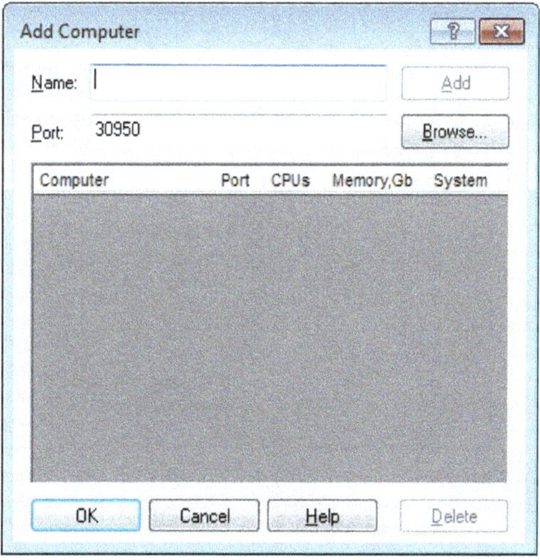

Figure-82. Add Computer dialog box

- Click in the **Port** edit box and enter the port number.
- Click on the **Browse** button from **Add Computer** dialog box. The **Browse for Computer** dialog box will be displayed; refer to Figure-83.

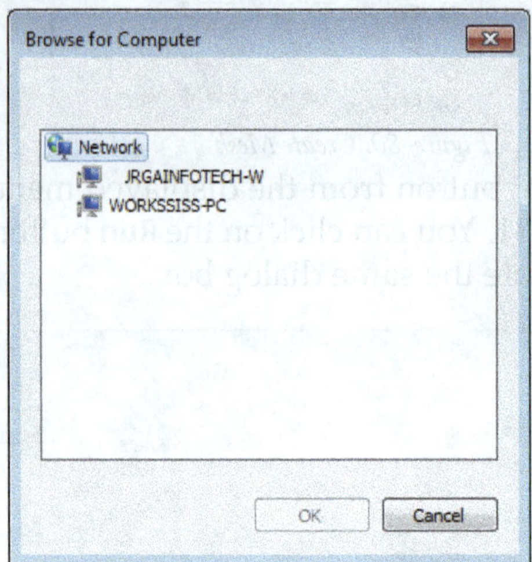

Figure-83. Browse for Computer dialog box

- Select the computer which you want to add and click on the **OK** button. The name of the selected computer will be added in the **Name** edit box.
- Click on the **Add** button from **Add Computer** dialog box. The computer will be added and listed under list section.
- After adding the computer, click on the **OK** button from **Add Computer** dialog box.
- Click on the **Use** drop-down from **CPU and Memory usage** section and select the number of processor used for calculation process.
- Select the **Load results** check box to automatically load results when the calculation is finished.
- Click on the **Batch Results** button from **Run** dialog box to define the plots, parameter tables, and reports to be created. The **Batch Result Processing** dialog box will be displayed; refer to Figure-84.

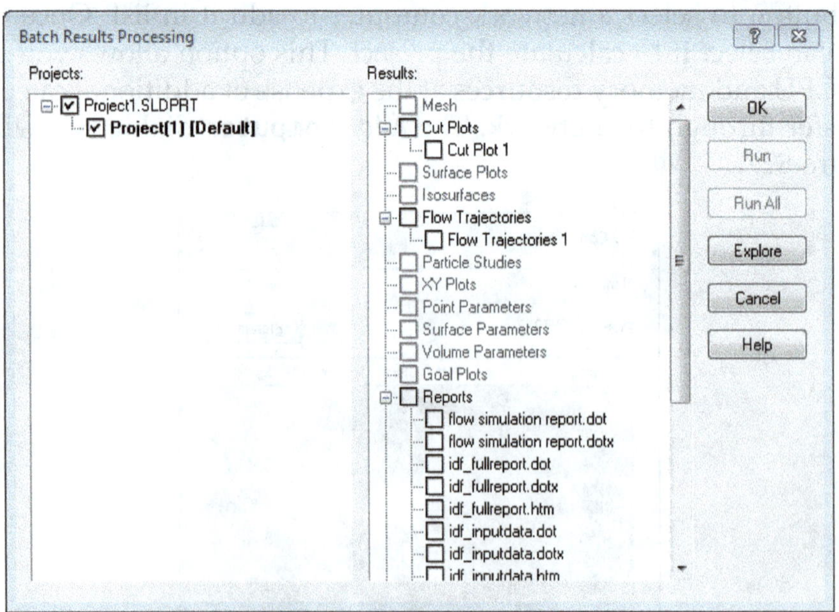

Figure-84. Batch Results Processing dialog box

- Select the required check box from **Results:** section and click on the **OK** button from **Basic Results Processing** dialog box.

- After specifying the parameters from **Run** dialog box, click on the **Run** button. The **Mesh Generation** dialog box will be displayed; refer to Figure-85.

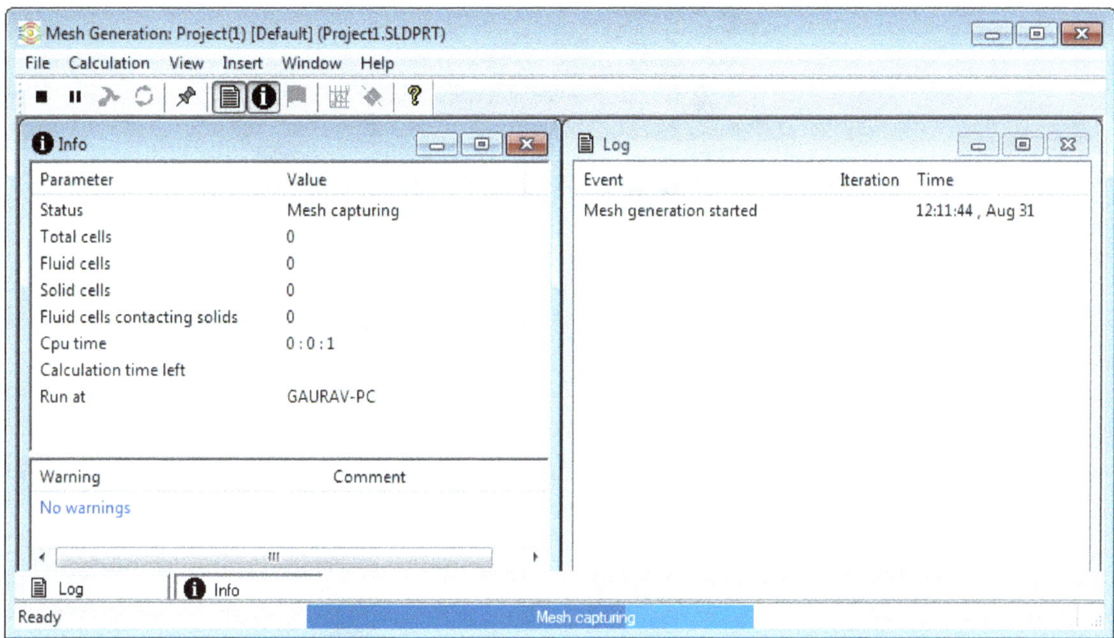

Figure-85. Mesh Generation dialog box

- The calculation is generated in **Mesh Generation** dialog box. During calculation, you can pause or stop the on-going calculation by clicking on the respective button from the **Toolbar** of **Mesh Generation** dialog box.
- All details regarding the calculation are displayed in the **Info** section.
- The information related to the generation of mesh is displayed in **Log** section of **Mesh Generation** dialog box.
- When **Mesh Generation finish normally** shows in bottom bar, click on the **Close** button from **Mesh Generation** dialog box to exit the dialog box.
- The analysis has been performed by the software and we are now ready to generate the results.

SETTING CALCULATION CONTROLS

The **Calculation Control Options** tool is used to set the calculation parameters related to running current analysis. The procedure to set the parameters is given next.

- Click on the **Calculation Control Options** tool from the **Flow Simulation** cascading menu in the **Tools** menu or right-click on **Input Data** node in the **Flow Simulation Analysis Tree** and select the **Calculation Control Options** tool from the shortcut menu; refer to Figure-86. The **Calculation Control Options** dialog box will be displayed; refer to Figure-87.

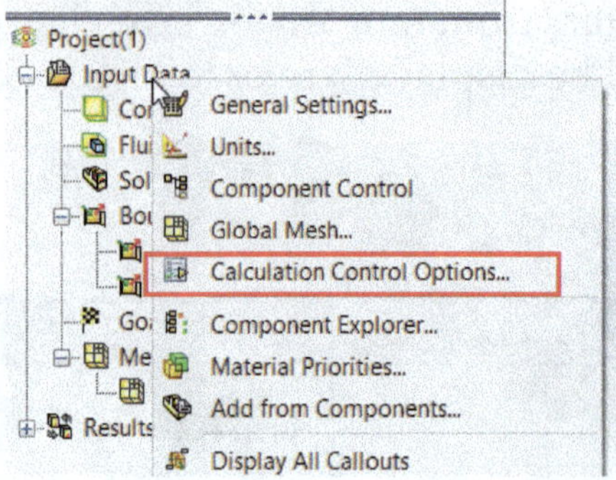

Figure-86. Calculation Control Options tool

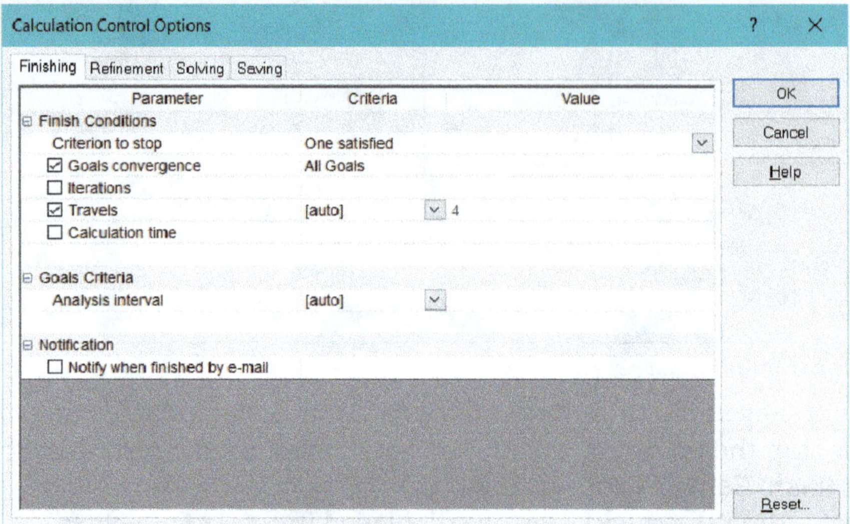

Figure-87. Calculation Control Options dialog box

Finishing Options

- Select the desired option in the **Finishing** tab of the dialog box to specify the conditions of defining limit after which the analysis will be assumed as completed.
- Select the **One satisfied** option from the **Criterion to stop** drop-down if you want to close the calculation when any one of the condition is satisfied. Select the **All satisfied** option from the **Criterion to stop** drop-down if you want all the conditions to be satisfied before analysis is assumed as completed.
- Select the **Goals convergence** check box to include goals convergence as condition for completion of analysis calculations.
- Select the **Iterations** check box if you want to specify the number of iterations after which the analysis will be assumed as completed.
- Select the **Travels** check box if you want to specify the number of fluid travels through the domain. By default, **auto** is selected for travel but if you want to specify desired value of travels then select the **manual** option from related drop-down and specify desired value of travels.
- Select the **Calculation time** check box to define the time period upto which the analysis can run.
- Select the desired option from the **Analysis interval** drop-down in the **Goals Criteria** node of the dialog box. By default, **auto** is selected in the drop-down so interval is defined automatically. Select the manual option and specify the desired value of interval in terms of travels, iterations, or physical time.
- Select the **Notify when finished by e-mail** check box if you want to be notified once analysis is complete.

Refinement Options

- Click on the **Refinement** tab to define the level of mesh refinement during calculation. The options in this tab work more like adaptive meshing during analysis. The options in the dialog box will be displayed; refer to Figure-88.

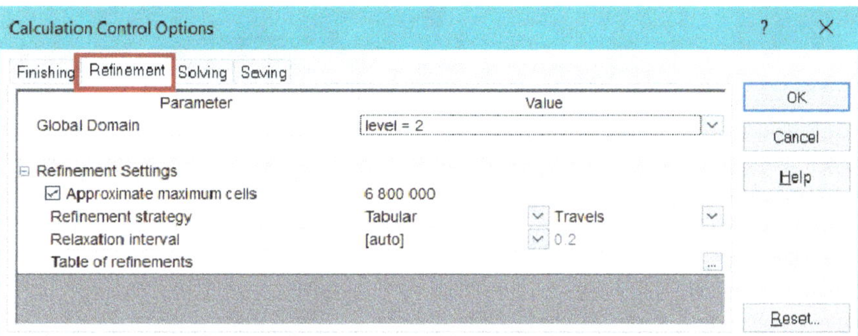

Figure-88. Refinement tab of dialog box

- Select the desired level of refinement in the **Global Domain** drop-down. During calculation if initial mesh is of level 2 and you have specified refinement level 3 then mesh can further split up to 2+3 = 5th level.
- If you have specified local meshing then set the level of refinement for local mesh in the same way.
- From **Refinement Settings** node, set the parameters related to refinement like strategy, relaxation interval, and so on. Select the **Approximate maximum cells** check box to define the approximate value up to which cells can be created after refinement.
- The options in the **Refinement strategy** drop-down are used to define the parameters for performing refinement during analysis. There are four options available in this drop-down; **Periodic**, **Tabular**, **Goal Convergence**, and **Manual only**. For each of these options, you can select iterations, travels or time period in the next drop-down. The selected option will be used as parameter for refinement strategy. Select the **Periodic** option if you want to specify start point at which refinement will begin and period after which refinement will advance to next level. Select the **Tabular** option from the drop-down if you want to create a table with value in term of iteration/travel/time period at which refinement will occur. Select the **Goal Convergence** option if you want to select goals for refinement. Whenever a goal is achieved, the next refinement will occur. Note that you can select **Delay** check box to define the delay time after which refinement will occur once goal convergence has occurred. Select the **Manual only** option from the drop-down if you want to activate refinement manually.
- Select the manual option from the **Relaxation interval** drop-down if you want to specify the value of relaxation interval. Relaxation interval is time/iteration/travel value which will be enforced after refinement and before completion of calculations. Specify the desired value of relaxation interval in the edit box next to **Relaxation interval** drop-down.
- You can specify refinement parameters in the same way for local mesh.

Solving Options

- Click on the **Solving** tab in the dialog box to display options related to analysis solution process.
- Select the **Scale factor** check box so that automatic time step of analysis will be multiplied if you are solving stationary problem. On selecting the check box, an edit box will be displayed to specify the scale value.

- Select the **Averaged** check box if you want to average any result parameter at specified interval. Select the desired parameter from the next drop-down to define interval parameter for averaging. Click on the **Browse** button for **Parameters** field. The **Customize Parameter List** dialog box will be displayed; refer to Figure-89. Select the desired option from the Averaging interval drop-down to define the interval for averaging.

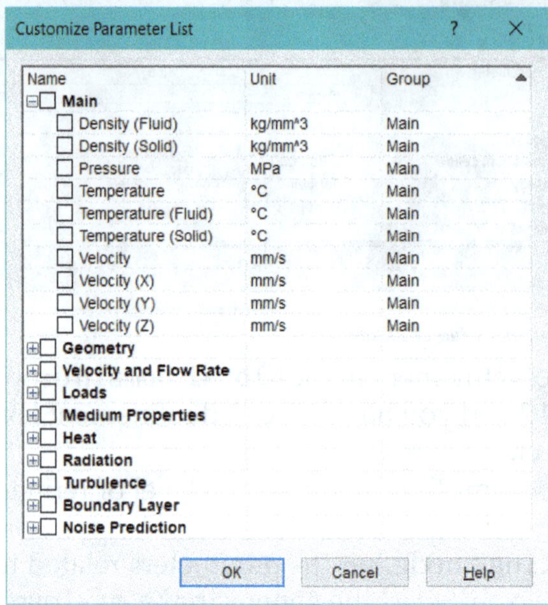

Figure-89. Customize Parameter List dialog box

- Select the **Calculate Local Mean Age (LMA)** check box to calculate average time τ for which fluid travel from inlet and remains in the fluid domain before exiting.
- Select the **Calculate Comfort Parameters** check box to enable calculations of parameters like Mean Radiant Temperature (MRT), Operative Temperature, Predicted Mean Vote (PMV), Predicted Percent Dissatisfied (PPD), Draught Rate, Draft Temperature, Air Diffusion Performance Index (ADPI), Contaminant Removal Effectiveness (CRE), Local Air Quality Index (LAQI), and Flow Angle.
- Select the desired option from the **Results Processing Speed-up Data** drop-down to manage how to save the data required for creating **Surface Plots**. The information about the model geometry and the computational mesh required for creating **Surface Plots** are saved in the *.gdb files in the project folder. Select the **On meshing** option from the drop-down if you want to save the data for results at mesh generation stage. Select the On loading results option from the drop-down if you want to save data at result loading stage for surface plots. Select the **Never** option from the drop-down if you do not want to save the data.
- Select desired option from the **Freezing strategy** drop-down to define strategy for saving CPU time by freezing values of all flow parameters, with the exception of fluid and solid temperatures and fluid substances concentrations (if several substances are considered), which converge more slowly than the other flow parameters, so the temperature and concentrations are calculated at each iteration. This option will be useful when a steady-state or time-dependent problem with substantial heat transfer and/or fluid substances propagation is solved. Set the parameters for freezing as discussed earlier.

Saving Options

- Click on the **Saving** tab in the dialog box to display save options during analysis. The dialog box will be displayed as shown in Figure-90.

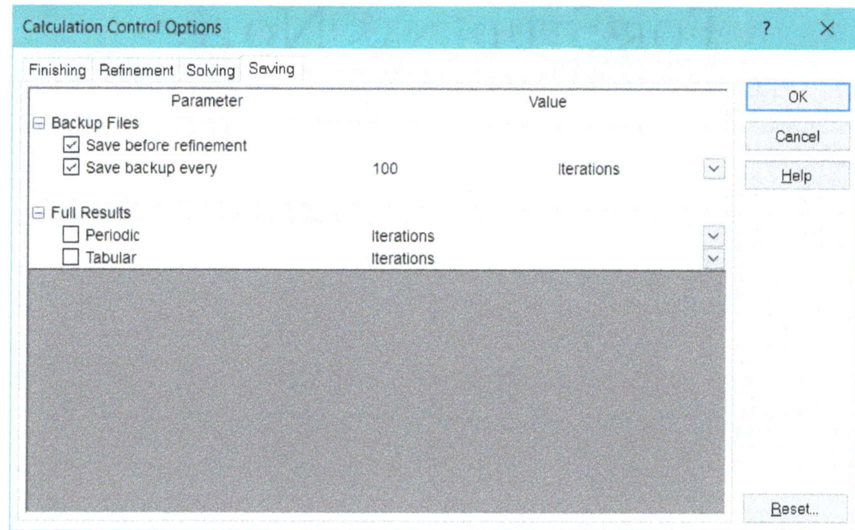

Figure-90. Saving tab

- Select the desired check boxes from **Backup Files** node in the dialog box to save backup copies of the analysis files. Select the **Save before refinement** check box to save the files before adaptive refinement occurs during the analysis. Select the **Save backup every** check box and specify the number of iterations or time at which backup copy will be saved.
- Similarly, you can set parameters in **Full Results** node for various parameters.
- Click on the **OK** button to apply the calculation control parameters.

The procedure of generating results will be discussed in the next chapter.

Self Assessment

Q1. Write down steps to create a project for simulation.

Q2. What is Fluid Sub-domains and why it is used?

Q3. Write down basic steps to apply inlet mass flow boundary conditions.

Q4. Write down the procedure of inserting Surface Goal in a project.

Q5- What is the use of Global Mesh?

Q6- Which button is used to view the basic mesh?

For Students Notes

Chapter 3

Analyzing and Generating Results of Analysis

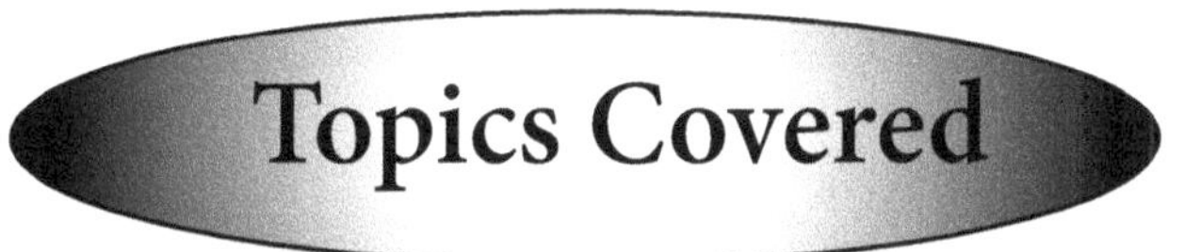

The major topics covered in this chapter are:

- ***Results***
- ***Cut Plots***
- ***Surface Plots***
- ***Isosurfaces***
- ***Flow Trajectories***
- ***Particle Studies***
- ***Point Parameters***
- ***Report***

INTRODUCTION

In the last chapter, we have learned the procedure of preparing data for analysis of a component. In this chapter, we will learn the procedure of generating and analyzing the reports.

GENERATING RESULTS

You can load the results after you have performed the analysis by using **Run** button. The procedure is given next.

- Click on the **Run** button from **CommandManager** to load the results. The full procedure to load the results has been discussed earlier in the last chapter.
- The results for the analysis are displayed under **Results** section of the **Flow Simulation Analysis**; refer to Figure-1.

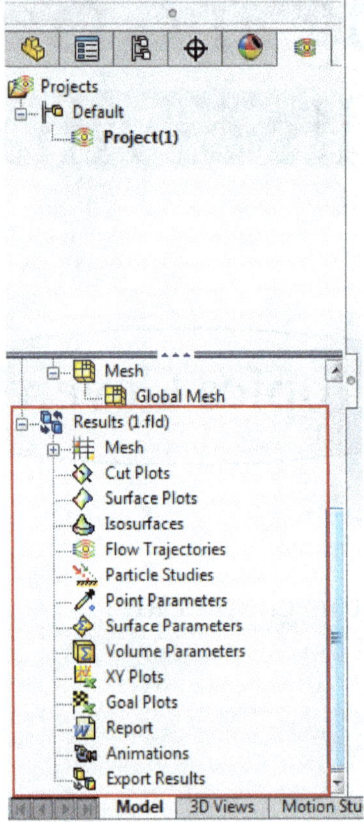

Figure-1. Results section

- The options of **Results** section are discussed next.

Mesh

The **Mesh** option is used to display the computational mesh cells and mesh related parameters at the calculation moment selected for getting the results. The procedure to generate mesh result is discussed next.

- Right-click on the **Mesh** option from **Results** design tree. The right-click shortcut menu will be displayed; refer to Figure-2.

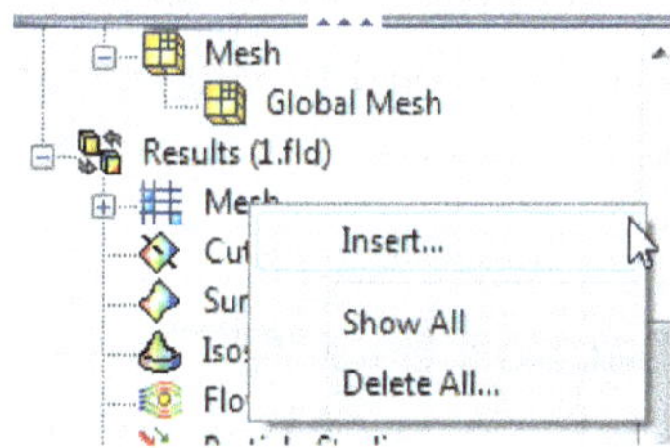

Figure-2. Mesh option for analysis

- Click on the **Insert** option from the shortcut menu. The **Mesh PropertyManager** will be displayed; refer to Figure-3.

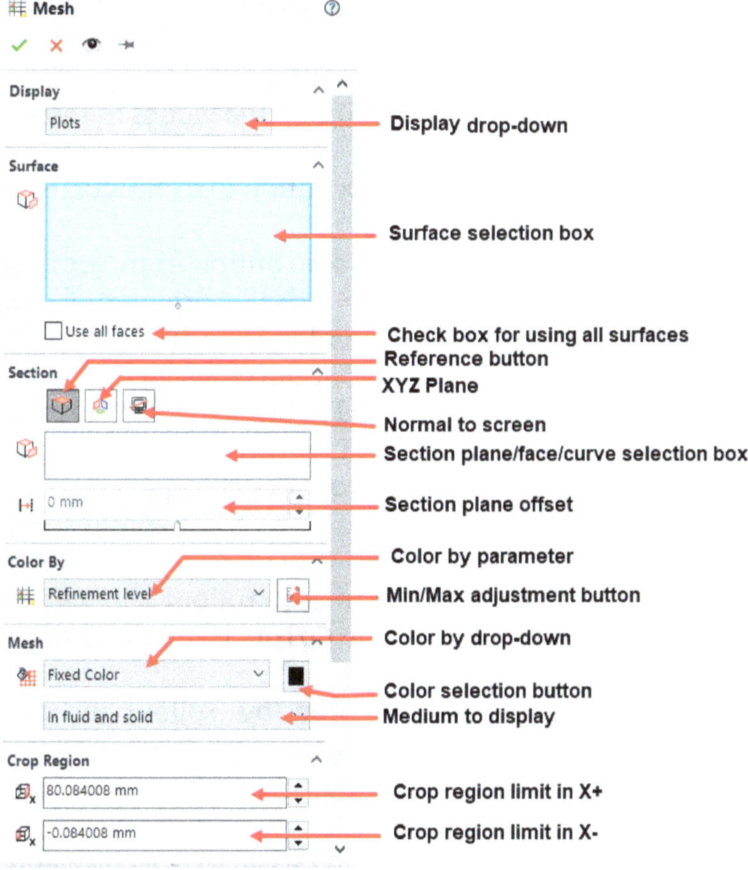

Figure-3. Mesh PropertyManager

Various parameters of the **PropertyManager** are discussed next.

Plots

- Click on the **Plots** option from the **Display** drop-down to display the computational mesh cells in a section view or on the selected model faces or surfaces.
- Click in the **Surface** selection box and select the faces of component for which you want to display the mesh distribution. Select the **Use all faces** check box to select all the face of model for displaying mesh. The **Cut with section** check box will be displayed. Select the **Cut with section** check box to cut the faces with selected section plane.

Section by Reference

- Select the **Reference** button from **Section** rollout to select a plane or planer face for section.

- Click on the **Section Plane or Planer Face** selection box and select the desired face, plane, or curve in which you want to display the results.
- Click in the **Offset** edit box and enter the desired value of offset. You can also move the slider to set the offset value.

Section by XYZ Planes

- Click on the **XYZ Planes** button from **Section** rollout to specify the plane.
- Select the desired radio button to use respective plane for section.
- Specify desired offset value in the **Offset** edit box.

Section by Plane Normal to Screen

- Click on the **Normal to Screen** button from **Section** rollout to define the section plane normal to the screen. After choosing this button, click at the desired point where you want the section plane to start and drag mouse in desired direction to set the orientation of the plane. You can also position the plane by specifying the values in respective edit box from **Section** section.
- After specifying the parameter, click on the **Fix** button to fix the current position of normal line.
- Select the **Normal to Screen, Vertically** radio button from **Section** rollout to define the section plane normal to the screen and vertically oriented. Click in the **Vertical Position** edit box from **Section** section and enter the desired value. You can also specify the location of vertical section line by clicking at the desired location in screen.
- Select the **Normal to Screen, Horizontally** button from **Section** rollout to define the section plane normal to the screen and horizontally oriented. Click in the **Horizontal Position** edit box from **Section** section of **PropertyManager** and enter the desired value to position plane. You can also specify the location of horizontal section line by clicking at the desired location in screen.
- Click in the **Parameter** drop-down of **Color by** rollout and select the **Fixed Color** option to apply the desired color. Click on the **Color** button and select the desired color to apply.
- Click in the **Parameter** drop-down from **Color by** rollout and select the **Curvature Criteria** option to color the surface and planes in accordance with their curvature. Click on the **Adjust Minimum/Maximum** button and specify the maximum and minimum values used for coloring.
- Click in the **Number of Levels** edit box from **Color by** section and enter the value to specify the levels up to which color can change. You can also set the value of number of levels by moving the **Number of Levels** slider.
- Similarly, specify the value of other options like **Tolerance Criteria** and **Refinement Criteria** of **Parameter** drop-down in **Color by** section.
- Click in the **Parameter** drop-down and select **None** option if you do not want to apply any color to faces/surfaces of mesh or make the meshing transparent.
- Click on the **Color by** drop-down from **Mesh** section and select the **Fixed Color** option to change the default mesh color. Click on the **Color** button next to drop-down and select the desired color.
- Select the desired option from the **Medium to Display** drop-down to define where mesh will be created. You can create mesh for fluids, solids, or fluids and solids both.
- Click in the desired edit box from **Crop Region** section and enter the desired values in the respective edit boxes to specify to crop the mesh visualization area.

Cells

- Select the **Cells** option from **Display** drop-down to display the computational mesh cells of the selected type as cuboids; refer to Figure-4.

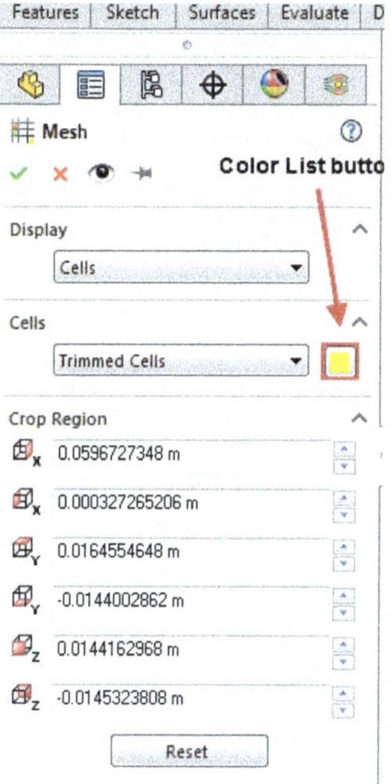

Figure-4. Cells option

- Click on the **Cells** drop-down from **Cells** section and select **Trimmed Cells** option to display trimmed cells of mesh if present. Select **Irregular Cells** option to display uneven cells of mesh.
- Click on the **Color List** button and select the desired color from the displayed list of colors. The other options of the **PropertyManager** are same as discussed earlier.

Channels

- Click on the **Channels** option from **Display** drop-down to display the model surfaces colored by the channels height on the model faces; refer to Figure-5.

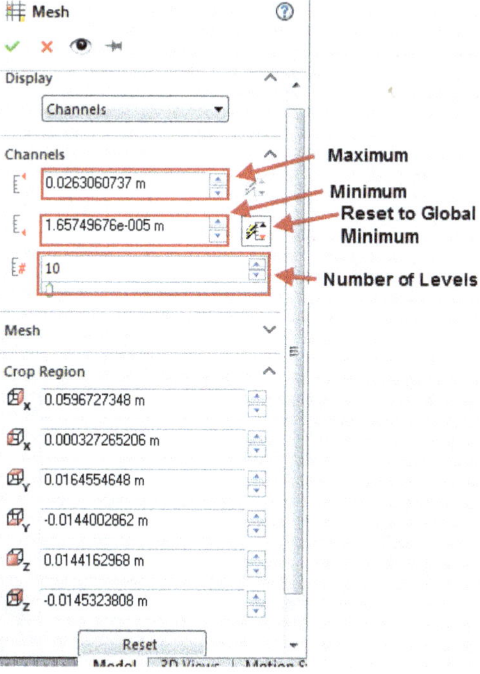

Figure-5. Channels options

- Click in the **Maximum** edit box from **Channels** section and enter the desired value.
- Click in the **Minimum** edit box from **Channels** section and enter the desired value.
- Click on the **Reset to Global Minimum:** button to reset the value.
- Click in **Number of Levels** edit box from **Channels** section and enter the value to set the number of levels required for coloring. You can also set the value by moving the **Number of Levels** slider.
- Other options of the **PropertyManager** are same as discussed earlier in last section.
- After specifying the parameters, click on the **OK** button. The mesh will be calculated and displayed on the component; refer to Figure-6.

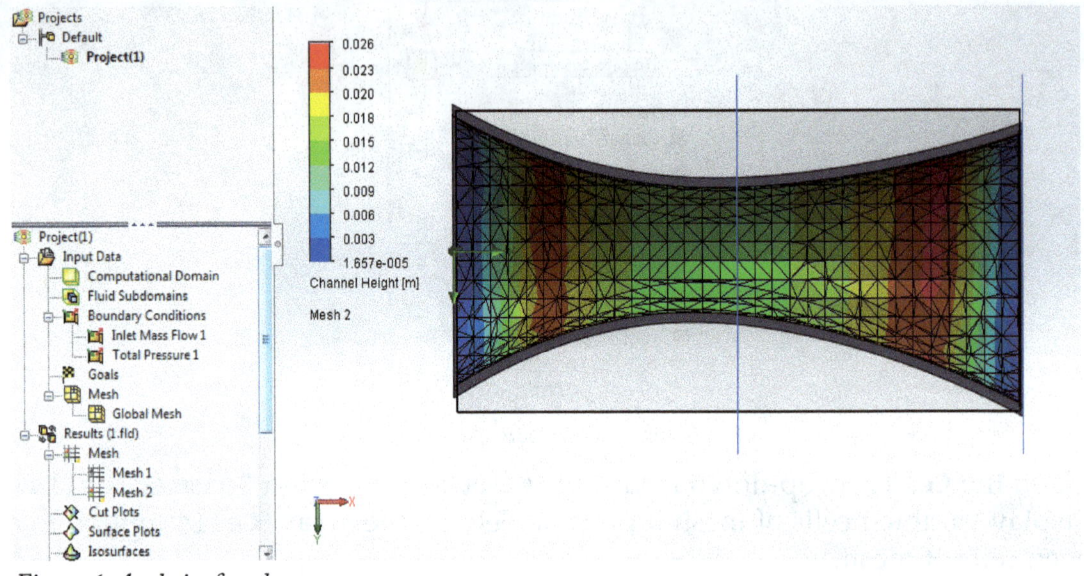

Figure-6. Analysis of mesh

Cut Plots

The **Cut Plots** option is used to display various plots fluid flow in the model. The procedure to generate a cut plot is discussed next.

- Right-click on the **Cut Plots** option from **Design Tree**. The right-click menu will be displayed; refer to Figure-7.
- Click on the **Insert** option from displayed menu. The **Cut Plot PropertyManager** will be displayed; refer to Figure-8.

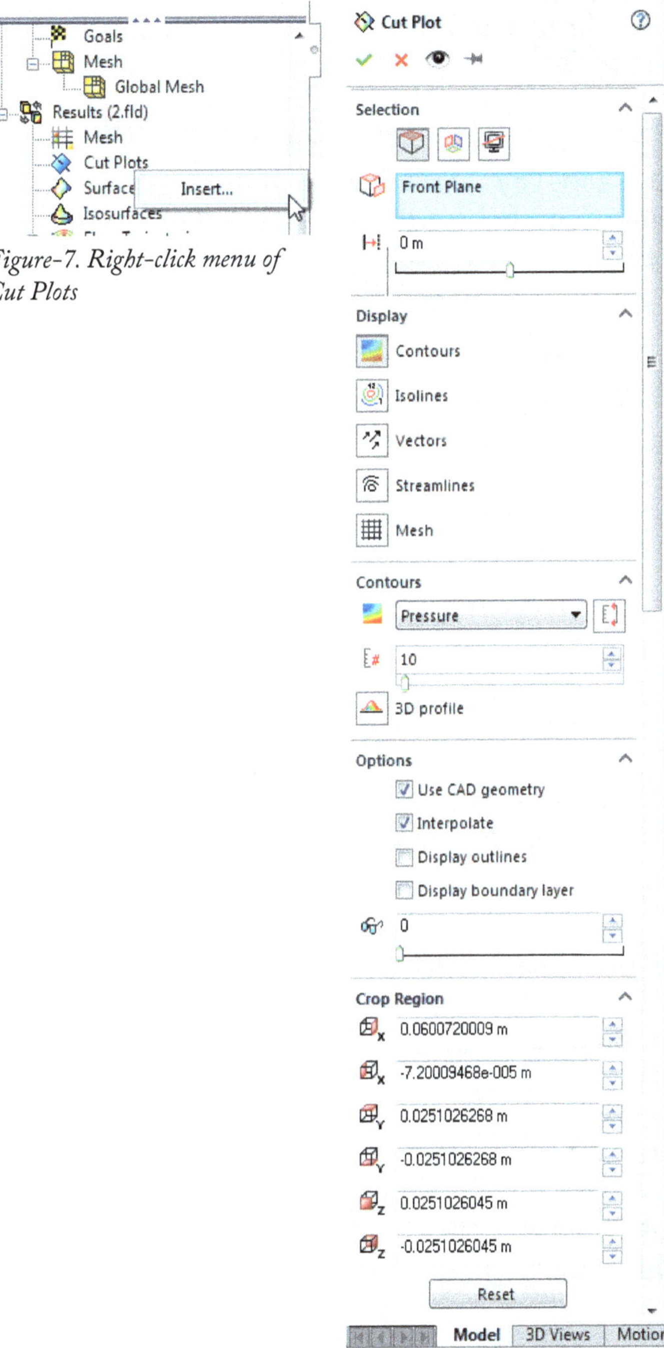

Figure-7. Right-click menu of Cut Plots

Figure-8. Cut Plots Property Manager

- The mid plane of the model is selected by default in the **Section Plane** selection box. If you want to select another plane then right-click in the **Section Plane** selection box and click on the **Clear Selections** button from the right-click menu; refer to Figure-9.

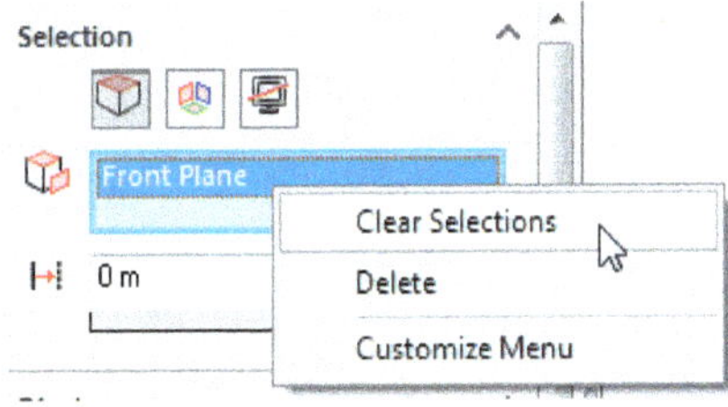

Figure-9. Clear Selection button

- Select the desired plane/face to use as section plane.
- Click in the **Offset** edit box and enter the desired value of offset. You can also set the value by moving the **Offset** slider.

Options

- Select the **Use CAD Geometry** check box from **Options** section if you want to display original CAD geometry otherwise the Flow Simulation-interpreted geometry will be displayed; refer to Figure-10.

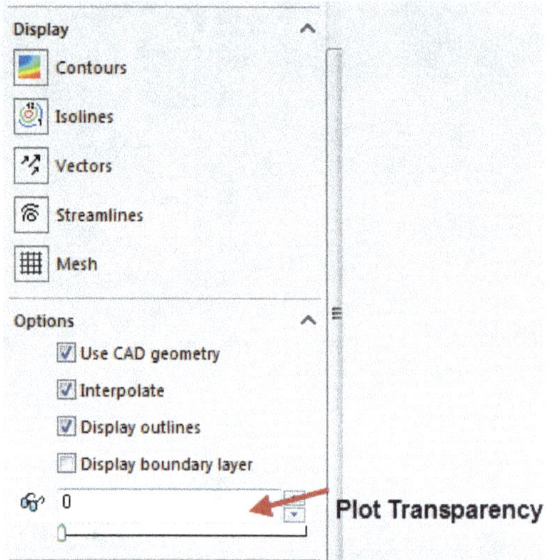

Figure-10. Options section of Cut Plots

- Select the **Interpolate** check box from **Options** rollout if you want to switch to interpolation and accelerate operations related to displaying results.
- Select the **Display Outlines** check box from **Options** rollout to display the plot area outlines.
- Select the **Display boundary layer** check box from **Options** rollout if you want to display boundary layers in the cut plots. But it requires additional computer resources to visualize the boundary layer.
- Click in the **Plot Transparency** edit box from **Options** rollout and enter the desired value to set the transparency for the plot ranging from 0 to 1. You can also set the transparency by moving the **Plot Transparency** slider.
- The options of **Crop region** rollout are same as discussed earlier in the last section.

Contours

- Click on the **Contours** button from **Display** rollout to display the distribution of parameter specified in **Parameters** drop-down of **Contours** rollout. The options of **Contours** rollout will be displayed; refer to Figure-11.

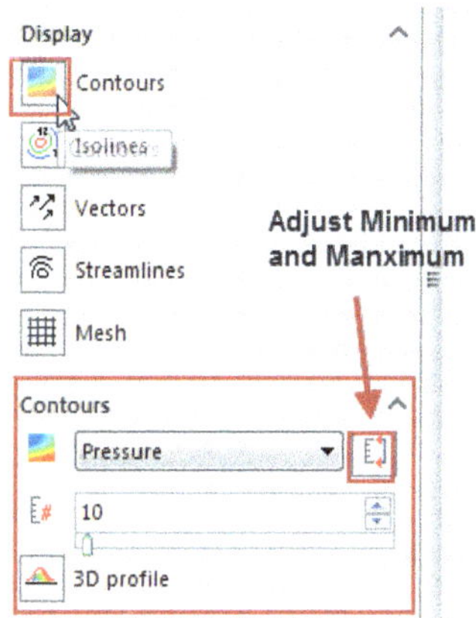

Figure-11. Contours Section

- Click in the **Parameter** drop-down from **Contours** section and select the parameter which you want to display in contours.
- Click on the **Adjust Minimum and Maximum** button to specify the maximum and minimum values.
- Click in the **Number of Levels** edit box from **Contours** section and enter the value to specify the number of level. You can specify up to 255 levels (Divisions). You can also set the value by moving the **Number of Levels** slider.
- Click on the **3D Profile** button from **Contours** section to represent the parameter distribution over a 3D profile. The option of 3D profile will be displayed; refer to Figure-12.

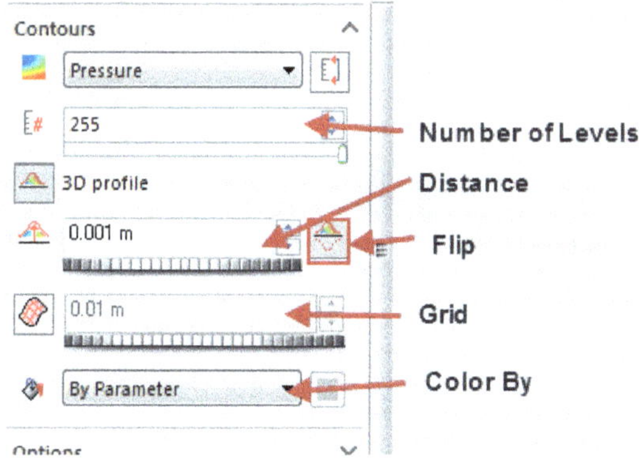

Figure-12. Options of 3D Profile button

- Click in the **Distance** edit box from **Contours** section to specify the value of distance between maximum parameter value and reference value. You can also move the **Distance** slider to set the value of distance.
- Click on the **Flip** button to change the direction of the 3D Profile elevation to the opposite one.
- Click on the **Grid** button from **Contours** section. The **Grid Step** edit box will be activated. Click in the **Grid Step** edit box to specify the distance between adjacent lines of grid projected in 3D Profile. You can also move the **Grid Step** slider to set the value.
- Select the **Fixed Color** button from **Color by** drop-down to display the 3D profile using fixed color. If this option is selected, the **Number of Levels** option is unavailable.
- Click on the **By Parameter** button from **Color By** drop-down if you want the 3D profile to be colored in accordance with the distribution of the parameter.

Isolines

The **Isolines** button is used to display the parameter over isolines and specify isolines options. The procedure to use this button is discussed next.

- Click on the **Isolines** button from **Display** section. The options of **Isolines** will be displayed below **Display** section; refer to Figure-13.

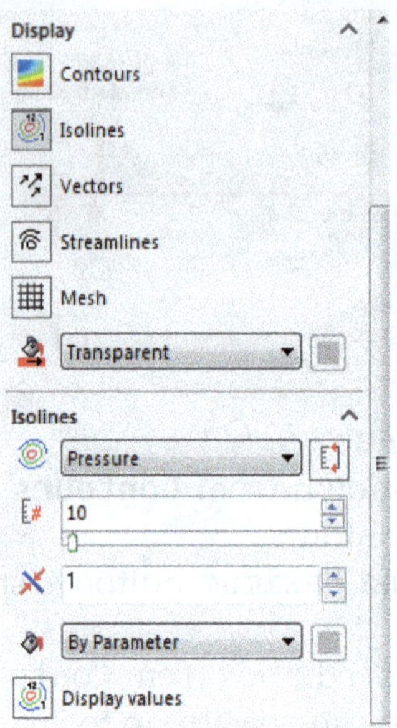

Figure-13. Options of Isolines from Cut Plot

- Click in the **Parameter** drop-down from **Isolines** section and select a parameter for which you want to display isolines.
- The other options of this section were discussed earlier.
- Click on the **Display Values** button of **Isolines** section to display the parameter values on the isolines.

Vectors

The **Vectors** button is used to display vectors which are used to visualize flow of fluid with vector arrows; refer to Figure-14.

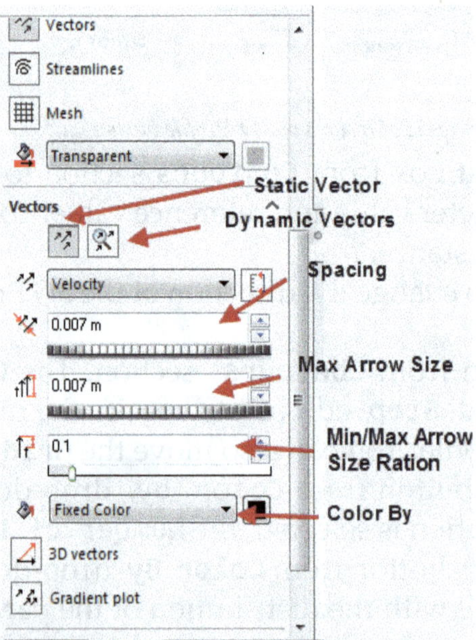

Figure-14. Options of Vectors button

- Click on the **Static Vectors** button from **Vectors** rollout to display the distribution of vector parameter as a static image.

- Click on the **Dynamic Vectors** button from **Vectors** section to update the parameter of distribution of a vector in a real time, as you manipulate the model.
- Click in the **Parameter** drop-down of **Vectors** section and select a parameter for which you want to display Vectors.
- Click in the **Spacing** edit box from **Vectors** section and enter the value to control the distance between the vector starting points. You can also move the slider to adjust or set the value of spacing. If the **Dynamic Vectors** mode is selected in place of **Static Vectors** then it is the distance between vectors along the dynamic trajectory.
- Click in the **Max Arrow Size** edit box from **Vectors** section and enter the value to specify the vector size corresponding to the maximum parameter's value. You can also move the slider to set the value of **Max Arrow Size**. If **Dynamic Vectors** mode is selected, the vector size is specified in pixels.
- Click in the **Min/Max Arrow Size Ratio** edit box from **Vectors** section and enter the value to control the vector size corresponding to the minimum value of selected parameter. The value of **Min/Max Arrow Size Ratio** is lie between 0 to 0.999. You can also set the value by moving the slider.
- Select the **Logarithmic scale** check box if you want to use logarithmic ratio for arrow size.
- Click in the **Color By** drop-down from **Vectors** section to select the way for coloring vectors.
- Click on the **3D Vectors** button from **Vectors** section to display spatial vectors otherwise it display vector projections into surface.
- Click on the **Gradient Plot** button from **Vectors** section to set the spacing between vectors based on gradient of vector (slope) otherwise spacing between all the vectors is constant.

Streamlines

The **Streamlines** button is used to specify display options for streamlines. It allows us to visualize field lines projected on the selected planes.

Evenly Spaced Lines Style

- Click on the **Streamlines** button from **Display** rollout of **Cut Plots PropertyManager**. The tools or option related to **Streamlines** button will be displayed; refer to Figure-15.

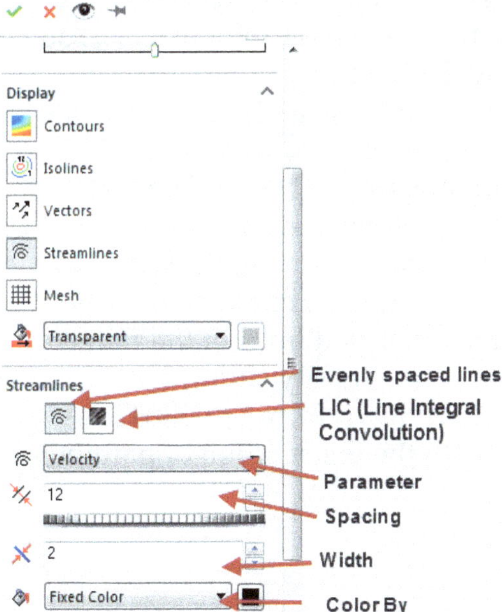

Figure-15. Options of Streamlines

- Click on the **Evenly Spaced Lines** button from **Streamlines** rollout to display a vector parameter as evenly spaced plane streamlines.
- Click in the **Parameter** drop-down from **Streamlines** rollout and select the required option which you want to visualize with the streamlines.
- Click in the **Spacing** edit box from **Streamlines** rollout and enter the value to control the distance between the lines starting points in the range from 0 to 128 pixels. You can also move the slider to adjust or set the value of spacing.
- Click in the **Width** edit box from **Streamlines** rollout and enter the value to set the width of displayed lines in the range from 2 to 16 pixels.
- Click on the **Color By** drop-down from **Streamlines** rollout and select the required parameter in accordance with the distribution of parameter.

LIC Style

- Click on the **LIC** button from **Streamlines** rollout to display a vector parameter using Line Integral Convolution (LIC) technique. It provides a global dense representation of the flow like wind-blown sand.

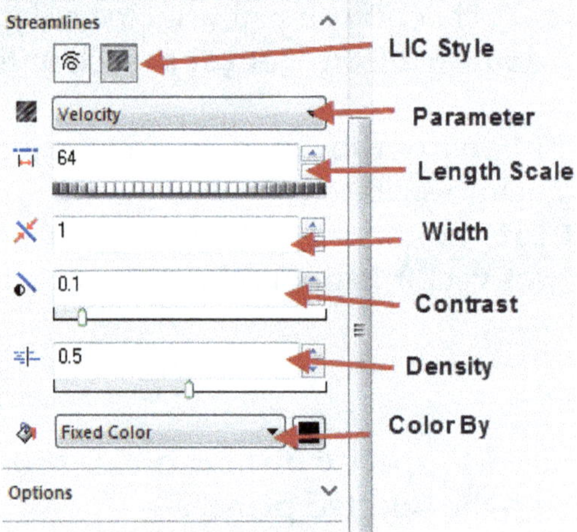

Figure-16. Options of LIC Style

- Click on the **Parameter** drop-down from **Streamlines** rollout and select the required option which you want to visualize with the streamlines.
- Click in the **Length Scale** edit box from **Streamlines** rollout to control the length of the displayed lines in the range from 4 to 128 pixels. You can also move the slider to set or adjust the value of **Length Scale**.
- Click in the **Width** edit box from **Streamlines** rollout and enter the value to set the width of displayed lines in the range from 1 to 10 pixels.
- Click in the **Contrast** edit box from **Streamlines** rollout and enter the value to set the difference in brightness between displayed lines and background in between 0 to 1. You can also set the value of contrast by moving the **Contrast** slider.
- Click in the **Density** edit box from **Streamlines** rollout and enter the value to control the line density scale of the displayed lines in the range from 0 to 1. You can also set the value of **Density** by moving the density slider.
- Click on the **Color By** drop-down from **Streamlines** rollout and select the required parameter in accordance with the distribution of parameter.

Mesh

The **Mesh** button is used to display the computational mesh in the Cut Plot if **Display Mesh** option is selected in **General Options**. The options of **Mesh** will be displayed; refer to Figure-17.

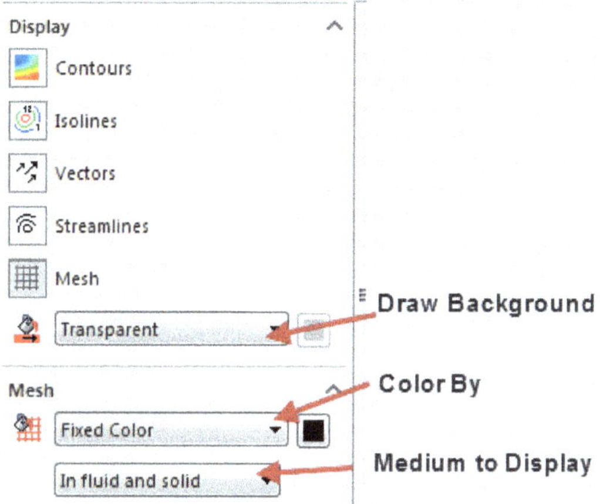

Figure-17. Options of Mesh

- Click on the **Draw Background** drop-down from **Display** section and select the required option to display the color background for the cut plot.
- Click on **Fixed button** from **Color By** drop-down of **Mesh** section and specify the a fixed color.
- Click in the **Medium to Display** drop-down from **Mesh** section to select a medium to display.
- After specifying the parameters of **Cut Plot**, click on the **OK** button. The model will be displayed with the selected options; refer to Figure-18.

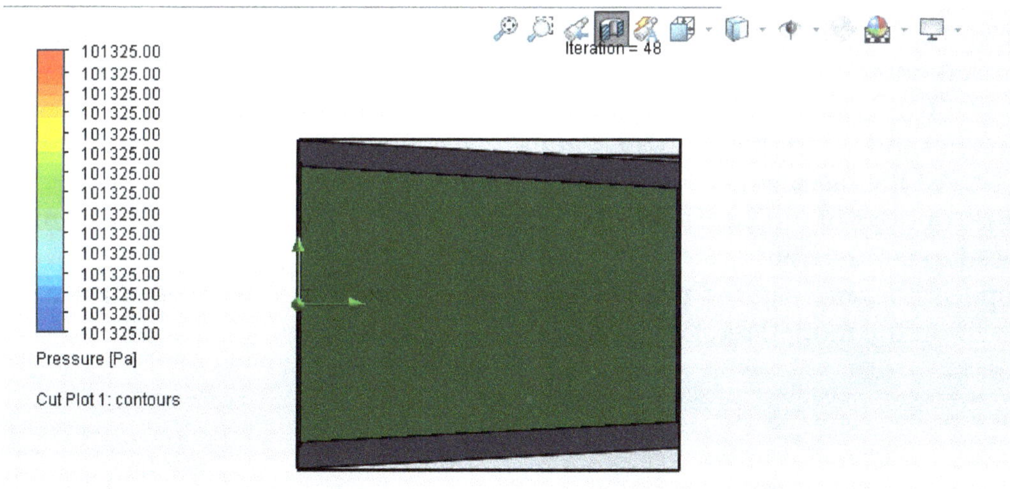

Figure-18. Cut Plot Contour

Surface Plots

The **Surface Plots** option is used to display parameter distribution on the selected plane, face, or surfaces.

- Right-click on the **Surface Plots** option from **Flow Simulation Analysis Tree**. The right-click menu will be displayed; refer to Figure-19.

- Click on the **Insert** button from displayed right-click menu. The **Surface Plot PropertyManager** will be displayed; refer to Figure-20.

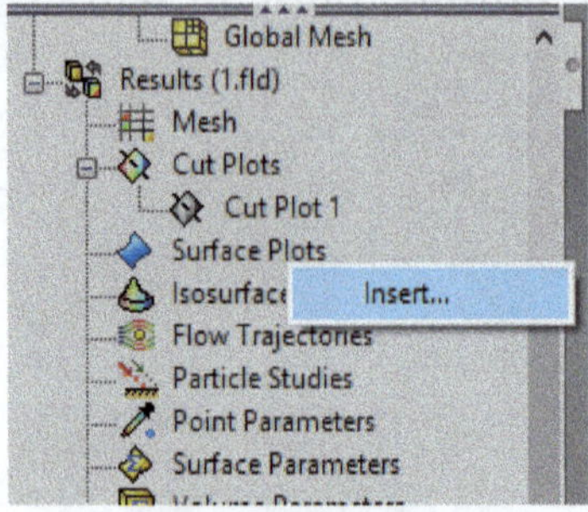

Figure-19. Right click menu of Surface Plots

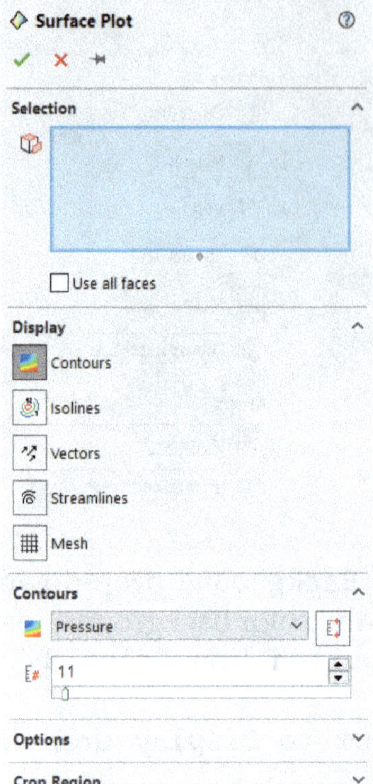

Figure-20. Surface Plot Property Manager

- The **Selection box** is activated by default. You need to select faces or surface of the model to display the parameter distribution. The selected faces will be displayed in the **Selection box**.
- If you want to select all the faces of a component then select the component from the selection list and select the **Use all faces** check box from the **Surface Plot Property Manager**. All the faces will be selected.
- The other options of the **Surface plot** are similar to the options of **Cut Plot** which are discussed in last section.
- After specifying the required parameters, click on the **OK** button. The surface plot will be displayed on the model; refer to Figure-21.

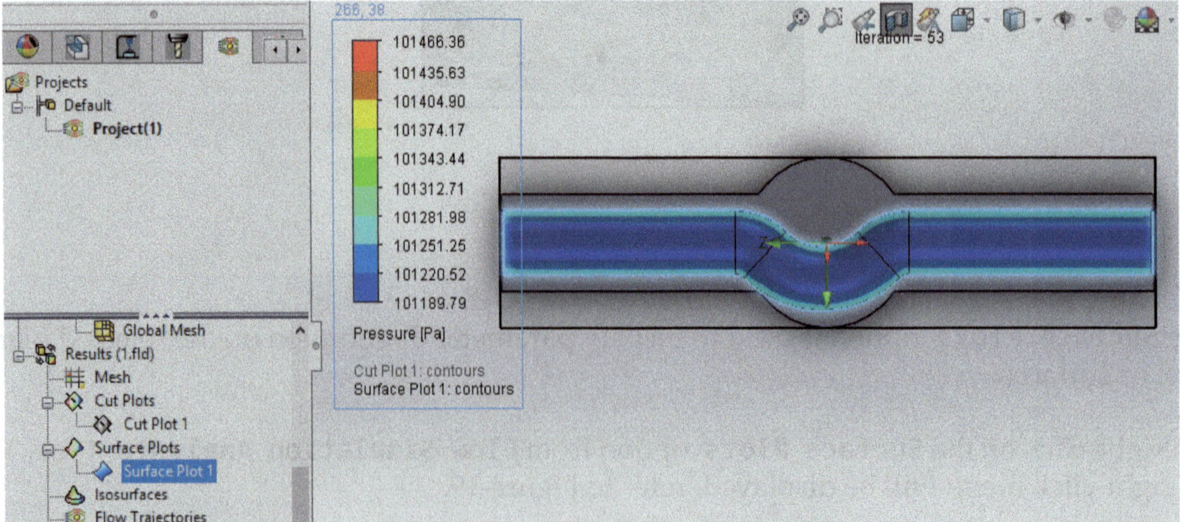

Figure-21. Displayed Surface Plot

Isosurfaces

The **Isosurfaces** option is used to display iso surfaces. Isosurfaces is a term defined as surface in which a parameter is constant. The Isosurface is fully defined by the parameter's value. If you want to check the span of 'x' parameter (like pressure) value in the fluid then this is the option to be used.

- Right-click on the **Isosurface** option from **Flow Simulation Analysis Tree**. The right-click menu will be displayed; refer to Figure-22.
- Click on the **Insert** button from the displayed menu. The **Isosurfaces PropertyManager** will be displayed; refer to Figure-23.

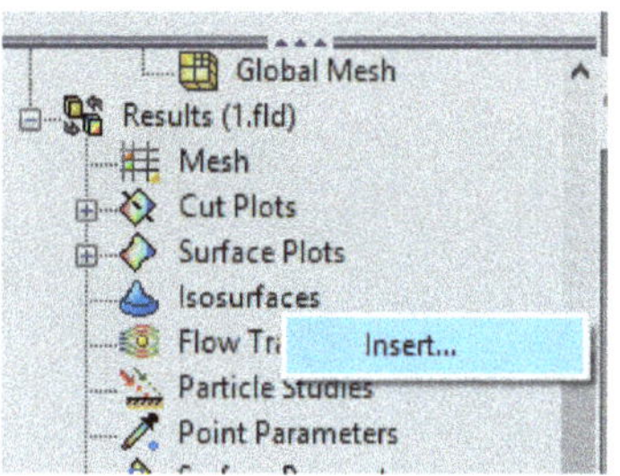

Figure-22. Right click menu of Isosurface option

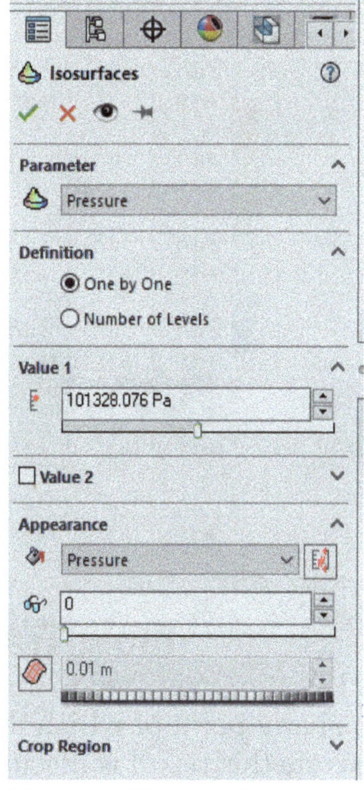

Figure-23. The Isosurfaces FeatureManager Design tree

- Click in the **Parameter** drop-down and select the required parameter for which you want to display isosurface.

One by One

- Select the **One by One** radio button from **Definition** section to specify each displayed value of the parameter separately.
- Click in the **Value 1** edit box and enter the desired value of the selected parameter.
- Select the **Value 2** edit box to create another isosurface. The **Value 2** edit box will be displayed. Enter the desired parameter. Similarly, you can set the value for **Value 3** and consecutive isosurfaces.

Number of Levels

- Select the **Number of Levels** radio button from **Definition** section to specify the range within which the parameter changes.
- Click in the **Minimum** and **Maximum** edit box from **Value** section and enter the desired values of the selected parameter.

- Click in the **Number of Levels** edit box from **Value** section to specify the number of levels for displaying selected parameter.
- Click on the **Color By** drop-down from the **Appearance** rollout and select the required parameter to specify the color in accordance with the distribution of parameter.
- Click in the **Plot Transparency** edit box and enter the desired value to set the plot transparency. The value of plot transparency lies between 0 to 1. You can also move the **Plot Transparency** slider to adjust the value.
- Click on the **Grid** button to activate the edit box. Click in the **Grid Step** edit box and enter the desired value to set the distance between adjacent lines of the grid projected on isosurfaces. You can also set the grid distance by moving the **Grid Step** slider.
- After specifying the parameters, click on the **OK** button. The chart will be displayed; refer to Figure-24.

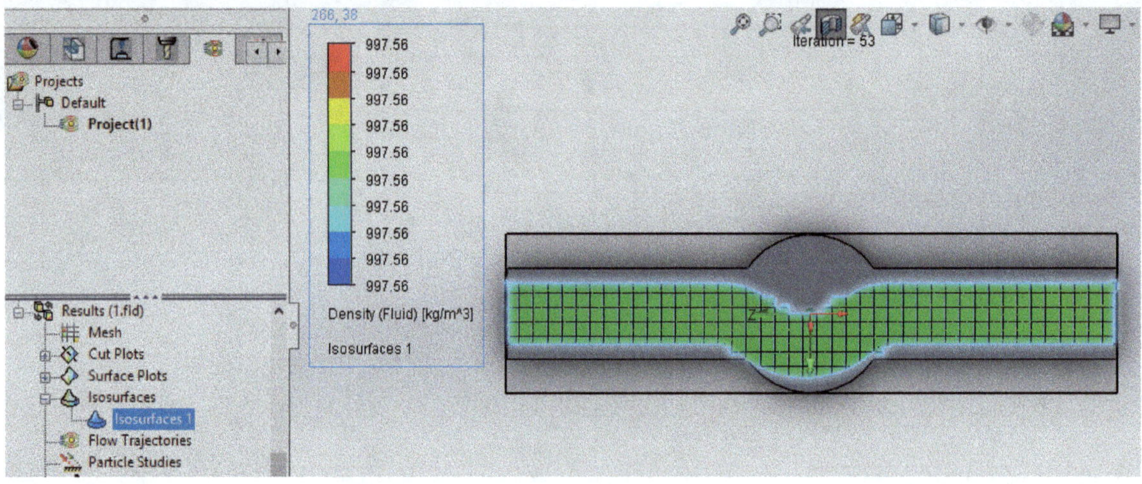

Figure-24. Chart of Isosurface option

Flow Trajectories

The **Flow Trajectories** option is used to display flow trajectories as flow streamlines. Flow streamline is a curve, at any point of which the flow velocity vector is tangent to that curve. The procedure to generate this result is discussed next.

- Right-click on the **Flow Trajectories** option from **Flow Simulation Analysis Tree**. The right-click menu will be displayed; refer to Figure-25.
- Click on the **Insert** option from the displayed menu. The **Flow Trajectories PropertyManager** will be displayed; refer to Figure-26.

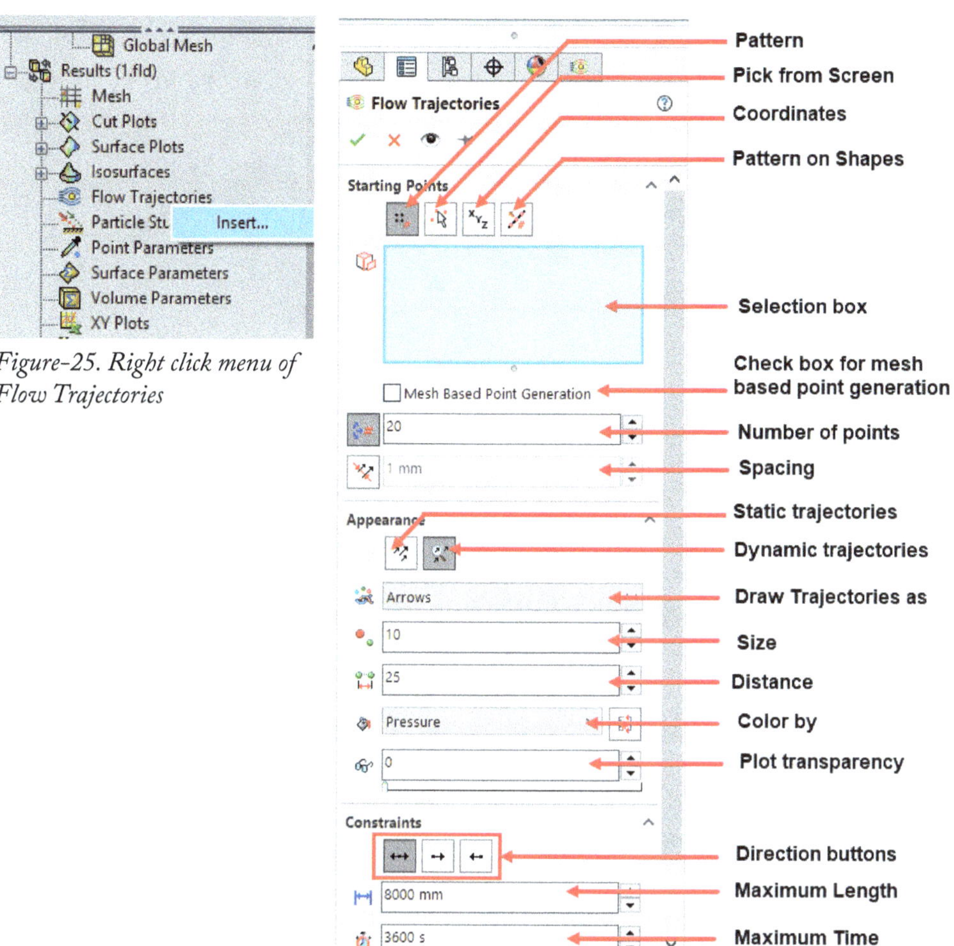

Figure-25. Right click menu of Flow Trajectories

Figure-26. Flow Trajectories PropertyManager

Pattern

- Click on the **Pattern** button from **Starting Points** section to evenly distribute the trajectories starting point over the selected planes, faces, edges, sketches, and curves.
- The selection box is active by default. You need to select the plane or face for flow trajectories.
- Select the **In plane** check box to obtain flow trajectories on the selected plane if you have selected a plane to display flow trajectories.
- Click in the **Offset** edit box and enter the desired vale to offset the selected plane.
- Click in the **Number of Points** edit box and enter the desired value to display or spacing defining the distance between the trajectories starting points.
- Click on the **Spacing** button to activate the **Spacing** edit box. Click in the **Spacing** edit box and enter the desired value to define the distance between the two trajectories starting point.

Pick from Screen

The **Pick from Screen** button is used to select starting points on a planer surface in the graphics area.

- The **Selection** box is active by default. You need to select the plane, face, or surface; refer to Figure-27. The only requirement for assemblies is that the selected plane should be perpendicular to fluid flow.

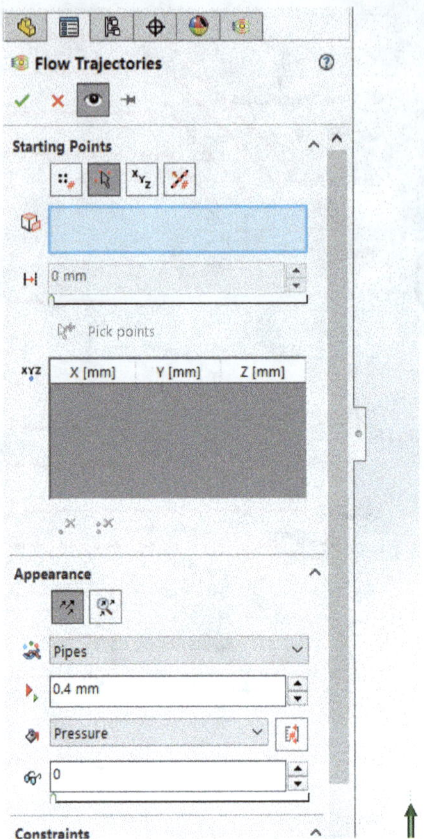

Figure-27. Options of Pick from Screen button

- Click in the **Offset** edit box from **Starting points** section and enter the desired value to offset starting point plane from the selected location. You can also move the slider to set the offset value.
- Click on the **Pick Point** button to allow the point selection and click on the selected plane or face from the graphics area to select a point on the plane. The selected point will be displayed in **Coordinates in Global coordinate system** box.
- If you want to delete the point then select the desired point from the table and click on the **Delete** button. If you want to delete all the selected points of table then click on the **Delete All** button.

Coordinates

- The **Coordinates** button is used to specify a set of starting points by creating a table with points coordinates.
- Click in the **X Coordinate** edit box and enter the desired coordinate of X for locating point.
- Click in the **Y Coordinate** edit box and enter the desired coordinate of Y for locating point.
- Click in the **Z Coordinate** edit box and enter the desired coordinate of Z for locating point.
- After specifying the coordinates of X,Y, and Z, click on the **Add Point** button. The point will be added and displayed in the point table; refer to Figure-28.

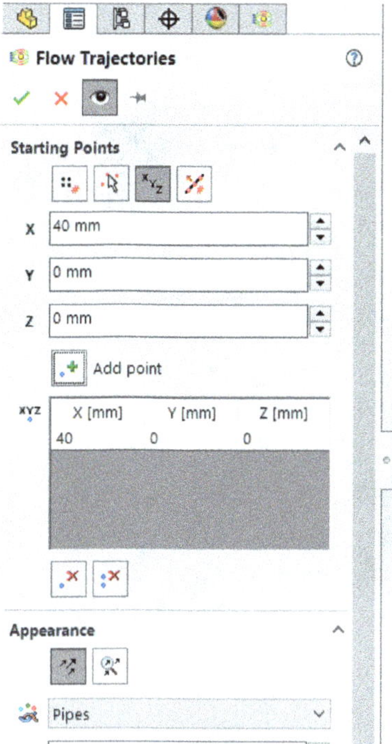

Figure-28. Specifying coordinates

Pattern on Shapes

- Click on the **Pattern on Shapes** button display the trajectory in the form of pattern of selected shape (line, rectangle, or sphere). The options in **PropertyManager** will be displayed as shown in Figure-29.
- Select the desired radio button to define shape on which flow trajectory will be created.
- Specify the desired value for number of points at which trajectory lines will be created; refer to Figure-30.

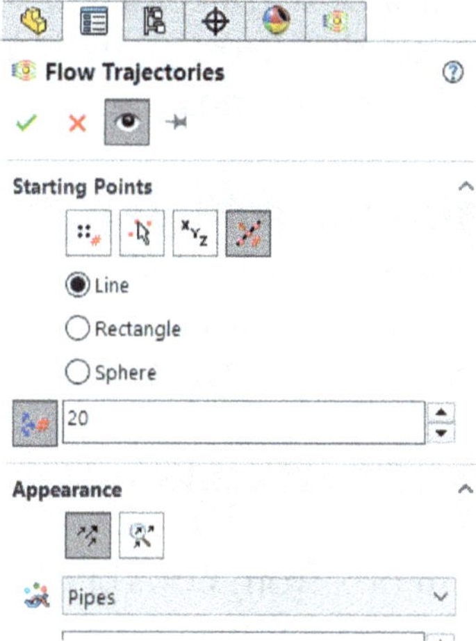

Figure-29. Pattern on shape options

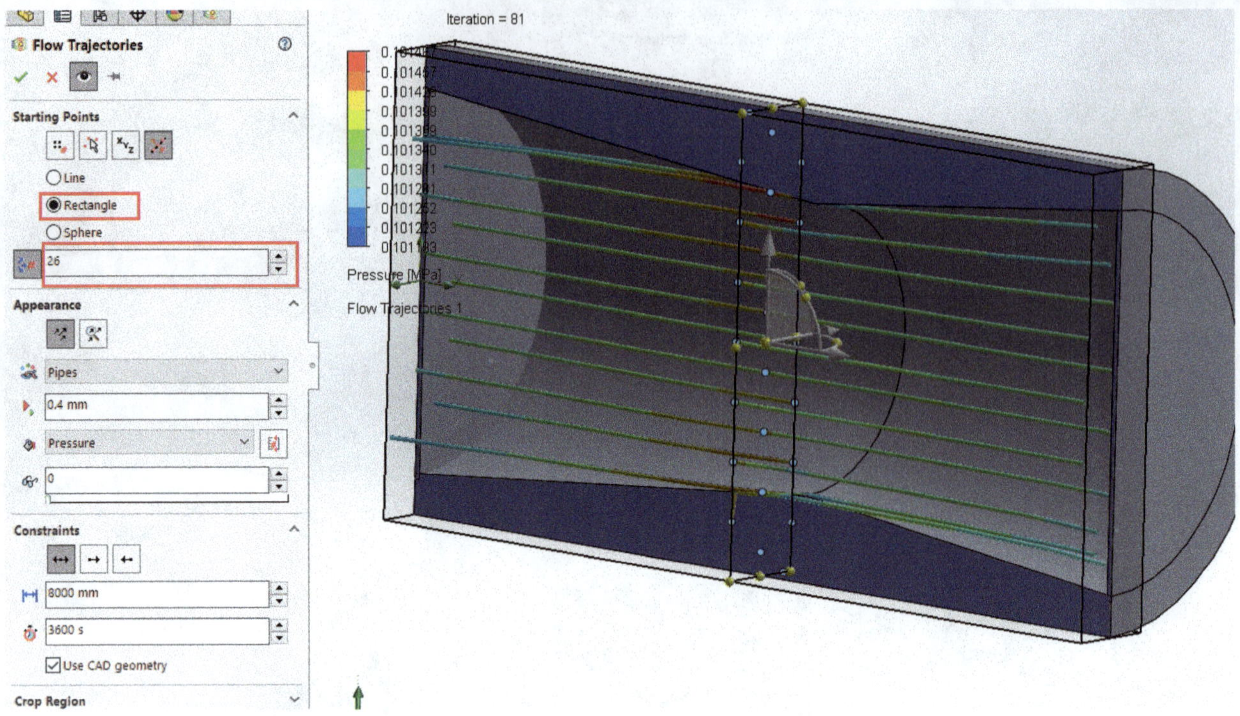

Figure-30. Flow trajectory on rectangle shape

Appearance rollout options

- Click on the **Draw Trajectories As** drop-down from **Appearance** rollout and select the option to specify the required trajectories.
- Click on the **Width** edit box from **Appearance** rollout and enter the value to specify the width of the selected option of **Draw Trajectories As** drop-down.
- Click in the **Color By** drop-down from **Appearance** rollout and select the required parameter to set the way of coloring flow trajectories.
- Click in the **Plot Transparency** edit box from **Appearance** rollout and enter the desired value to set the transparency of the trajectories.

Constraints rollout options

- Click on the **Both** button from **Constraints** rollout to display a whole flow streamline passing through the selected point from star end.
- Click on the **Forward** button from **Constraints** rollout to display the forward direction of streamline as starting from this point.
- Click on the **Backward** button from **Constraints** rollout to display the backward direction of streamline as ending from this point.
- Click in the **Maximum Length** edit box from **Constraints** rollout and enter the value of length of the trajectory to the specified value.
- Click in the **Maximum Time** edit box from **Constraints** rollout and enter the value to stop the trajectory travelling time of the flow has reached the maximum value.
- Select the **Use CAD geometry** check box from **Constraints** rollout to do not use the flow simulation interpreted geometry for the plot.
- Other parameters of this options have been discussed earlier.
- After specifying the parameters, click on the **OK** button from **Flow Trajectories PropertyManager**. The **Progress** dialog box will be displayed in which processing and calculations are displayed.

- After completion of calculations, the flow trajectories will be displayed on the model; refer to Figure-31.

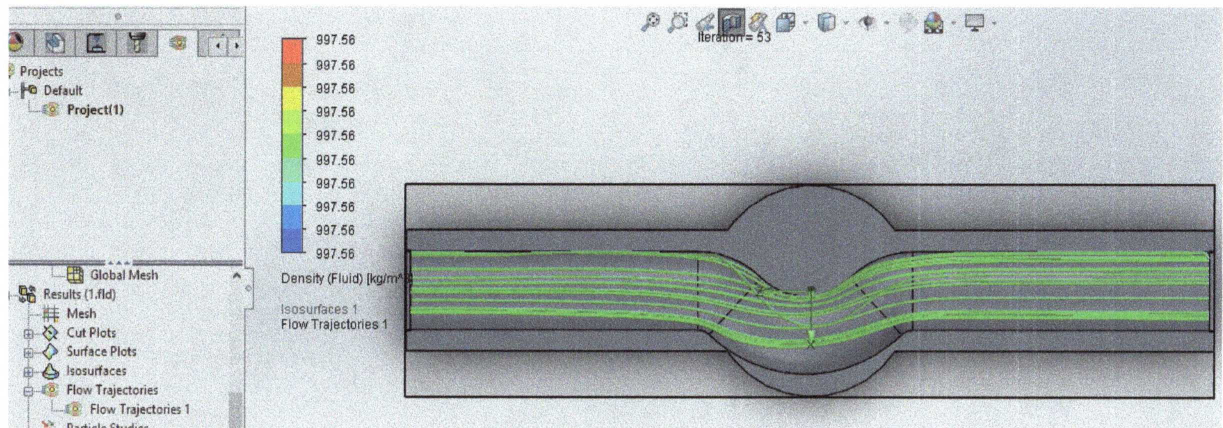

Figure-31. Flow Trajectories chart

Particle Studies

The **Particle Studies** option is used to display the trajectories of physical particles and obtain various information about the behaviors of particle. It also includes the effect of wall erosion or particles material accumulation as the result of particles interaction with walls.

- Right-click on the **Particle Studies** option from **Flow Simulation Analysis Tree**. The right-click shortcut menu will be displayed; refer to Figure-32.

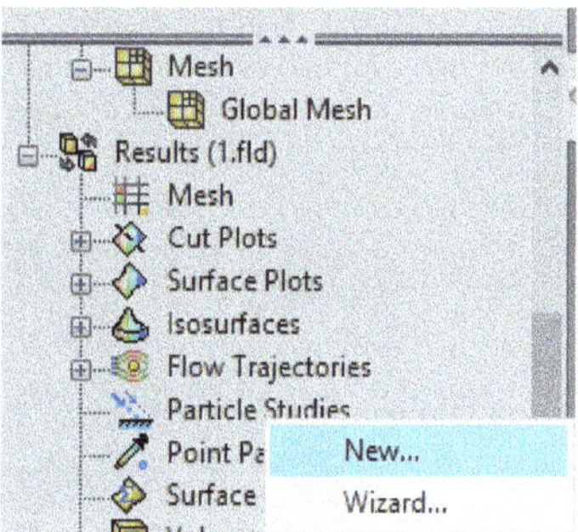

Figure-32. Right-click menu of Particle Studies

- Click on the **New** button from displayed right-click menu to start the particle study by using **PropertyManager**.
- Click on the **Wizard** button from displayed right-click menu to start the particle study by using wizard.

Creating a particle study by using wizard

Wizard is an easy way of creating particle study. The particle study is generally used to specify the particle injections, defining the physical settings, specifying the wall conditions, and defining calculation settings. After calculation you can see the trajectories of particle and obtain statistical information about particles. The procedure to use this is discussed next.

- Right-click on the **Particle Studies** option from **Flow Simulation Analysis tree**. The right-click menu will be displayed.
- Click on the **Wizard** button from displayed menu. The **Welcome** page of wizard will be displayed in the **PropertyManager**; refer to Figure-33.
- Click in the **Name** edit box and specify the desired name of particle study.
- After specifying the name, click on the **Next** button. The **Injection 1 Property Manager** will be displayed on the left in the screen; refer to Figure-34.

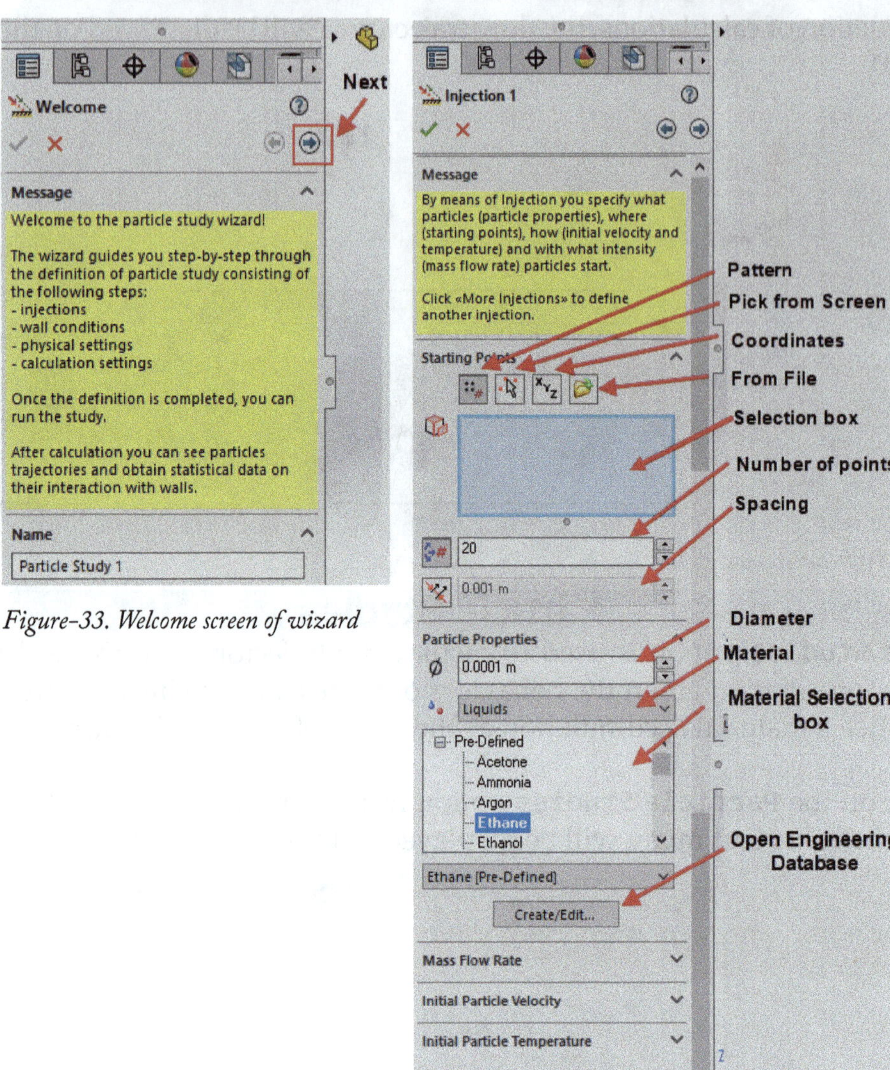

Figure-33. Welcome screen of wizard

Figure-34. Injection 1

Injection means a group of particles of the same material with initial conditions like velocity, diameter, temperature etc.

- The **Selection box** is active by default. You need to select the face or planer surface from where the particles will start flowing.
- Click in the **Offset** edit box and enter the desired offset value. You can also set the value by moving the **Offset** slider.
- Click in the **Number of Points** edit box and enter the desired value to define the number of particle points.
- Click on the **Spacing** button to specify spacing between consecutive particles, the **Spacing** edit box will become active. Click in the **Spacing** edit box and enter the desired value to define the distance between the two trajectories starting point
- Click in the **Diameter** edit box from **Particle Properties** section and enter the desired value of initial particle diameter. Since the considered particles are of constant mass so the volume of particle is variable.
- Click in the **Materials** drop-down from **Particle Properties** section and select the material type viz. Liquids or Solids.
- Click on the desired material from the **Material Selection** box to select.

- Click on the **Create/Edit** button to create a new material or edit the properties of the selected material.
- Click in the **Mass Flow Rate** edit box and enter the desired value to specify the mass flow rate of particles. The value of **Mass Flow Rate** defines the number of particles that are injected through the opening of a unit time. If you have used a file to define the injection, the mass flow rate should be specified for each particle trajectory.
- Select the **Relative** radio button of the **Initial Particle Velocity** section to specify the initial particle velocity vector at the particle starting point and specify the value in **Velocity Magnitude** edit box for specifying the Velocity vector.
- Select the **Absolute** radio button of the **Initial Particle Velocity** section to specify the initial particle velocity and specify the particle velocity vector in the respective edit boxes; refer to Figure-35.
- Select the **Relative** radio button from **Initial Particle Temperature** section to specify the initial particle temperature. The specified temperature is a difference between the particle and the fluid temperatures at the particle start point. The negative value corresponds to the lower temperature of the particle with respect to the fluid temperature.
- Click in the **Temperature** edit box and specify the value of initial particle temperature.
- Select the **Absolute** radio button from **Initial Particle Temperature** section to specify the initial particle temperature and enter the specific value in **Temperature** edit box.
- Click on the **More Injections** button to add more injections. The **Injection 2 PropertyManager** will be displayed.
- Click on the **Delete Injection** button to delete the recently created injection.
- After specifying the parameters, click on the **Next** button of **Injection PropertyManager**. The **Physical Settings PropertyManager** will be displayed; refer to Figure-36.

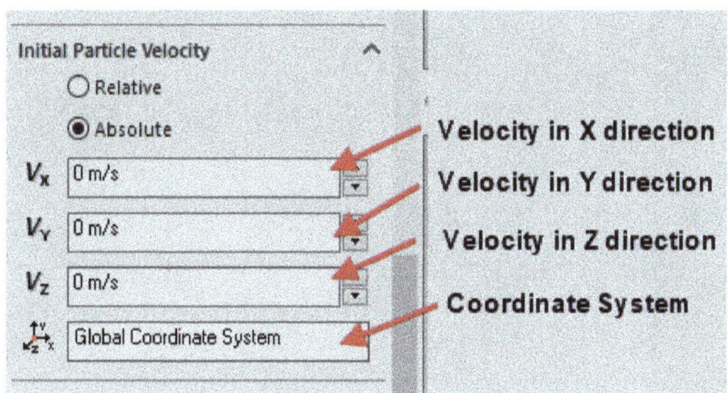

Figure-35. Options of Absolute radio button

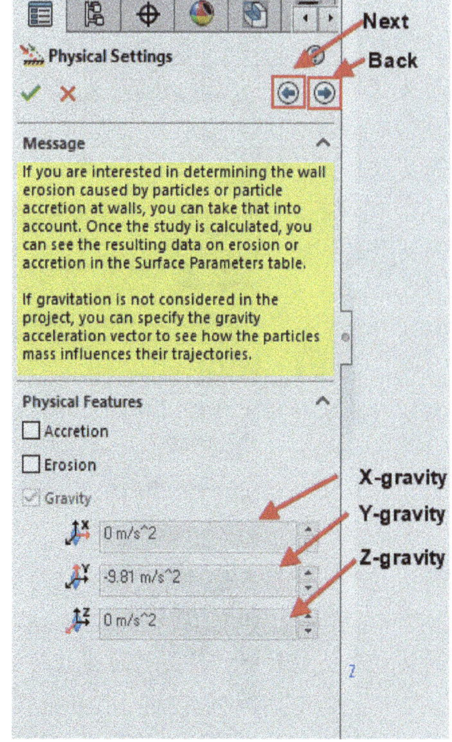

Figure-36. Options of Physical Settings

- The **Physical Setting** is used to see the physical features for details in the particle study.
- Select the **Accretion** check box of **Physical Features** section to specify mass accumulation rate in the particle study.
- Select the **Erosion** check box of **Physical Features** section to see the erosion of model wall in the particle study.

- In our case, the **Gravity** check box of **Physical Setting Property Manager** is disabled because the gravity check box is not selected at the start of **Project Wizard-Analysis Type**; refer to Figure-37.

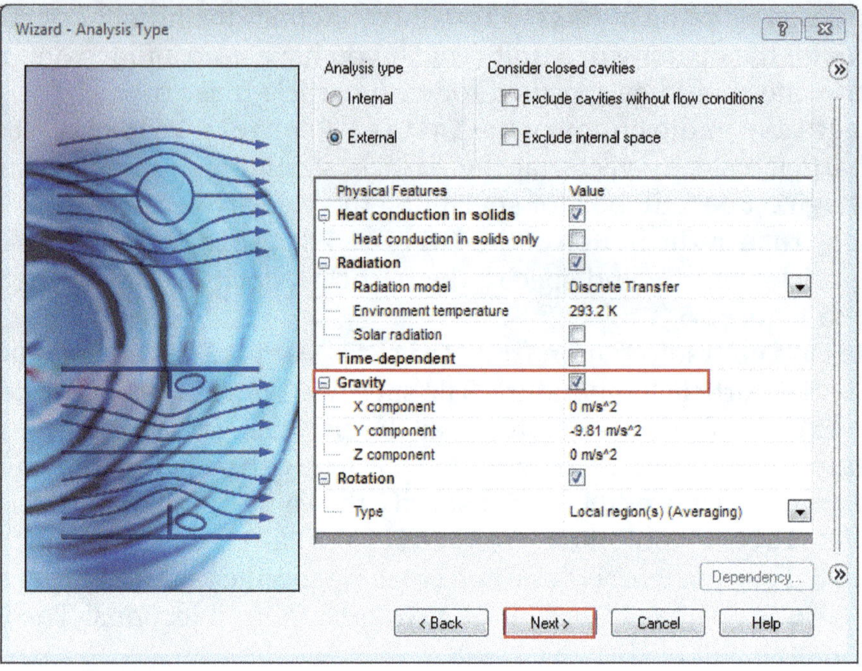

Figure-37. Updated wizard-Analysis Type dialog box

- After specifying the parameters of physical properties, click on the **Next** button. The **Default Wall Condition PropertyManager** will be displayed; refer to Figure-38.

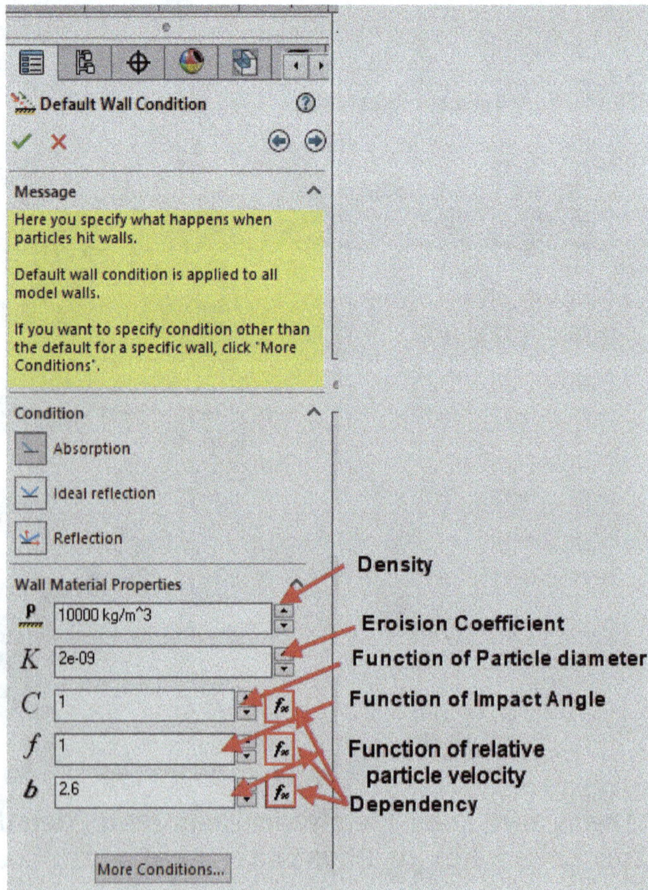

Figure-38. Default Wall Condition

- The **Default Wall Condition** is applied to all wall of the model unless it is redefined for a specific wall. It allows us to specify the particle behavior when particles meet a wall.
- Click on the **Absorption** button from **Condition** section to specify the absorption of particles through walls. This is typical for liquid particles.
- Click on the **Ideal Reflection** button from **Condition** section to specify the reflection of particles by the wall.
- Click on the **Reflection** button from **Condition** section to specify the normal and tangential restitution coefficients. The normal and tangential restitution coefficients are the ratio of absolute values of normal and tangential velocity. Enter the required value in respective edit box.
- The options of **Wall Material Properties** will be displayed if the **Erosion** check box was selected in **Physical Setting Property Manager**.
- Click in the **Density** edit box from **Wall Material Properties** section and enter the desired value of wall material density.
- Click in the **Erosion Coefficient** edit box from **Wall Material Properties** section and enter the value of coefficient. The Erosion Coefficient can be used as a proportionality constant if other parameters are specified in a system of units which are different from the units of project system. The default value is $2x10^{-9}$.
- Click in the **Function of Particle Diameter** edit box from **Wall Material Properties** and enter the value of particle diameter which depends on erosion.
- Click on the **Dependency** button to specify the table dependency on the diameter. The **Dependency** dialog box will be displayed which was discussed earlier.
- Click in the **Function of Impact** edit box from **Wall Material Properties** section and enter the value to define the dependency of erosion from the particle impact angle.
- Click in the **Function of Relative Particle Velocity** edit box from the **Wall Material Properties** section to define the dependency of erosion from the particle relative velocity. The default value is **2.6**.
- Click on the **More Conditions** button to customize the wall boundary condition for a specific wall after specifying the Default Wall Condition.
- After specifying the parameters of **Default Wall Condition**, click on the **Next** button. The **Calculation Settings Property Manager** will be displayed; refer to Figure-39.
- The **Calculation Setting PropertyManager** allow us to specify the criteria for terminating calculation of particles trajectories and parameters that you want to be available for results processing and display options.
- Select the **Trajectories and statistics** radio button from **Results Savings** section to visualize the particle trajectories and obtain statistical information about their interaction with the walls after calculation.
- Select the **Statistics only** radio button from **Results Saving** section to save statistical information on results processing after the finishing of particle study. It is recommended that use this radio button when you need to study a huge number of trajectories. This **Default Appearance** rollout is disabled when the **Statistics** radio button is selected.
- The other sections like **Default Appearance**, **Constraints**, and **Computational Region** are same as discussed earlier.
- After specifying the parameters, click on the **Next** button. The **Run PropertyManager** will be displayed; refer to Figure-40.
- The particle study definition is completed here. Now you need to click on the **Run** button to calculate the study. Click on the **Run** button from **Run PropertyManager**. The processing will start.
- After the finishing of calculations, the **Wall Conditions** and **Injections** feature will be added in **Particle Studies** node of results; refer to Figure-41.

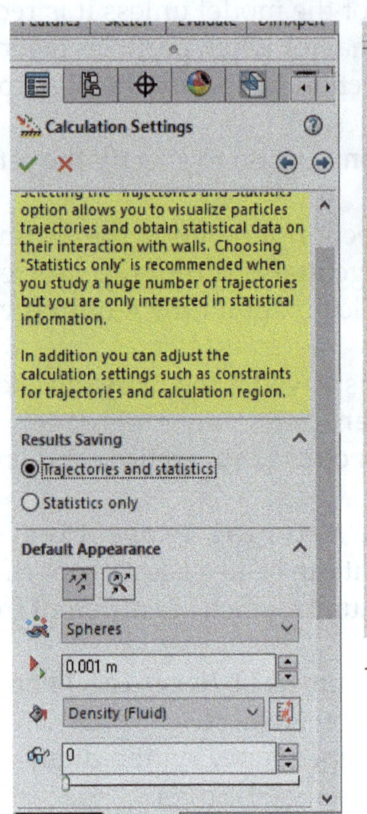

Figure-39. Calculation Settings Property Manager

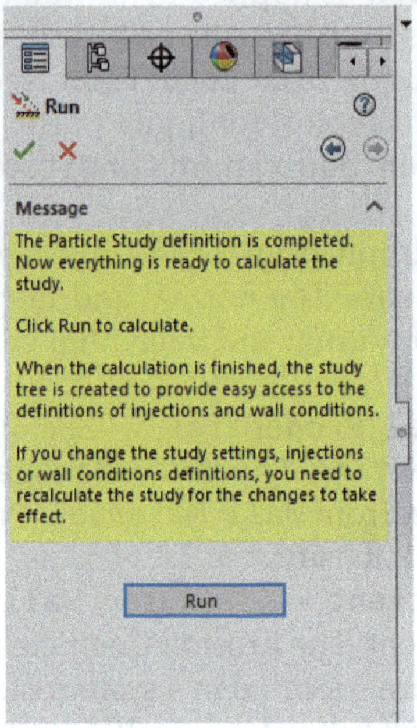

Figure-40. Run Property Manager

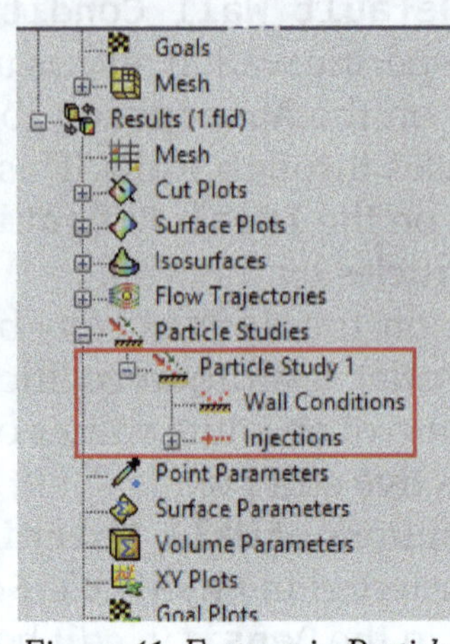

Figure-41. Features in Particle Study 1

Manual Setting of Particle Study Wall Condition

- Right-click on the **Wall Condition 1** feature from **Results Design tree**. The right-click menu will be displayed; refer to Figure-42.

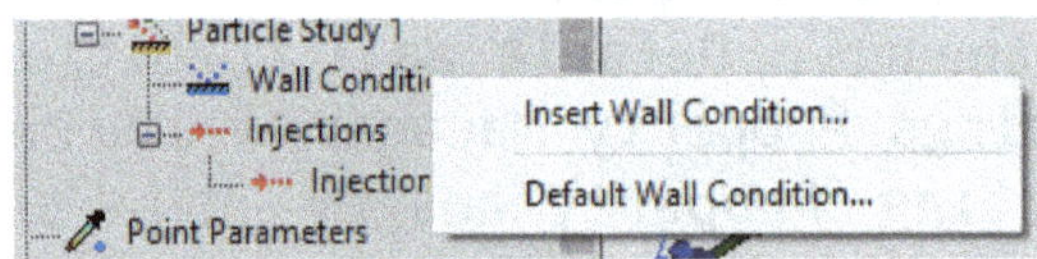

Figure-42. Right-click menu of Wall Conditions

- Click on the **Insert Wall Condition** button from the displayed right-click menu if you want to add a new wall condition different from what was specified at the beginning. The **Wall Condition 1 PropertyManager** will be displayed; refer to Figure-43

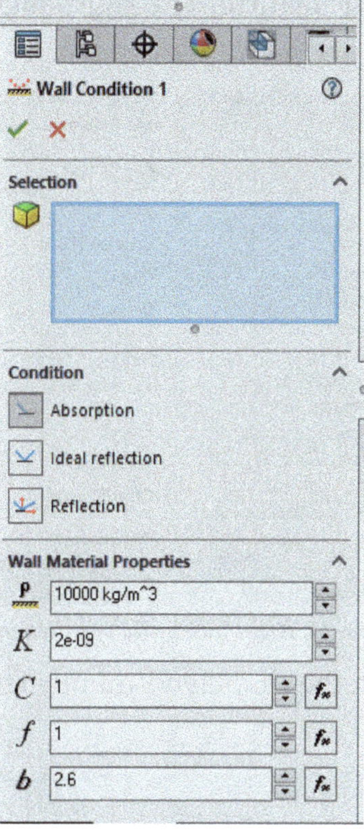

Figure-43. Wall Condition 1 Property Manager

- The **Selection box** is active by default. You need to click on wall of the model to select; refer to Figure-44. The selected wall will be displayed in the selection box.

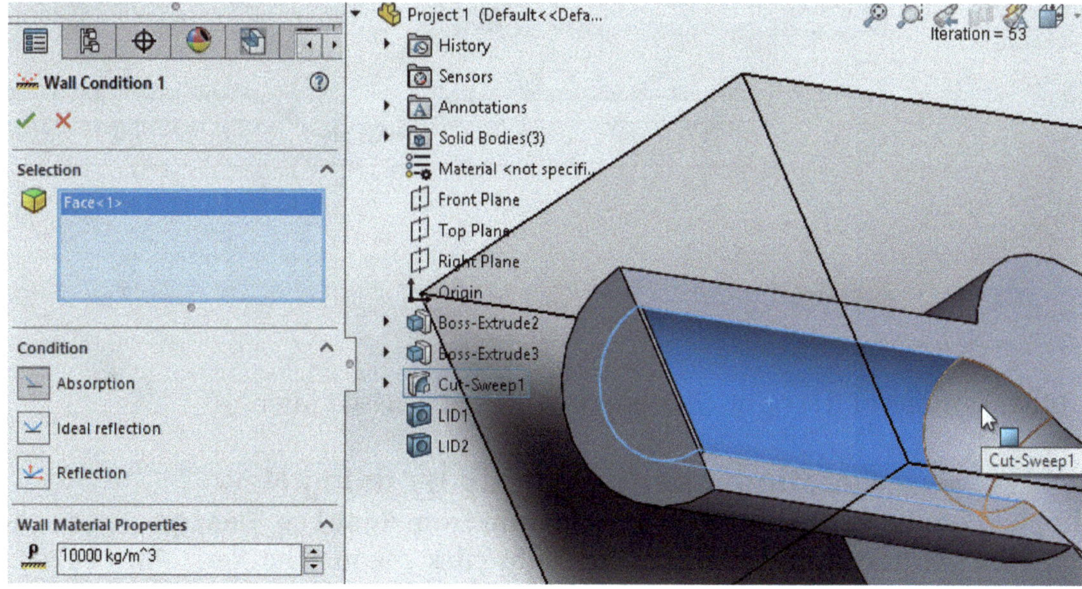

Figure-44. Selecting wall

- Now, select the conditions as per requirement and specify the parameters in the **Wall Material Properties** section of the wall condition.
- After specifying the parameters, click on the **OK** button from **Wall Condition 1 PropertyManager**. The **Wall Condition 1** will be added in the **Results Design Tree**.
- Click on the **Default Wall Conditions** button from the displayed right-click menu of the **Wall Conditions** option to use the default parameters stated in the **Particle Study** process.

- Right-click on the **Injection 1** feature. The right-click menu will be displayed; refer to Figure-45.

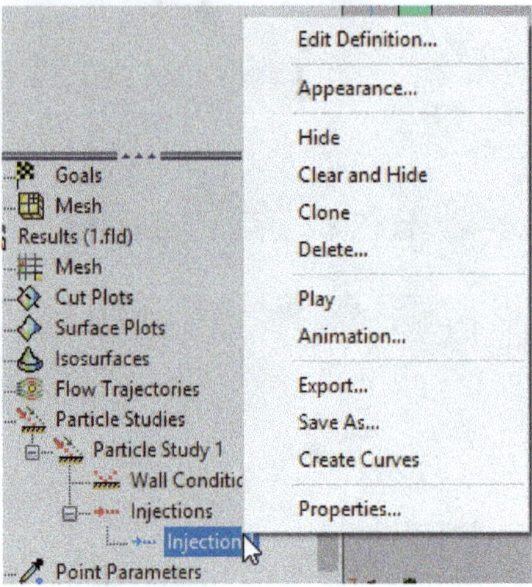

Figure-45. Right click menu of Injection

- Click on the **Show** button from the displayed menu. The animation on the model will be displayed; refer to Figure-46.

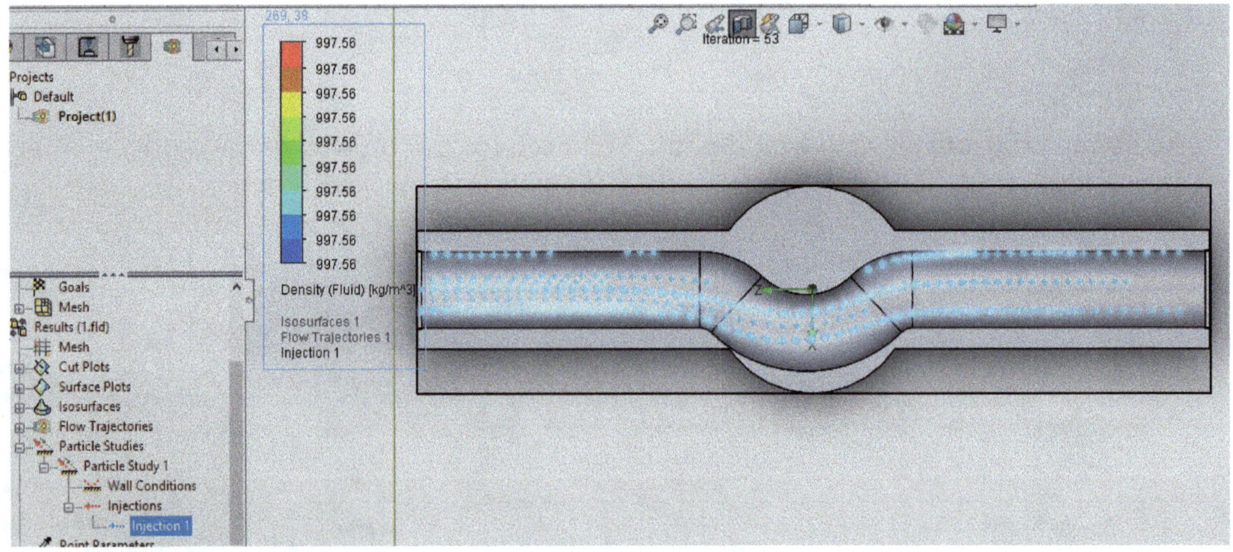

Figure-46. Showing Injection 1

- The other options of the right-click menu will be discussed later.

Creating a particle study by using New

- Right-click on the **Particle Studies** feature from **Results Design Tree** and click on the **New** button from the displayed right-click menu. The **New Particle Study PropertyManager** will be displayed; refer to Figure-47.

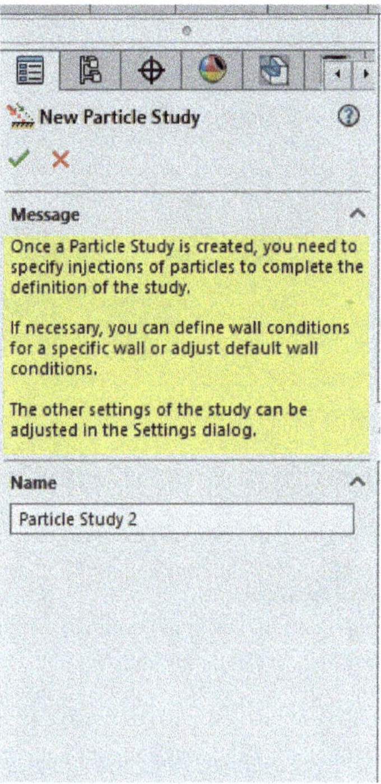

Figure-47. The New Particle Study Property Manager

- Click in the **Name** edit box and enter the desired name of particle study.
- Click on the **OK** button from **New Particle Study Property Manager**. The **Particle Study 2** will be created and added in **Particle Studies** node if **Particle Study 1** is already there; refer to Figure-48.

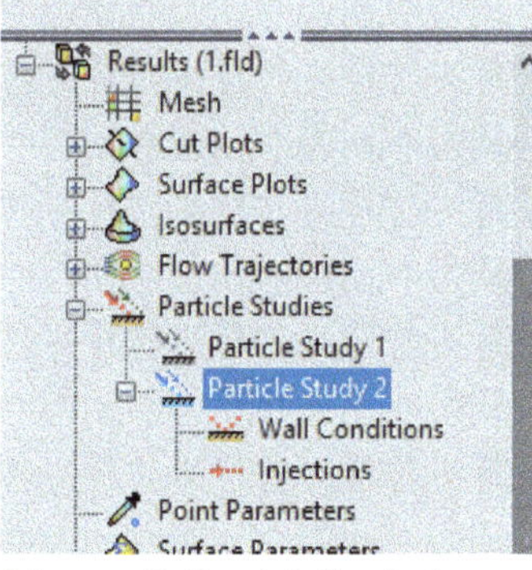

Figure-48. Particle Study 2

- Specify both the conditions of the particle study separately for the particle study as discussed earlier.

Point Parameters

The **Point Parameter** feature allow us to display parameter values at a specified points inside the **Computational Domain**. These points of interest can be specified by their coordinates or by reference geometrical features. The procedure to use this feature is discussed next.

- Right-click on the **Point Parameters** feature from **Results Design Tree**. The right-click shortcut menu will be displayed; refer to Figure-49.

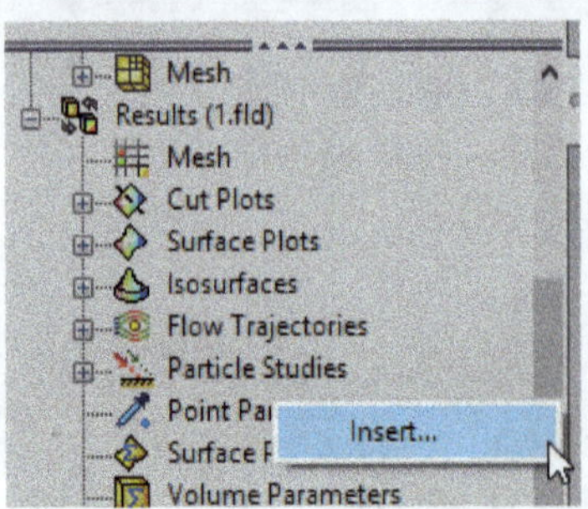

Figure-49. Right-click menu of Point Parameter

- Click on the **Insert** option from the displayed menu. The **Point Parameters PropertyManager** will be displayed; refer to Figure-50.

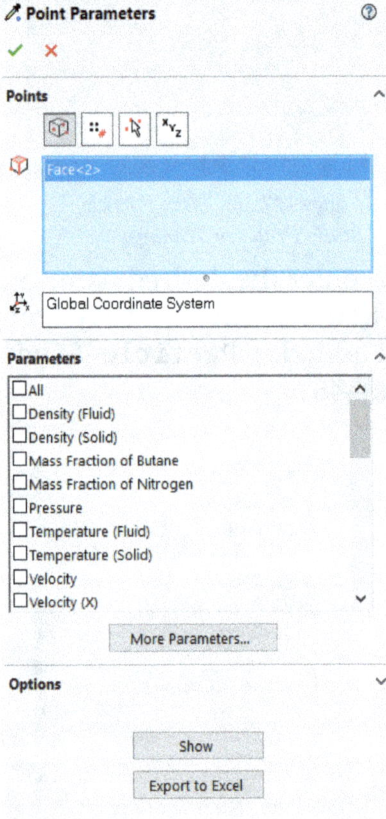

Figure-50. The Point Parameter Property Manager

- The **Selection box** is active by default. You need to click on the points, vertices, planer faces, or edges to check the parameters on selected location.
- Select the required parameter check box from **Parameters** section to evaluate the selected parameter.
- Click on the **More Parameters** button from **Parameters** section to change the list of parameters available for selection. The **Customize Parameter List** will be displayed; refer to Figure-51.

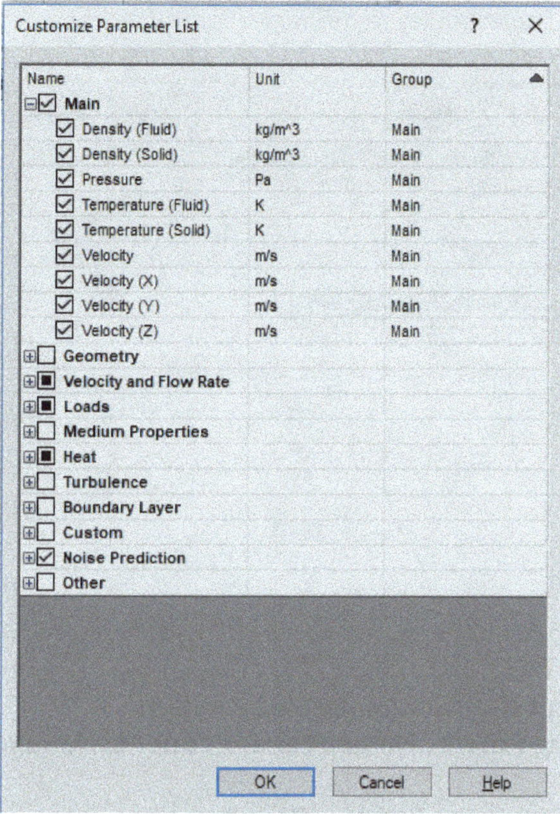

Figure-51. Customize Parameter List

- Select the check box of desired parameter to be checked from the dialog box and click on the **OK** button. The selected parameters will be added in the **Parameters** section.
- Click on the **Show** button from **Options** section to calculate and show data. The **Point Parameters 1** tab will be displayed along with calculated parameters; refer to Figure-52.

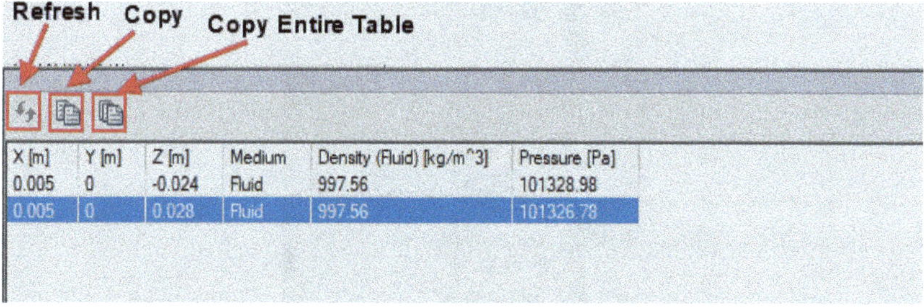

Figure-52. Point Parameter tab

- Click on the **Refresh** button to recalculate the displayed data.
- Click on the **Copy** button to copy the selected value. You can also copy values by using right-click shortcut menu. To do so, right-click on the particular value and click on **Copy** option.
- Click on the **Copy Entire Table** button to copy the values of entire table or we can say to copy the entire table.
- Click on the **Close** button of **Point Parameters 1** tab to close the tab window. The **Point Parameters Property Manager** will be displayed again.
- Click on the **Export to Excel** button from **Options** tab to calculate the parameter and export the data to the Microsoft Excel. The Microsoft Excel window will be displayed along with the calculated data; refer to Figure-53.

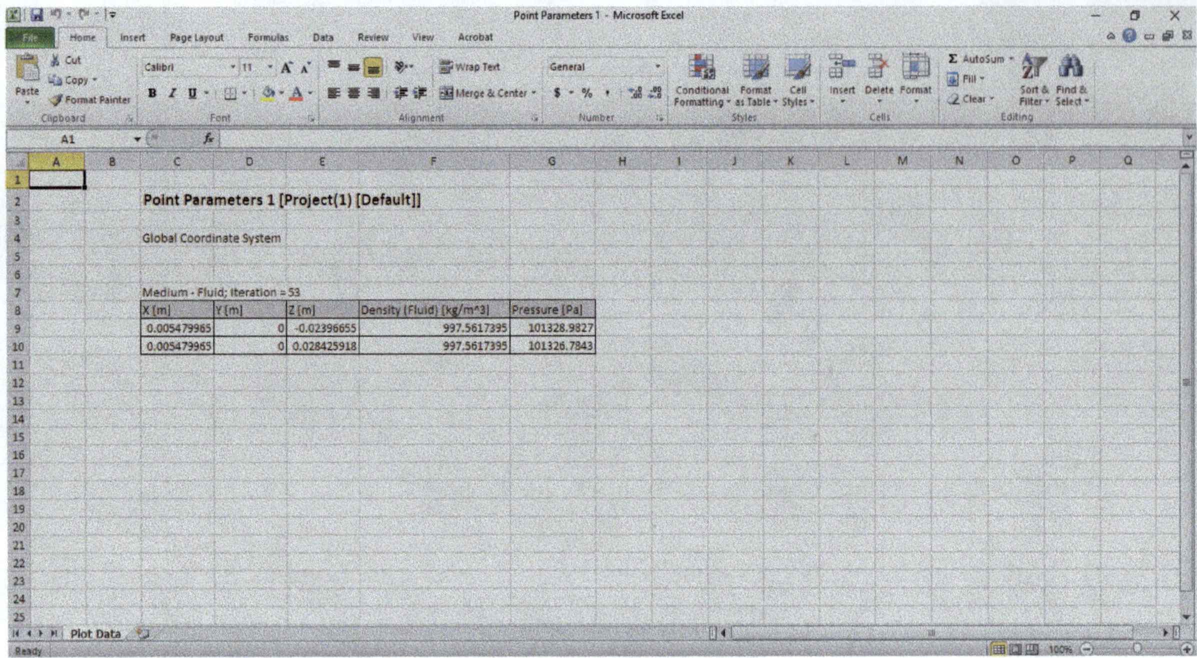

Figure-53. Window of Microsoft Excel

- Save the Microsoft Excel sheet to the desired location.
- Other parameters of the **Point Parameters PropertyManager** are same as discussed earlier.
- Click on the **OK** button from **Point Parameters PropertyManager**. The **Point Parameters 1** feature will be added in the **Design Tree**; refer to Figure-54 and the study of point parameters will be displayed on the model; refer to Figure-55.

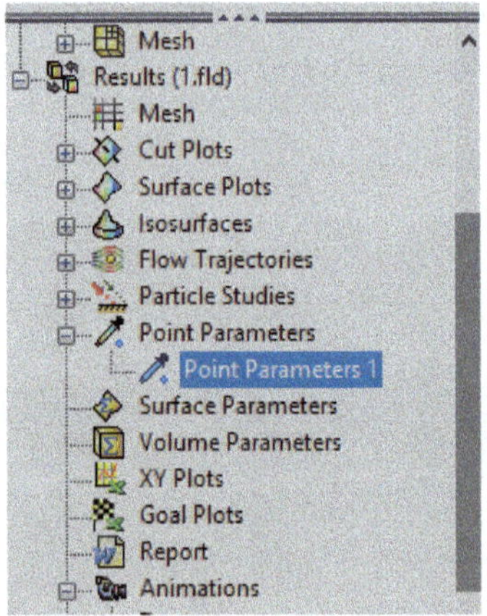

Figure-54. Point Parameters 1 feature

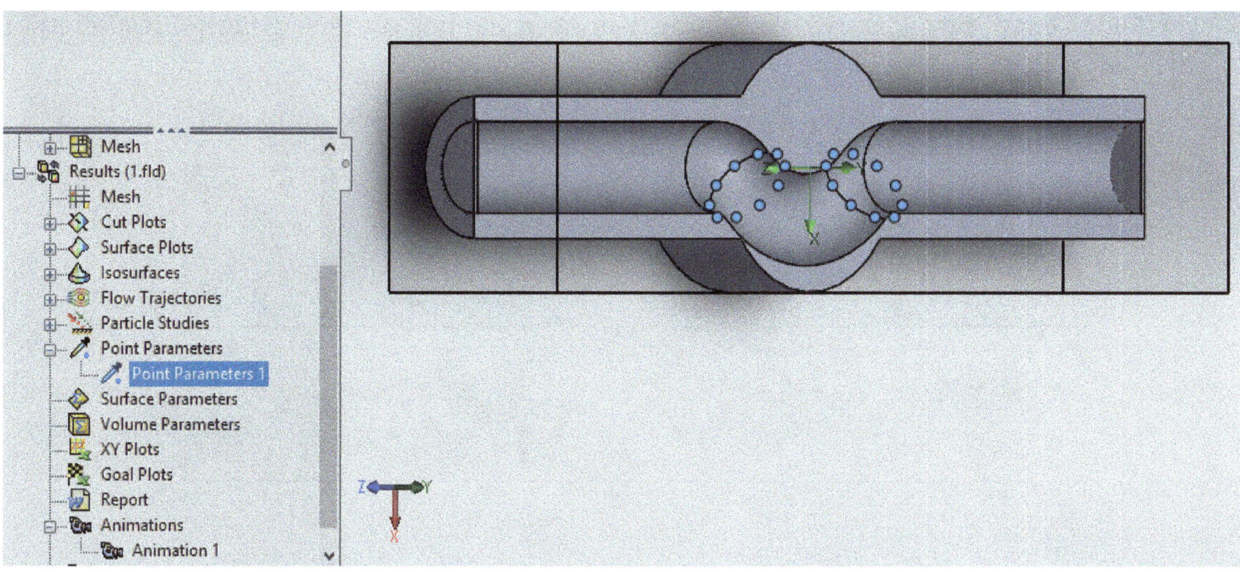

Figure-55. Point Parameter study

Surface Parameters

The **Surface Parameter** is used to display parameter values like minimum, maximum, average, and integral which are calculated over the specified surface. The procedure to use this is discussed next.

- Click on the **Insert** button from right-click menu of **Surface Parameters** feature in the **Results Design Tree**. The **Surface Parameters PropertyManager** will be displayed; refer to Figure-56.

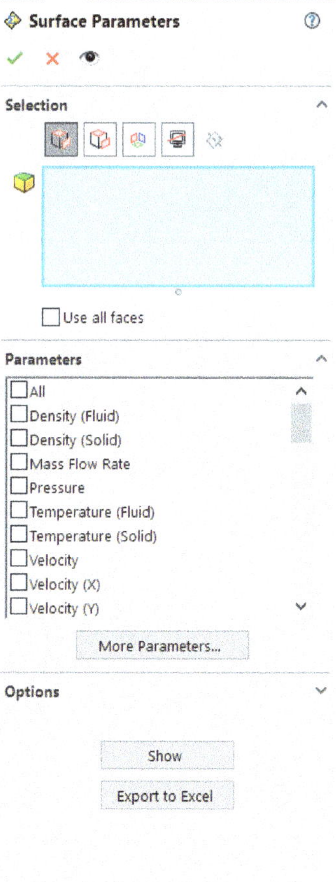

Figure-56. Surface Parameter Property Manager

- The **Selection box** is active by default. You need to click on the surface of model to be selected; refer to Figure-57.

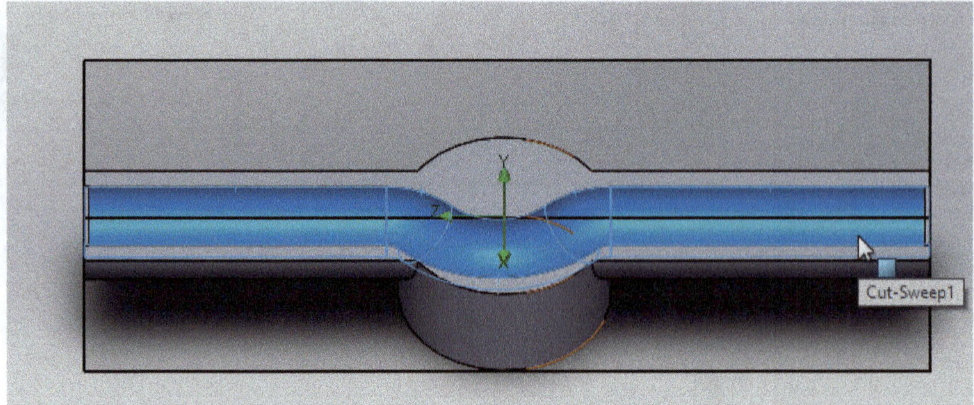

Figure-57. Selecting surface of model

- Select the desired surface. The selected surface will be displayed in **Selection box**.
- The other parameters of **Surface Parameters PropertyManager** are same discussed earlier.
- After specifying the parameters, click on the **OK** button from **Surface Parameters Property Manager**. The **Surface Parameter 1** feature will be added in **Design Tree**.
- Click on the **Surface Parameter 1** feature from **Design Tree** to see the surface parameter on the model; refer to Figure-58.

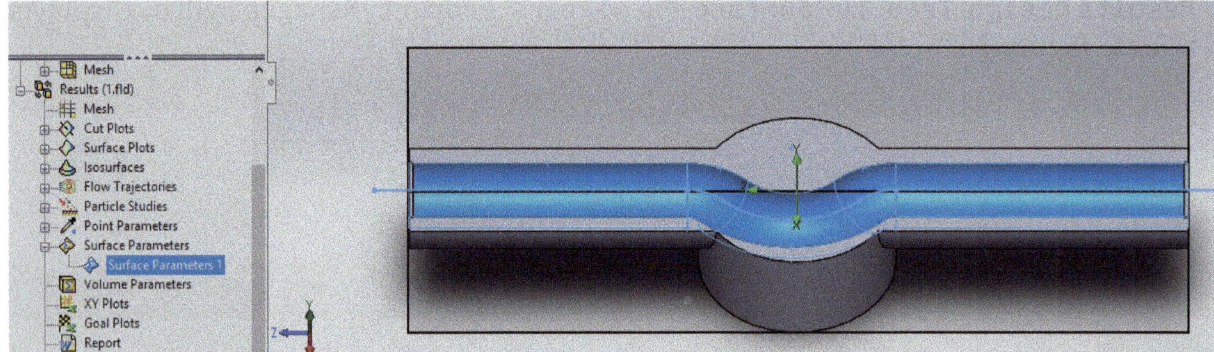

Figure-58. Surface Parameter on the model

Volume Parameters

The Volume Parameters is used to display parameter values like minimum, maximum, average, bulk average, and integral of the physical properties which are calculated within the specified volumes like subassembly components in assemblies as well as bodies in multi-body parts within the computational domain. The procedure to use this is discussed next.

- Click on the **Insert** button from right-click menu of **Volume Parameters** feature **Design Tree**. The **Volume Parameters PropertyManager** will be displayed; refer to Figure-59.

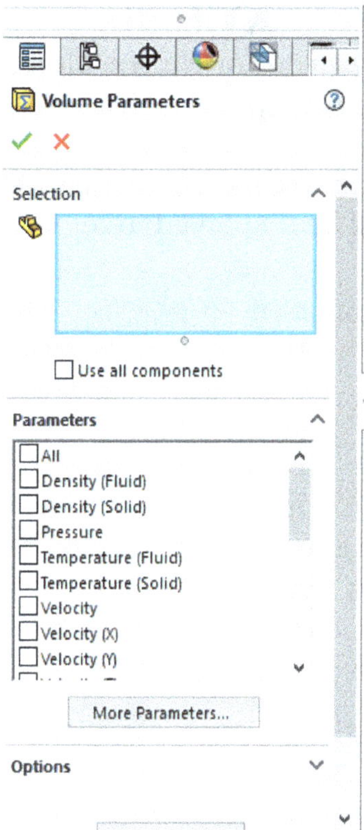

Figure-59. Volume Parameters Property Manager

- The **Selection box** is active by default. You need to click on the volume of model to selected; refer to Figure-60.

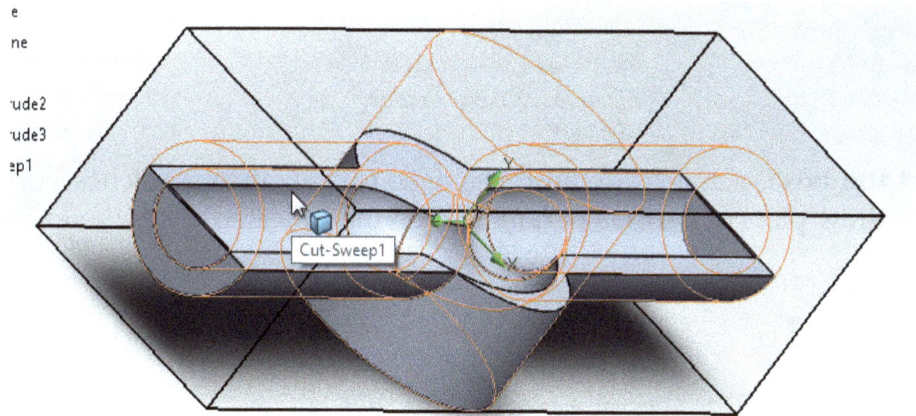

Figure-60. Selecting Volume

- Click at the desired part. The selected volume will be displayed in the selection box.
- Select the required parameter for study from **Parameters** section.
- The other parameters of **Volume Parameters Property Manager** are same as discussed earlier.
- After specifying the parameters, click on the **OK** button from **Volume Parameters Property Manager**. The **Volume Parameter 1** feature will be added in **Design tree**.

XY Plots

The **XY Plots** is used to create sketches or curves, along which you can see the parameter distribution. The procedure to use this option is discussed next.

- Click on the **Insert** button from right-click menu of **XY Plots** in **Design Tree**. The **XY Plots PropertyManager** will be displayed; refer to Figure-61.

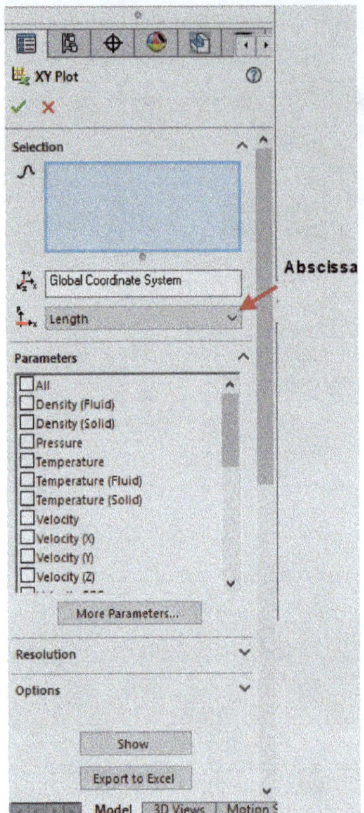

Figure-61. XY Plot Property Manager

- The **Selection** box is active by default. You need to click on the sketches, curves, or edges of model to draw plots of various parameters on that selected entity; refer to Figure-62.

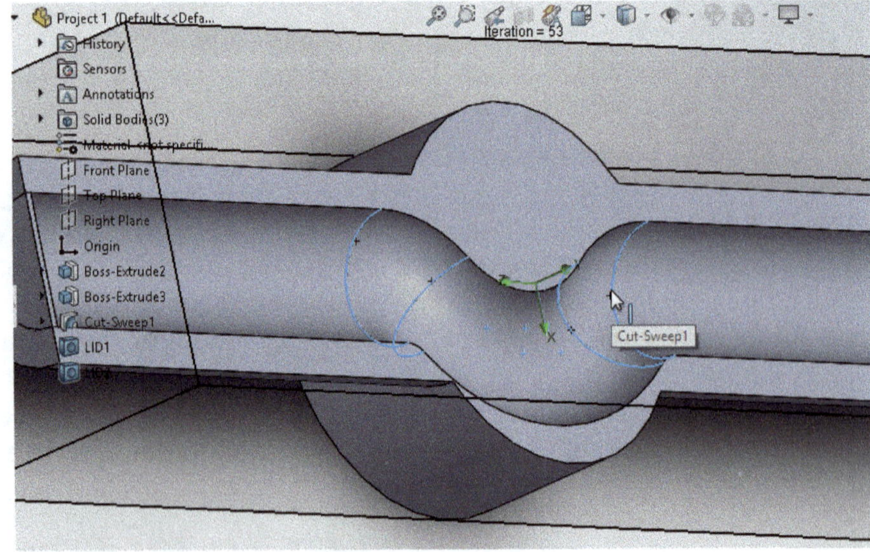

Figure-62. Selection of edges of XY Plots

- Select the **Length** button of **Abscissa** drop-down from **Selection** section to select the length of curve as chart's abscissa.
- Select the **Model X,Y, or Z** button of **Abscissa** drop-down from **Selection** section to select the Model X,Y, or Z axes as chart abscissa.
- The other parameters of this feature as same as discussed earlier.
- After specifying the parameters, click on the **OK** button from **XY Plots Property Manager**. The **XY Plot 1** feature will be added in the **Design Tree**.
- Right-click on the recently created **XY Plot 1**. The right-click menu will be displayed; refer to Figure-63.

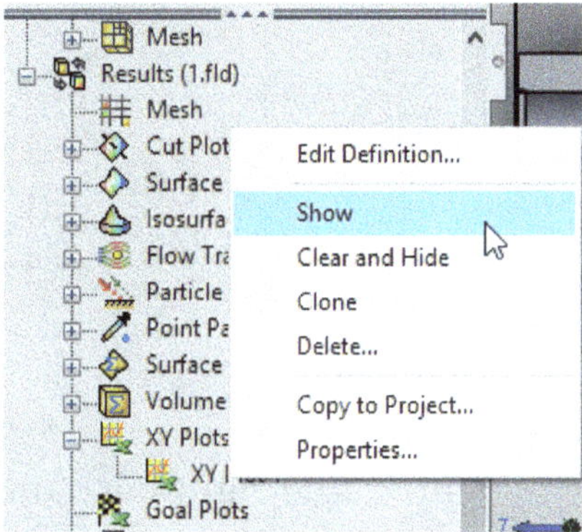

Figure-63. Right click menu of XY Plots 1

- Click on the **Show** button to view the study of selected parameters. The chart will be displayed along with the model; refer to Figure-64.

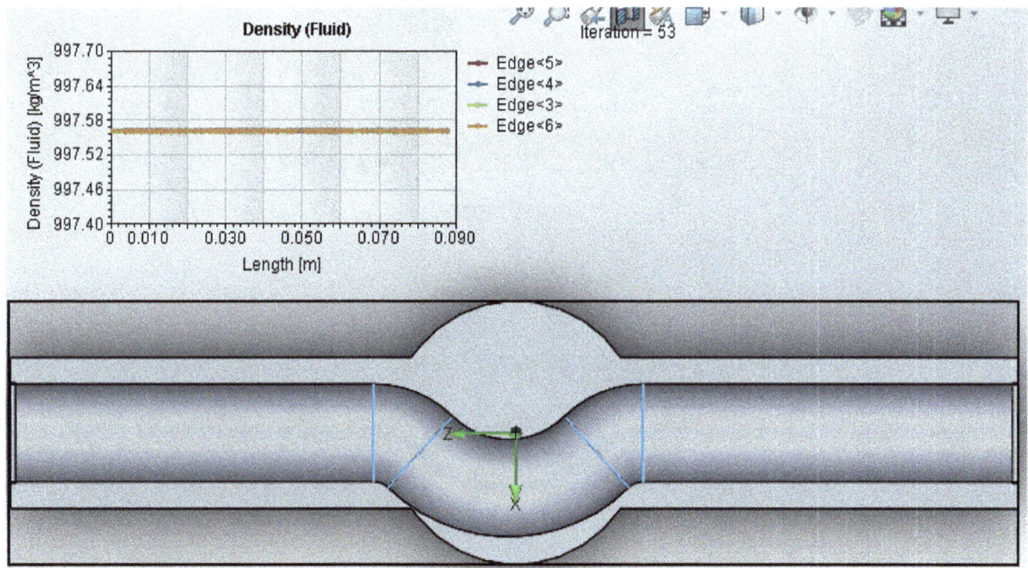

Figure-64. Chart of XY Plots

- Click on the **Hide** button from the right-click menu of **XY Plots 1**. The chart of the model will be hidden.

Goal Plots

The **Goal Plots** is used to study goal changes in the course of calculation. The procedure to use this feature is discussed next.

- To insert the **Goal Plots**, you should have at least one goal added in the project. You can add goals from the **Tools > Flow Simulation > Insert >** menu.

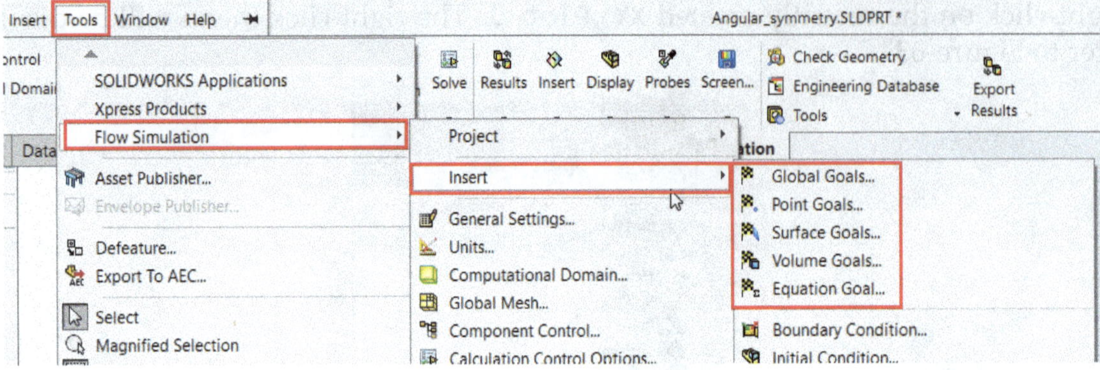

Figure-65. Options to define goals

- Right-click on the **Goal Plots** feature. The right-click menu will be displayed.
- Click on the **Insert** button from the displayed menu. The **Goal Plot PropertyManager** will be displayed; refer to Figure-66.

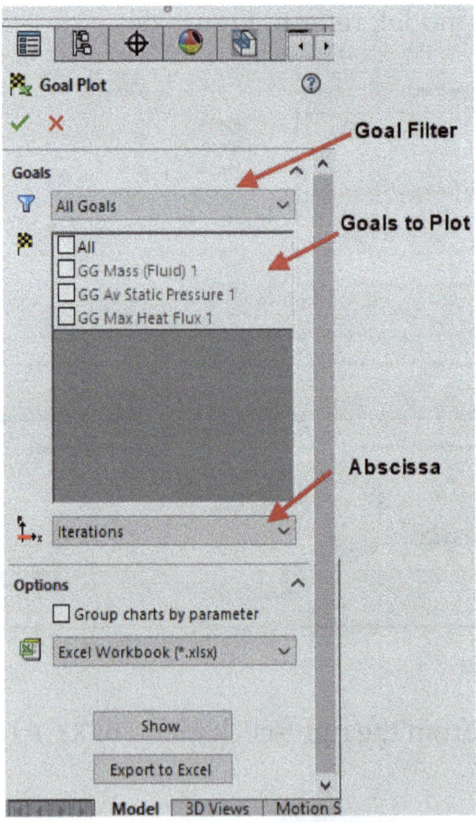

Figure-66. Goal Plot Property Manager

- Click on the **Goal Filter** drop-down from **Goals** section if you want to make available only the specific physical parameter goals in the **Goals to Plot** selection box which were selected by you in **Input Data** section.
- Select the check box of required parameters from **Goals to Plot** selection box to select for study.
- Select the **CPU time** button from **Abscissa** drop-down to specify the chart abscissa as overall CPU time in seconds.
- Select the **Iteration** button from **Abscissa** drop-down to specify iteration number as the chart abscissa.
- Select the **Travels** button from **Abscissa** drop-down to specify number of travels as the chart abscissa.
- Select the **Group Charts by parameter** check box from **Options** rollout to plot all the selected goals for the same parameter on the same charts.
- The other options of this feature are same as discussed earlier.
- After specifying the parameters, click on the **OK** button from **Goal Plot PropertyManager**. The **Goal Plot 1** feature will be added under **Goal Plots**.
- Right-click on the **Goal Plot 1** feature. The right-click menu will be displayed. Click on the **Show** button from the displayed menu. The chart will be displayed along with the model; refer to Figure-67.

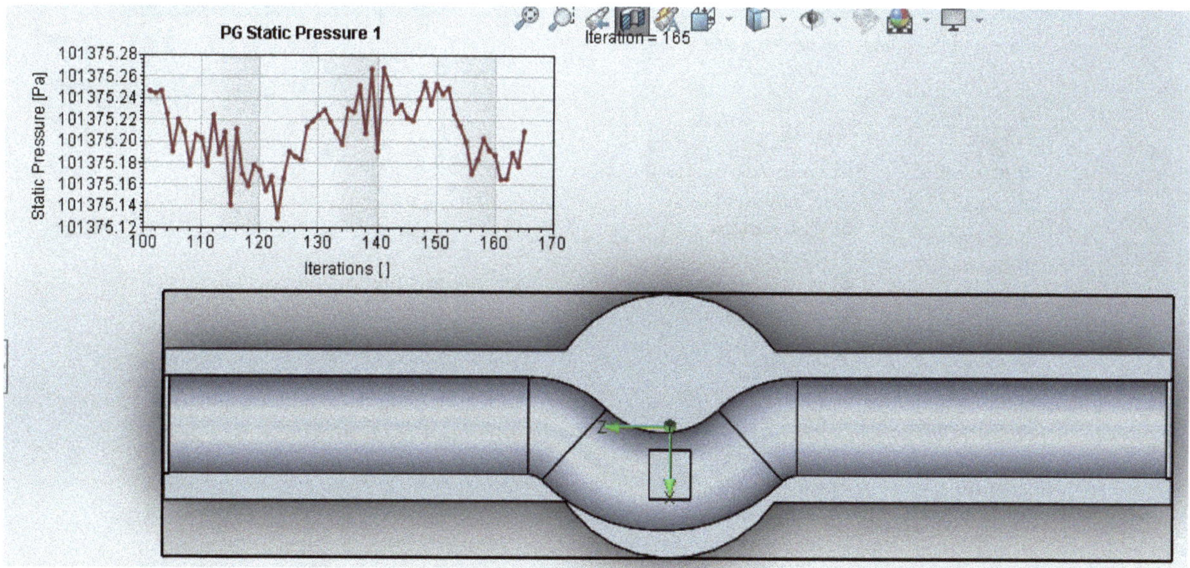

Figure-67. Chart of Goal Plot 1

Similarly, you can create flux plots. Flux plots display heat fluxes between model components and project features as a network chart. Also, you can save the chart as an image or export data in to an Excel workbook

Report

The **Report** option is used to generate analysis report in Microsoft Word and HTML file format. The procedure to use this option is discussed next.

- Right-click on the **Report** option and click on the **Create** button from the displayed right-click menu. The **Report** dialog box will be displayed; refer to Figure-68.

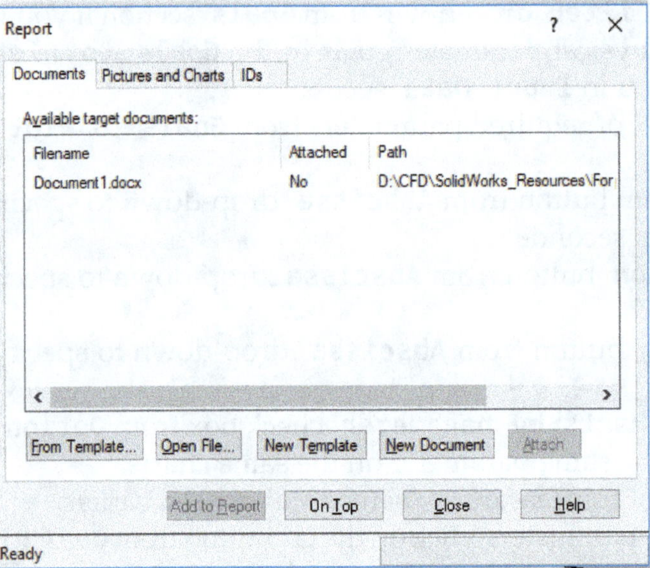

Figure-68. Report dialog box

- Click on the **From Template** button from **Report** dialog box to select a predefined report templates. The **Open** dialog box will be displayed; refer to Figure-69.

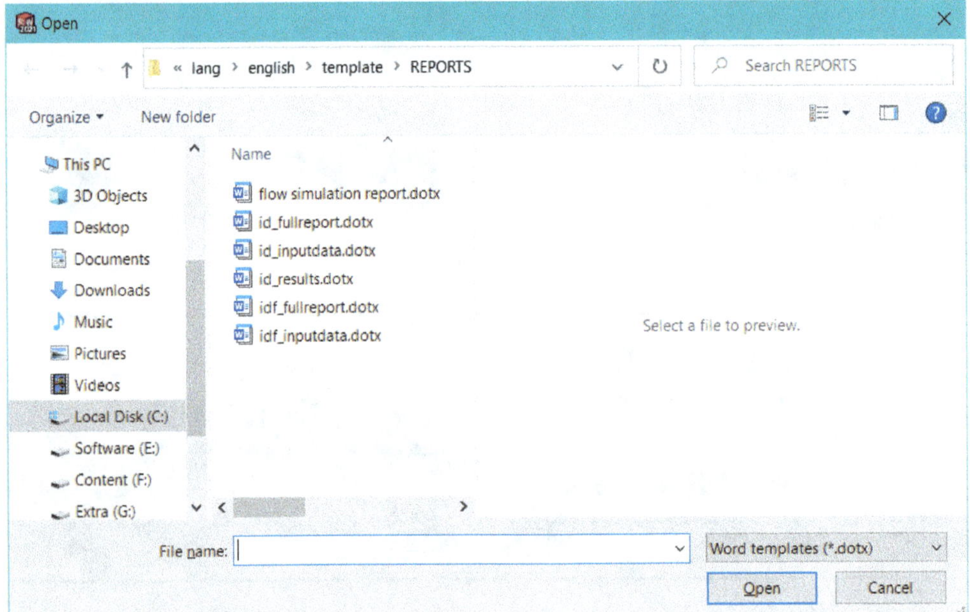

Figure-69. Open dialog box for selecting template

The following standard templates were available in the **Open** dialog box:

- **id fullreport** - This template is used to generates a text report including all available information about the current project in a brief form (all "short" IDs are used)
- **id inputdata** - This template is used to generate a text report of the project input data in a brief form.
- **id results** - This template is used to generates a text report of only the results information in a brief form.
- **idf fullreport** - This template is used to generates a text report including all available information about the current project in a full form.
- **idf inputdata** - This template is used to generates a text report of the project's input data information in a full form.

- Click on the desired template and click on the **Open** button from **Open** dialog box. The report will be generated and displayed in the **Microsoft Word** file; refer to Figure-70.

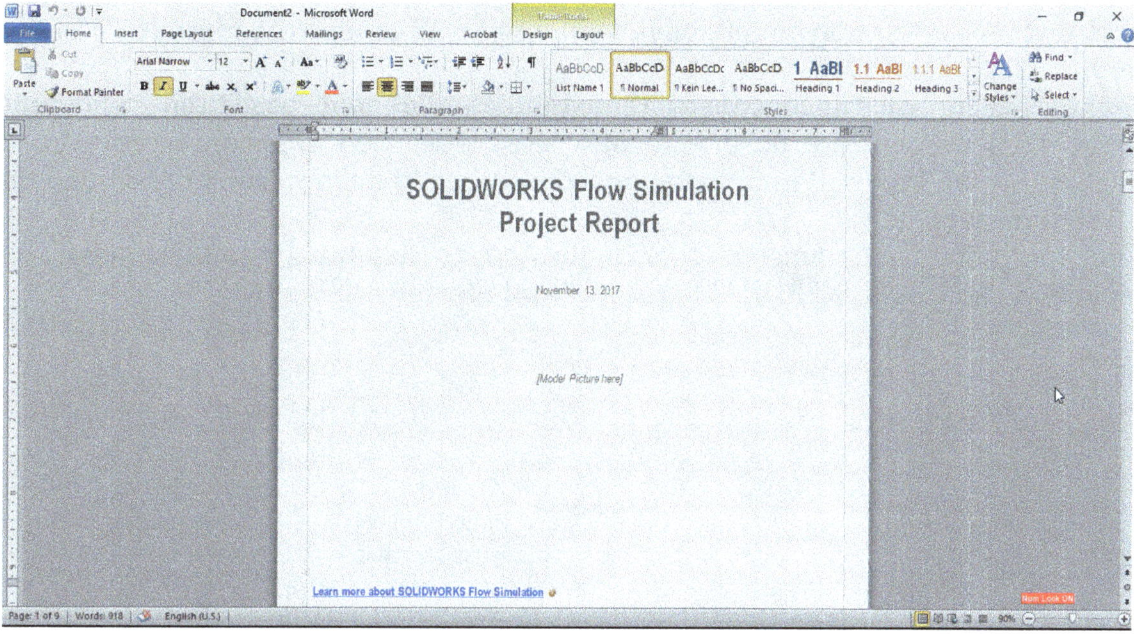

Figure-70. Flow simulation report

- The flow simulation report will be added in **Available target Documents** section.
- Click on the **Open File** button from **Report** dialog box to view the previously created report. The **Open** dialog box will be displayed.
- Click on the report to select and click on the **Open** button. The report will be displayed in **Available target Documents** section.
- Click on the **New Template** button from **Report** dialog box to create a new template. The **Template 1** file will be added in **Available target Documents** section.

Adding parameters to New Template

- Click on the **Template 1** file from **Available target Documents** section and click on **IDs** tab. The **IDs** tab will be displayed; refer to Figure-71.

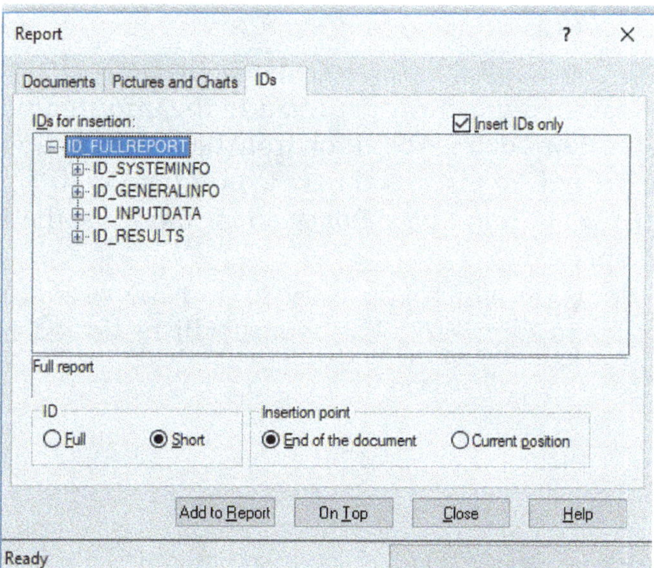

Figure-71. IDs tab

- From the **IDs for insertion** list, select an ID you want to add into the custom template.
- Click on the **Full** radio button for generating a full detail of report caption like Operating System: Windows 10 Service Pack 1 (Build 7601).
- Click on the **Short** radio button for generating a small caption of report like Windows 10 Service Pack 1 (Build 7601).
- Click on the **End of the document** radio button or **Current position** radio button to specify the location inside the document where the selected ID should be added.
- Click on the **On Top** button from **Report** dialog box to define a cursor position within a word document.
- After specifying the parameters, click on the **Add to Report** button. The report will be generated and displayed in the Microsoft Word file; refer to Figure-72.

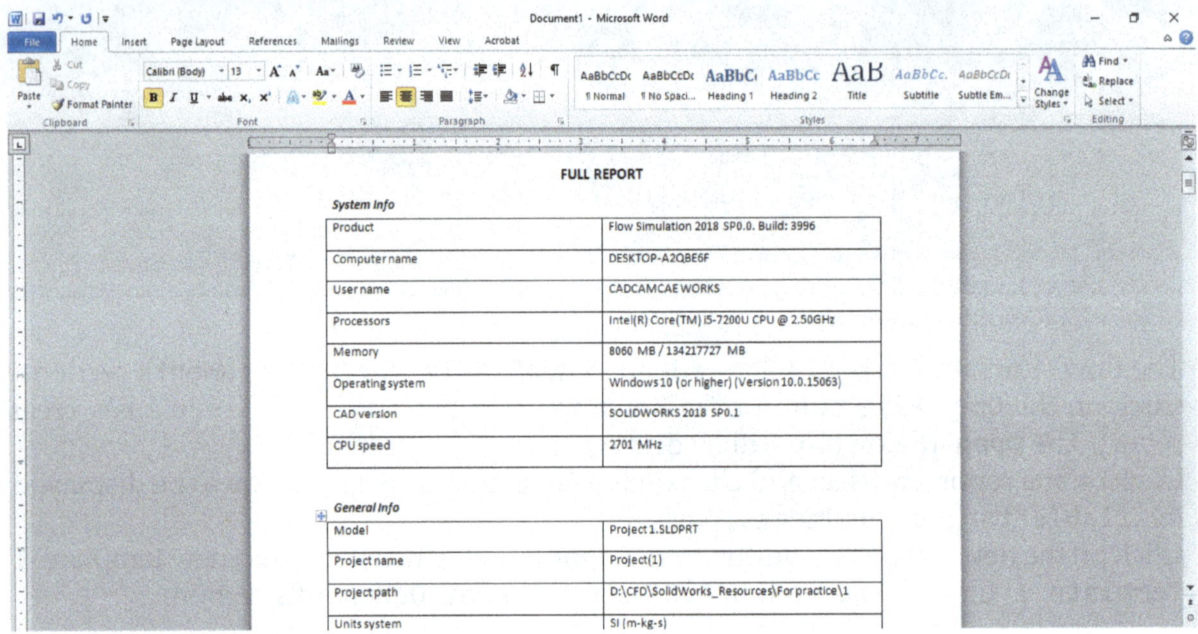

FULL REPORT

System Info

Product	Flow Simulation 2018 SP0.0. Build: 3996
Computer name	DESKTOP-A2QBE6F
User name	CADCAMCAE WORKS
Processors	Intel(R) Core(TM) i5-7200U CPU @ 2.50GHz
Memory	8060 MB / 134217727 MB
Operating system	Windows 10 (or higher) (Version 10.0.15063)
CAD version	SOLIDWORKS 2018 SP0.1
CPU speed	2701 MHz

General Info

Model	Project1.SLDPRT
Project name	Project(1)
Project path	D:\CFD\SolidWorks_Resources\For practice\1
Units system	SI (m-kg-s)

Figure-72. Generated Report

- The pages of a report depend on the study you have done in the flow simulation. The number of pages can go up to 20 pages per report.
- Save the report to the desired location and close the report.

Adding the data to the existing or new document

- The **New Document** button is generally used to add the information or data in the existing document. Click on the **New Document** button from **Documents** tab to add a new document and then follow the procedure discussed next to add data in it.
- Click on the **Open File** button of the **Documents** tab from the **Report** tab. The **Open** dialog box will be displayed.
- Select the existing report on which you want to add the new information and click on the **Open** button; refer to Figure-73. The report will be added in **Available target documents** section.

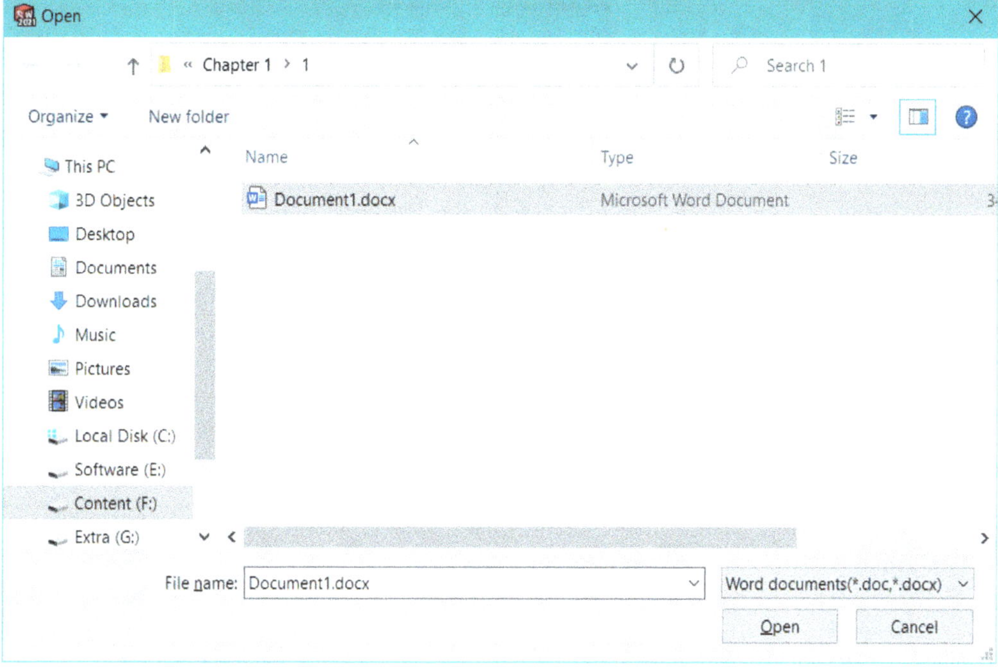

Figure-73. Open dialog box

- Click on the report to which you want to add a current information from **Available target documents** section and click on the **Pictures and Charts** tab of **Report** dialog box. The **Pictures and Charts** tab will be displayed; refer to Figure-74.

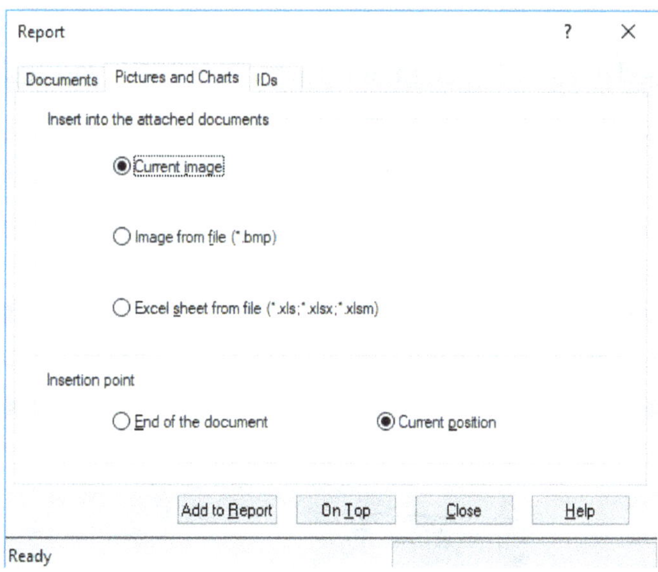

Figure-74. Pictures and Charts tab

- Click on the **Current image** radio button to insert the current image of model in the report.
- Click on the **Image from file (*.bmp)** radio button to add an image from file to the report.
- Click on the **Excel Sheet from file (*.xls,*xlsx,*xlsm)** radio button to add a excel file to the file.
- Select the **End of Document** or **Current Position** radio button as required to specify the location.
- After specifying the parameters, click on the **Add to Report** button. The new data or information will be added in existing report; refer to Figure-75.

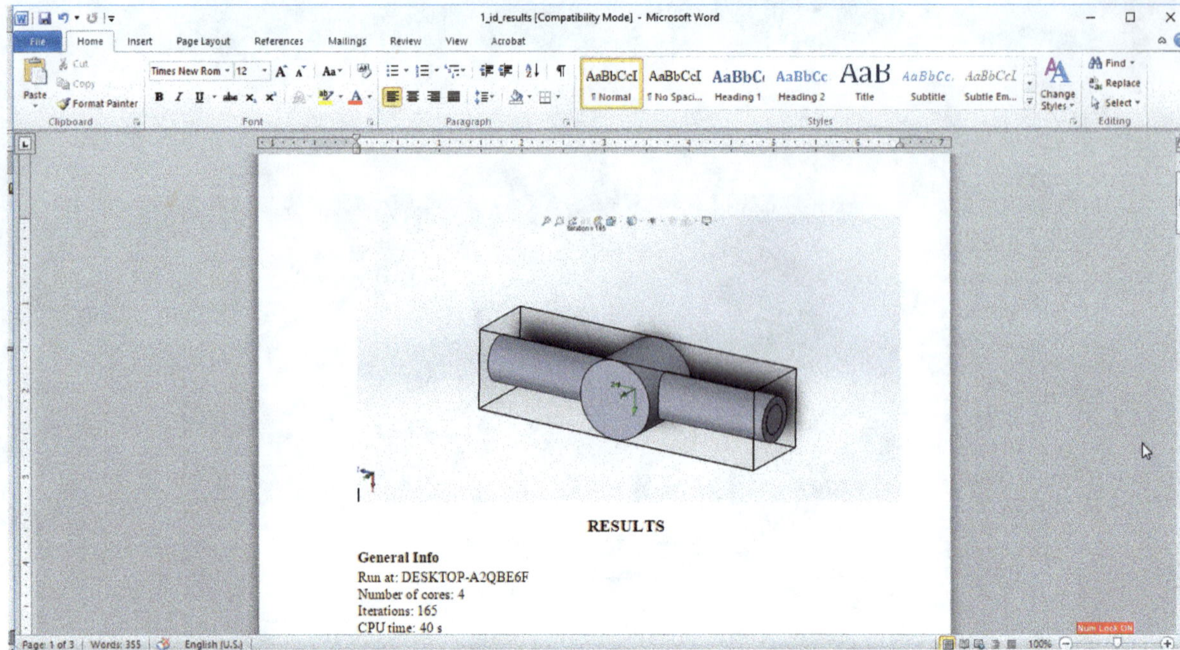

Figure-75. Added current data on report

- Save the Microsoft office file to the desired location.
- If you want to add the existing data to the new document then in the **Reports** dialog box, click on the **New Document** button. The **New Document** will be added in **Available target documents** section.
- Click on the **New document** and repeat the procedure of adding information to the existing file which is discussed earlier. The data or information will be added to the new document.

Animations

The **Animation** option is used to make visualized calculation results which are even more informational and easy to understand. The procedure to use this is discussed next.

- Click on the **Insert** button from right-click menu of **Animations** feature. The **Animation** tab will be displayed along with the model; refer to Figure-76.

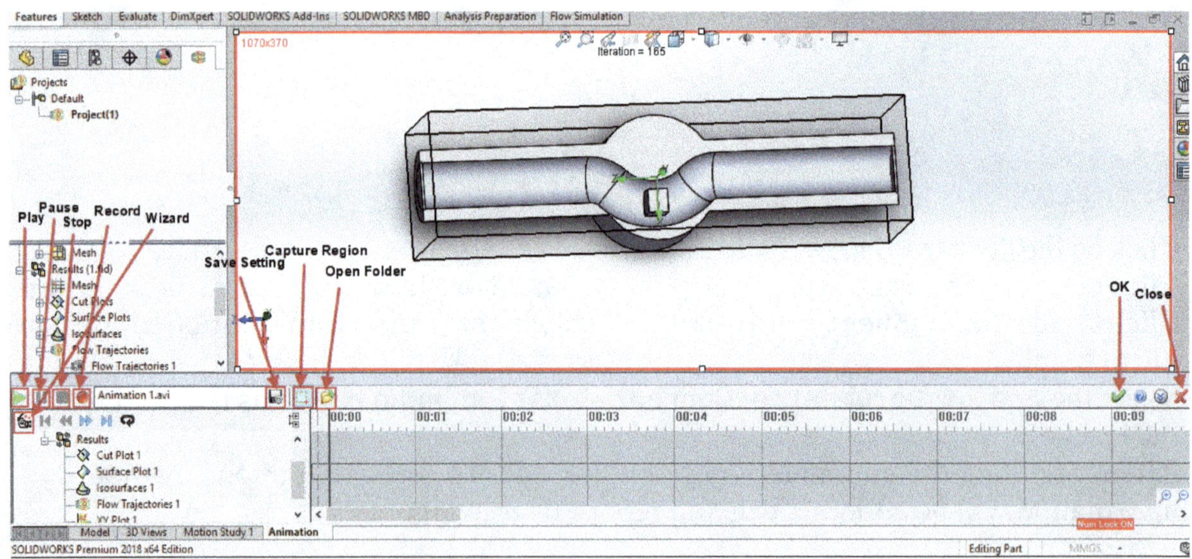

Figure-76. Animation tab

- The **Record** button is used to record the animation file in a specified format.
- The **Save Setting** button is used to select the output form and quality of animation. The output can be in the form of a video clip or series of images.
- The **Capture Region** button is used to specify an area of screen to capture the video. The recorded animation will be limited to the selected region. The capture region is shown as a red frame in the graphics area.

Note: Selecting a region does not affect preview and it plays the animation in the entire area.

- Click on the **Open Folder** button to open the project directory.
- Click on the **File Name** edit box and enter the desired name of animation file which is to be created.
- Click on the **Wizard** button from **Animation** tab to create animation in step by step process. The **Animation Wizard** dialog box will be displayed; refer to Figure-77.

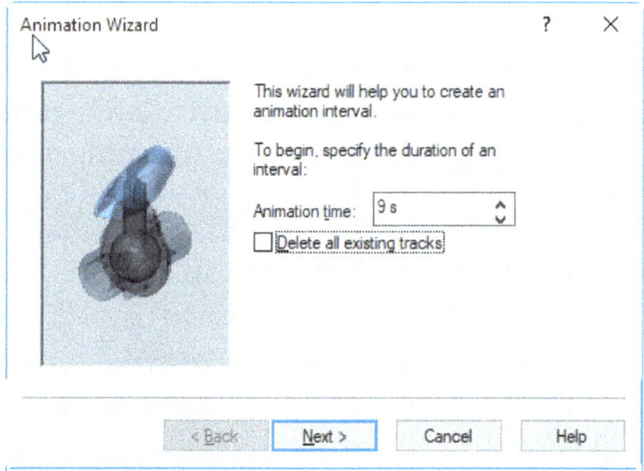

Figure-77. Animation Wizard dialog box

- Click in the **Animation time** edit box from **Animation Wizard** dialog box and enter the desired time for which you want to play the animation clip.
- Select the **Delete all existing tracks** check box from **Animation Wizard** dialog box to delete all the previous tracks or animation file.
- After specifying the required parameters, click on the **Next** button from **Animation Wizard** dialog box. The second page of **Animation Wizard** dialog box will be displayed; refer to Figure-78.
- Select the **Rotate Model** check box to specify whether the view model should be animated or note while performing animation.
- Click on the **Next** button from the second page of **Animation Wizard** dialog box if **Rotate model** check box is selected. The third page of **Animation Wizard** dialog box will be displayed; refer to Figure-79.

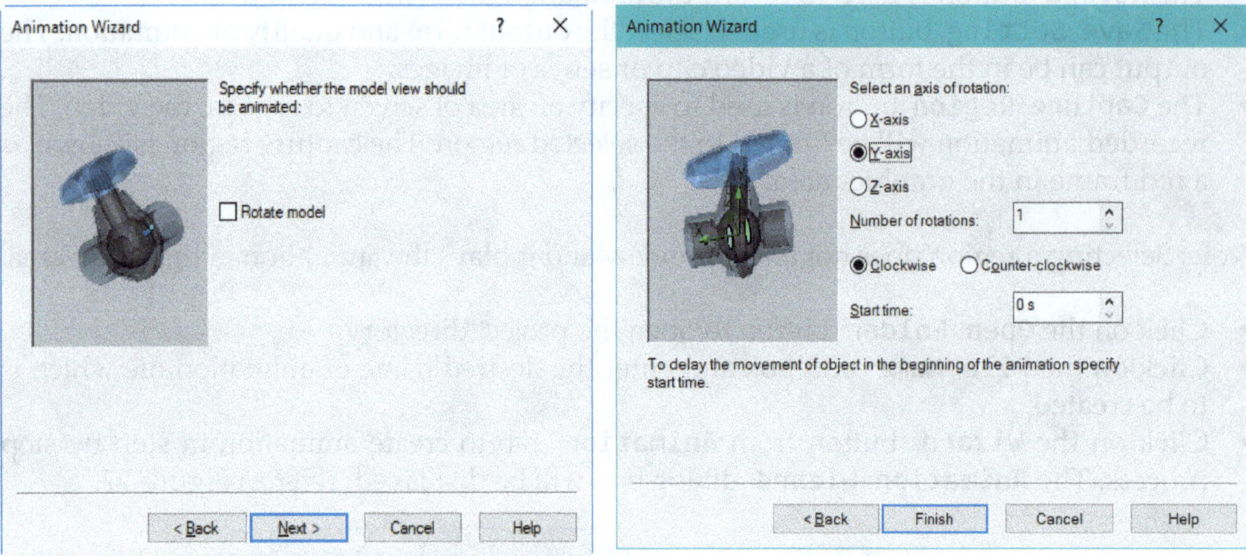

Figure-78. Second page of Animation Wizard dialog box

Figure-79. Third page of Animation Wizard

- Select the desired radio buttons from the dialog box to define rotational axis and direction of rotation. Specify the number of rotation and starting time in various edit boxes.
- Select the **Scenario** radio button to see how the visualized parameter distribution, flow and particle trajectories change from one results file to another in a series of results files.
- After specifying the parameters, click on the **Finish** button. The wizard will be created.

Adding a control point

- Hover the cursor on the created study in the **Design tree**. A rectangular box will be displayed in the Time line; refer to Figure-80.

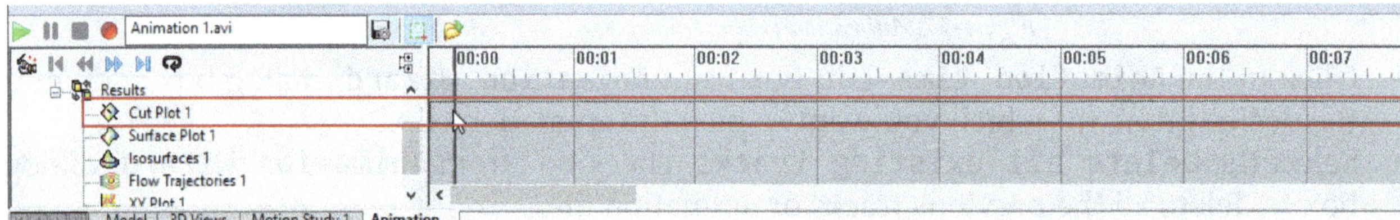

Figure-80. Timeline for adding control points

- In this specific box, we need to add control points of animation for the respected study.
- Right-click in the rectangular area and click on the **Insert Control Point** button from the displayed right click menu. The two control points will be displayed in which one is fixed and other moves with the cursor; refer to Figure-81.

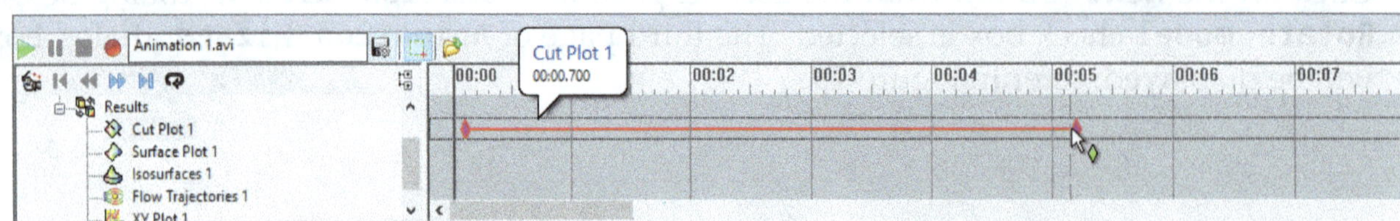

Figure-81. Specifying distance between control points

- Specify the distance between control points by moving the cursor and left-click on the **Timeline** location where you want to place the control point. The control point will be placed.
- To play the animation of **Cut Plot**, click on the **Play** button. The animation clip starts and will be stopped automatically when it completes the time; refer to Figure-82

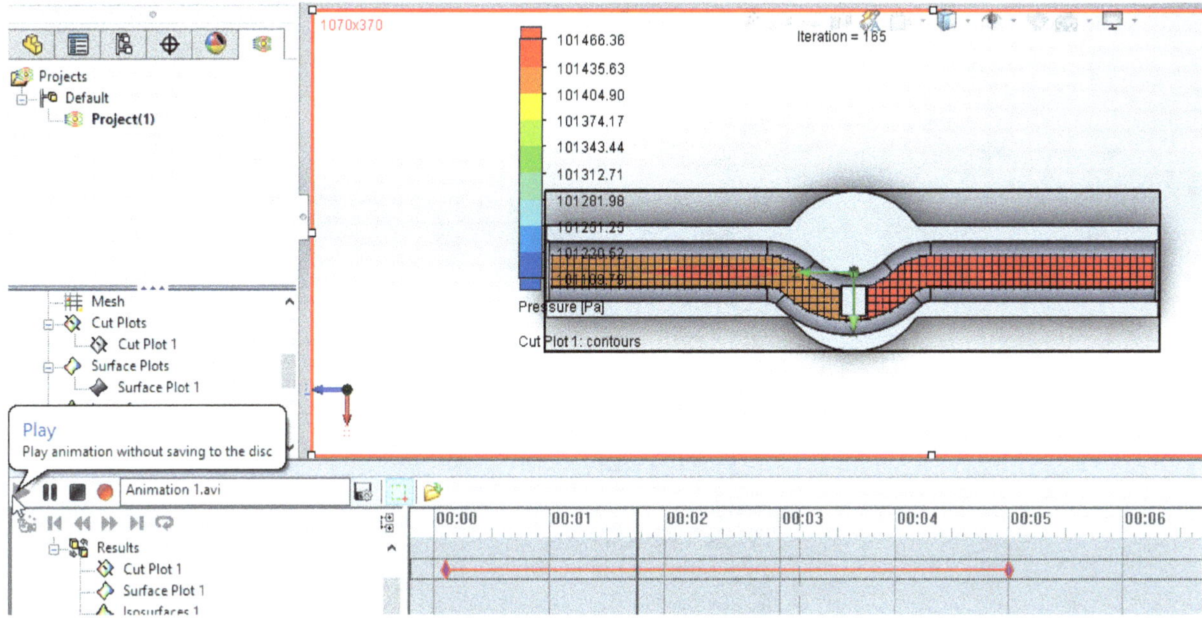

Figure-82. Playing animation

- Similarly, you can add more control points to the other feature and can play the animation together; refer to Figure-83.

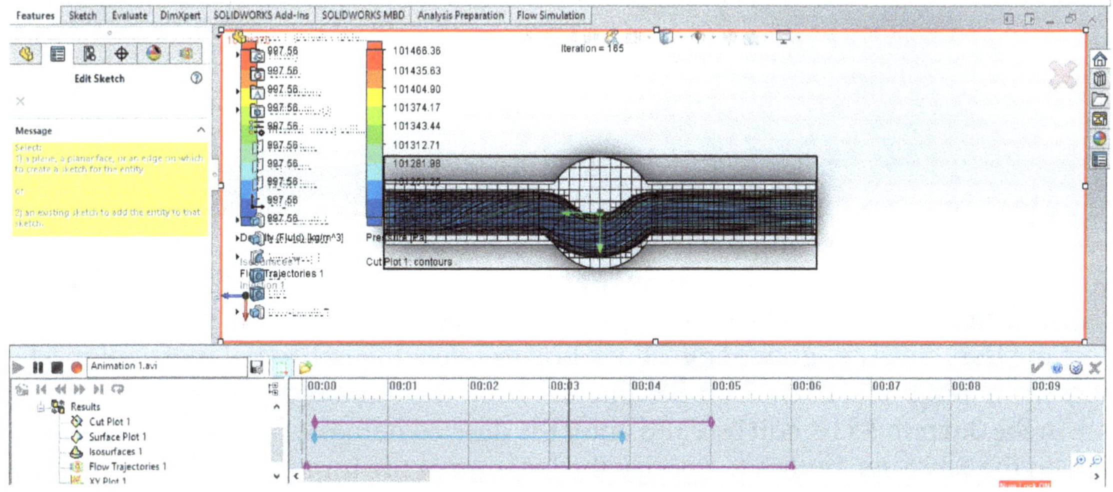

Figure-83. Playing animation together

- After creating the animation clips, click on the **OK** button from **Animation** tab. The **Animation 1** feature will be added in design tree.

Export Results

The **Export Results** is used to export the physical data or parameters values obtained in a selected mesh cells to an ASCII text file. The procedure to use this is discussed next.

- Click on the **Insert** button from the right-click shortcut menu of the **Export Results** option. The **Export Results Property Manager** will be displayed; refer to Figure-84.

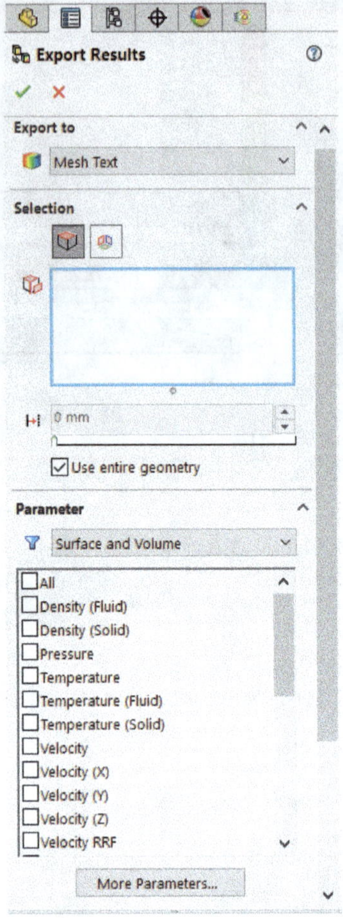

Figure-84. Mesh Text Property Manager

- The **Selection box** is active by default. You need to click on the plane or face of the model to select. On selecting the required face, the **Offset** edit box will be activated.
- Click in the **Offset** edit box from **Selection** section and enter the value for offset the selected face or plane. You can also move the **Offset** slider to set the value.
- Select the **Use entire geometry** check box to select the entire geometry of model .
- Select the check boxes of desired parameters from the **Parameter** rollout.
- Click in the **Output File** edit box and enter the desired name of file which is to be exported.
- Click on the **Save As** button to specify the location of file for saving. The **Save As** dialog box will be displayed.
- Specify the location and name in **Save As** dialog box and click on the **OK** button. The location will be specified.
- After specifying the parameters, click on the **Export** button. The calculation and exportation of file started.
- When the exportation is completed, the mesh text will be added under the **Export Results** node in the **Results** feature.
- Right-click on the **Mesh Text 1** node and select the **Edit Definition** option from shortcut menu. The **Mesh Text 1 Property Manager** will be displayed.
- Modify the parameters as needed and click on the **OK** button. The **Mesh Text 1** feature will be added in **Design Tree**. The exported file will be displayed in the saving folder.
- Open the exported file saving folder and double-click on the required file. The file will be opened in Notepad; refer to Figure-85.

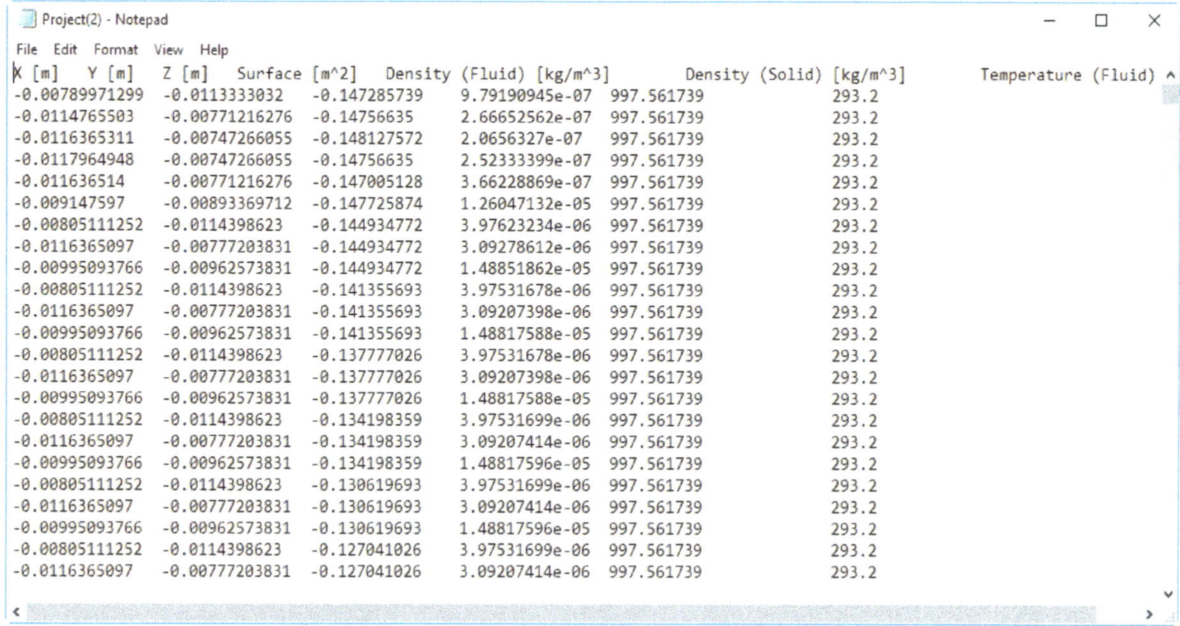

Project(2) - Notepad

File Edit Format View Help

X [m]	Y [m]	Z [m]	Surface [m^2]	Density (Fluid) [kg/m^3]	Density (Solid) [kg/m^3]	Temperature (Fluid)
-0.00789971299	-0.0113333032	-0.147285739	9.79190945e-07	997.561739	293.2	
-0.0114765503	-0.00771216276	-0.14756635	2.66652562e-07	997.561739	293.2	
-0.0116365311	-0.00747266055	-0.148127572	2.0656327e-07	997.561739	293.2	
-0.0117964948	-0.00747266055	-0.14756635	2.52333399e-07	997.561739	293.2	
-0.011636514	-0.00771216276	-0.147005128	3.66228869e-07	997.561739	293.2	
-0.009147597	-0.00893369712	-0.147725874	1.26047132e-05	997.561739	293.2	
-0.00805111252	-0.0114398623	-0.144934772	3.97623234e-06	997.561739	293.2	
-0.0116365097	-0.00777203831	-0.144934772	3.09278612e-06	997.561739	293.2	
-0.00995093766	-0.00962573831	-0.144934772	1.48851862e-05	997.561739	293.2	
-0.00805111252	-0.0114398623	-0.141355693	3.97531678e-06	997.561739	293.2	
-0.0116365097	-0.00777203831	-0.141355693	3.09207398e-06	997.561739	293.2	
-0.00995093766	-0.00962573831	-0.141355693	1.48817588e-05	997.561739	293.2	
-0.00805111252	-0.0114398623	-0.137777026	3.97531678e-06	997.561739	293.2	
-0.0116365097	-0.00777203831	-0.137777026	3.09207398e-06	997.561739	293.2	
-0.00995093766	-0.00962573831	-0.137777026	1.48817588e-05	997.561739	293.2	
-0.00805111252	-0.0114398623	-0.134198359	3.97531699e-06	997.561739	293.2	
-0.0116365097	-0.00777203831	-0.134198359	3.09207414e-06	997.561739	293.2	
-0.00995093766	-0.00962573831	-0.134198359	1.48817596e-05	997.561739	293.2	
-0.00805111252	-0.0114398623	-0.130619693	3.97531699e-06	997.561739	293.2	
-0.0116365097	-0.00777203831	-0.130619693	3.09207414e-06	997.561739	293.2	
-0.00995093766	-0.00962573831	-0.130619693	1.48817596e-05	997.561739	293.2	
-0.00805111252	-0.0114398623	-0.127041026	3.97531699e-06	997.561739	293.2	
-0.0116365097	-0.00777203831	-0.127041026	3.09207414e-06	997.561739	293.2	

Figure-85. Saved exported file

SELF-ASSESSMENT

Q1. The .... option is used to display the computational mesh cells and mesh related parameters at the calculation moment selected for getting the results.

Q2. Which of the following plots is used to section the model and check various result parameters in the form of contours?

(a) Cut Plot
(b) Isosurface
(c) Particle Study
(d) XY Plot

Q3. Which of the following plots is used to section the model and check various result parameters over selected face in the form of contours?

(a) Cut Plot
(b) Isosurface
(c) Surface Plot
(d) XY Plot

Q4. Which of the following plots is used to display result parameter of same value over the surface in the form of curves?

(a) Cut Plot
(b) Isosurface
(c) Surface Plot
(d) XY Plot

Q5. Which of the following plot is used to display fluid flow in the form of flow lines?

(a) Flow Trajectories
(b) Isosurface
(c) Surface Plot
(d) XY Plot

Q6. What is the use of Particle Studies in analysis result?

Q7. The .... analysis result is used to display the trajectories of physical particles and obtain various information about the behaviors of particle.

Q8. Which of the following plots is used to check result parameter at specified location?

(a) Surface Parameters
(b) Particle Parameters
(c) Particle Studies
(d) XY Plot

For Student Notes

FOR STUDENT NOTES

Chapter 4

Practical and Practice

The major topics covered in this chapter are:

- ***Practical 1***
- ***Practical 2***
- ***Practice 1***
- ***Practice 2***

In the previous chapters, we have learned various tools and techniques of Flow Simulation analysis on the model. In this chapter, we will perform the analyses on the models whose part files are available in the respective folder. The part files are available on www.cadcamcaeworks.com. You can download these part files or assembly files for reference.

PRACTICAL 1

In this tutorial, we will consider flow in a section of pipe whose diameter is decreasing from inlet of fluid to the exit of fluid. The part file will be available in resource folder.

Starting SolidWorks Flow Simulation

- Double-click on the **Tutorial 1** part file from the **Tutorials** folder. The SOLIDWORKS window will be displayed along with the part file; refer to Figure-1.

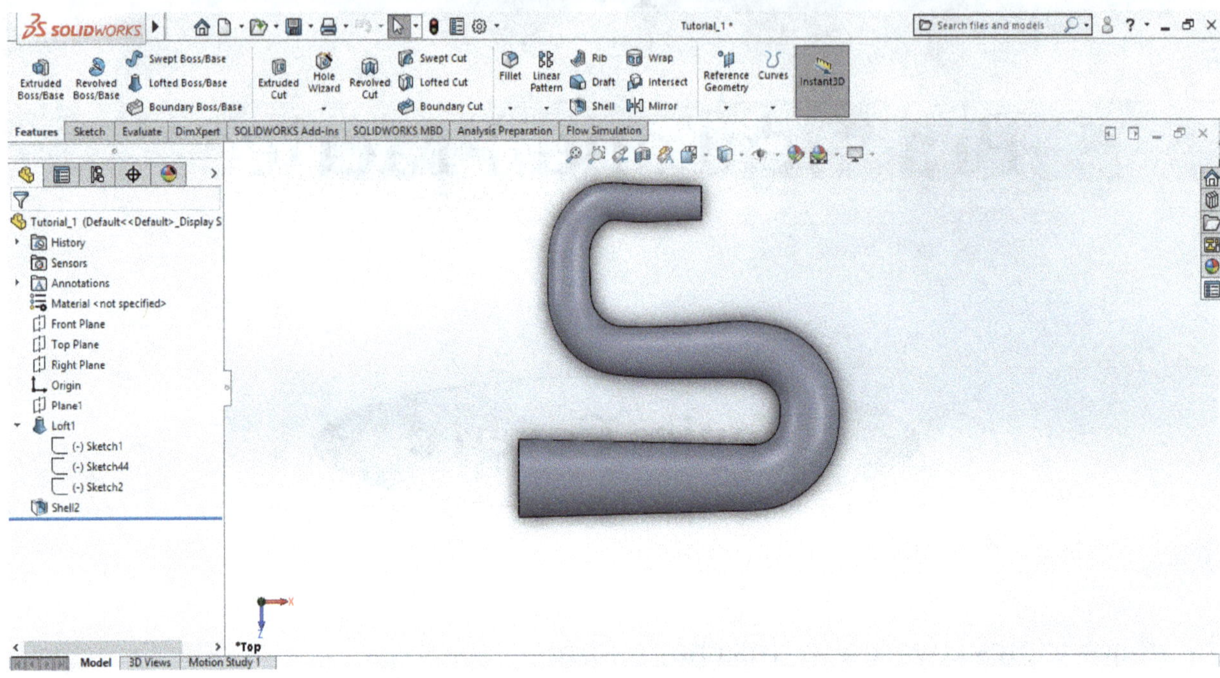

Figure-1. Welcome screen Tutorial

- There is an another method to open the part file. For this, open the SOLIDWORKS software by double-click on the SOLIDWORKS icon. The SOLIDWORKS window will be displayed.
- Click on the **Open** button from **Open** menu bar. The **Open** dialog box will be displayed; refer to Figure-2.

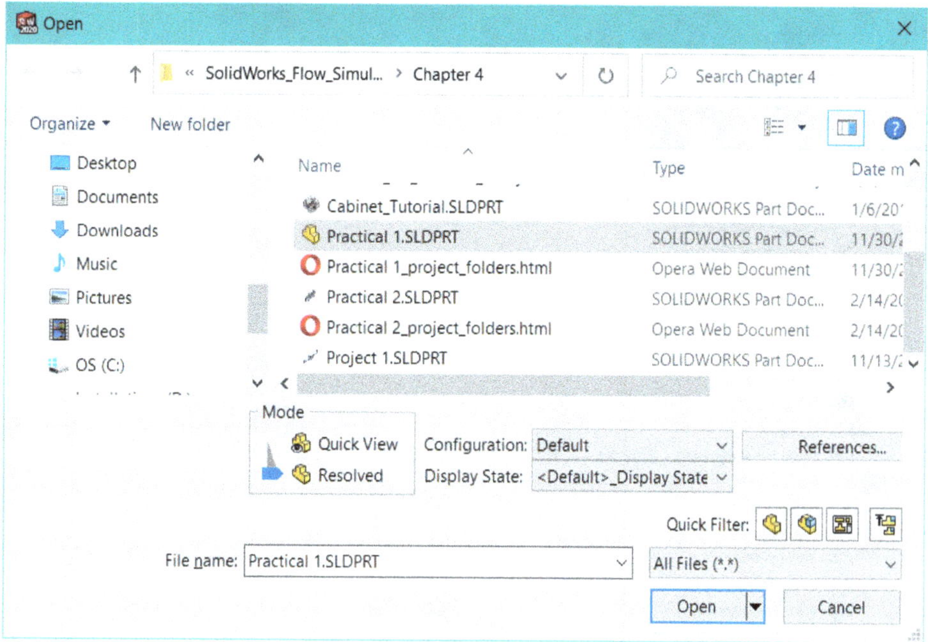

Figure-2. The Open dialog box for opening part files

- Select the **Practical 1** file and click on the **Open** button. The part file will be opened in SOLIDWORKS software.

Starting SOLIDWORKS Flow Simulation

- Click on the **Add-Ins** button from the **Tools** menu. The **Add-Ins** dialog box will be displayed.
- Select the **SOLIDWORKS Flow Simulation 2021** check box under **Active Add-ins** column and click on the **OK** button. The Flow Simulation environment will become active.
- Click on the **Flow Simulation** tab from **CommandManager**. The tools of **Flow Simulation** will be displayed; refer to Figure-3.

Figure-3. Tools of Flow Simulation

- Click on the **Wizard** button from **Flow Simulation** tab. The **Wizard-Project Name** dialog box will be displayed; refer to Figure-4.

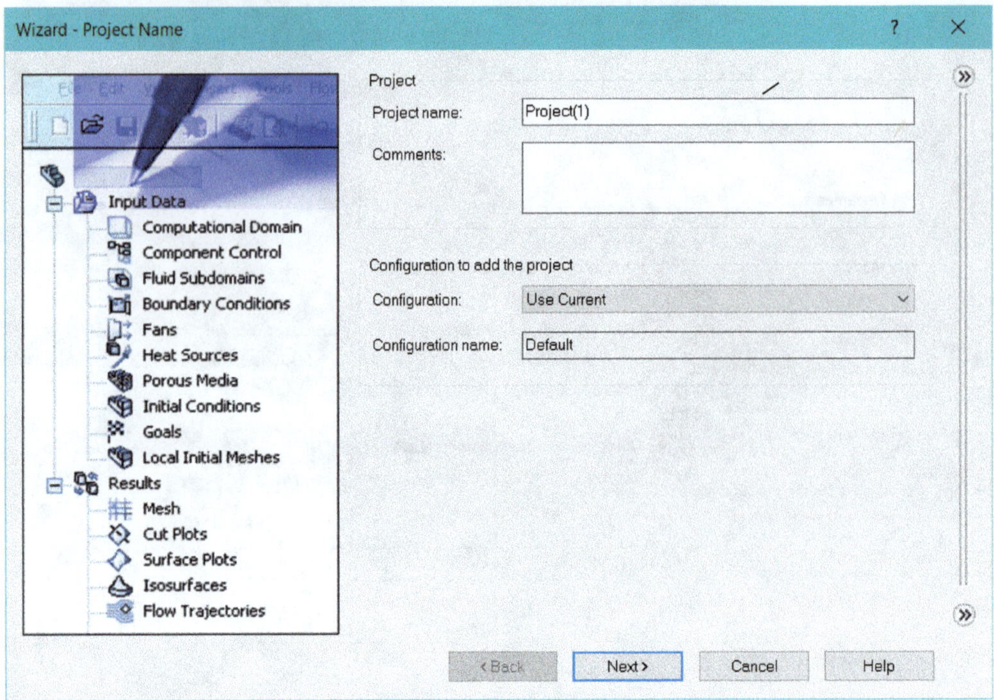

Figure-4. Wizard-Project Name dialog box

- Click in the **Project name** edit box and enter the name as Tutorial 1. Click on the **Next** button. The **Wizard- Unit System** dialog box will be displayed; refer to Figure-5.

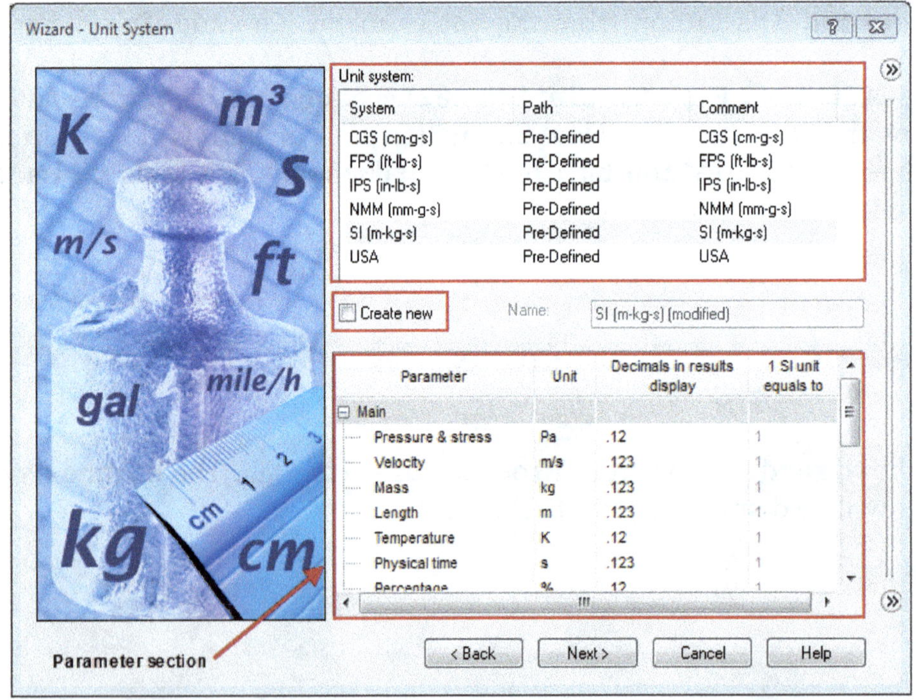

Figure-5. Wizard-Unit Name dialog box

- Select the **SI(m-kg-s)** unit system from **Unit System** section and click on the **Next** button. The **Wizard-Analysis Type** dialog box will be displayed; refer to Figure-6.

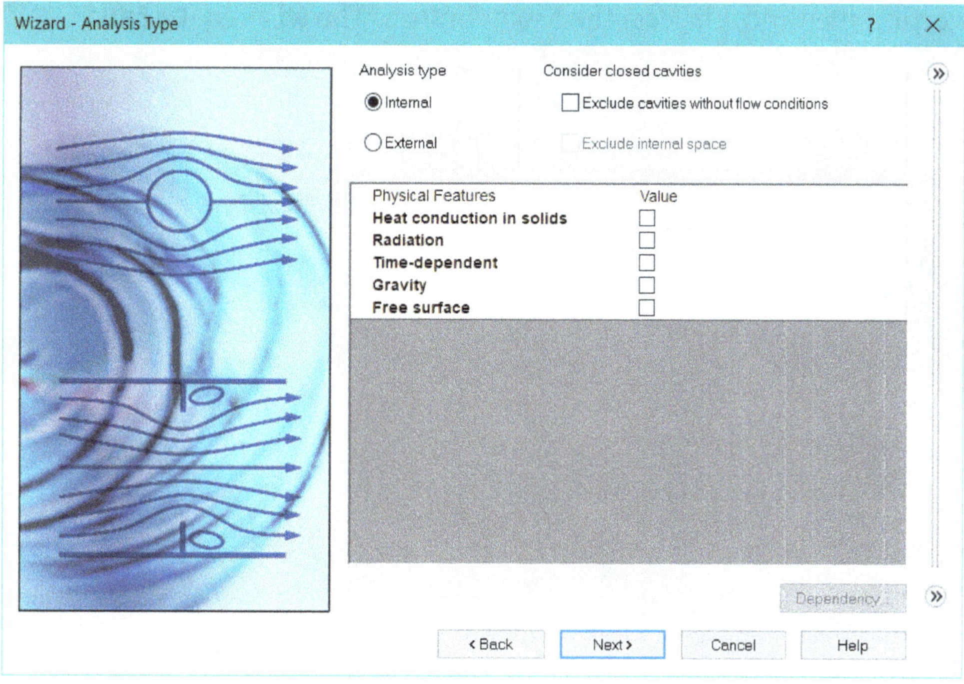

Figure-6. Wizard Analysis Type dialog box

- Select the **Internal** radio button from **Analysis** type section for internal analysis of model.
- Select the **Heat conduction in solids**, **Time-dependent**, and **Gravity** check boxes from **Physical Features** section.
- After specifying the parameters, click on the **Next** button. The **Wizard-Default Fluid** dialog box will be displayed; refer to Figure-7.

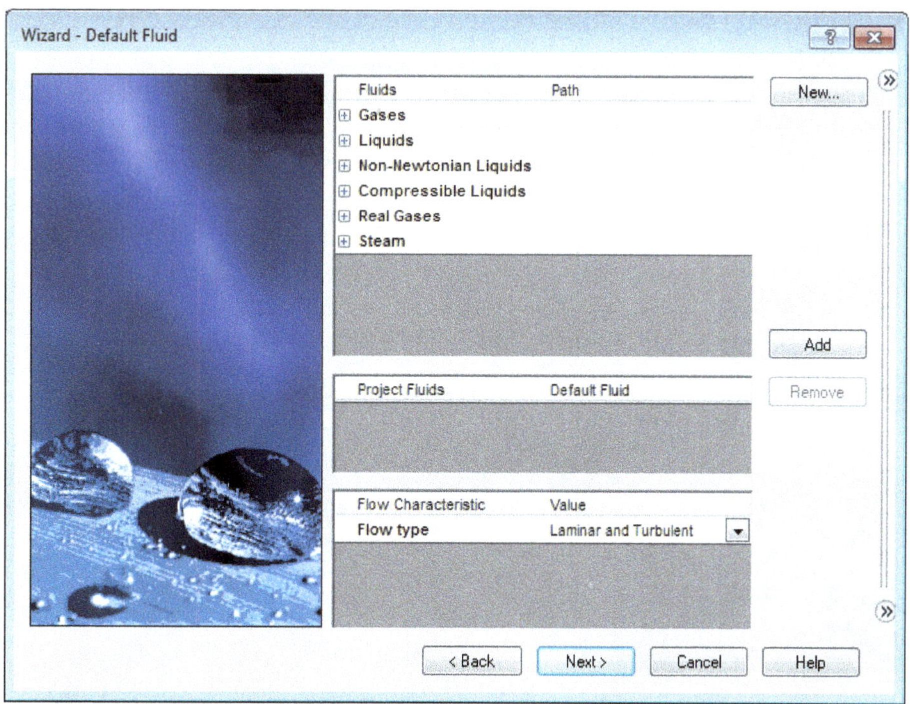

Figure-7. The Wizard-Default Fluid dialog box

- Select the fluid as **Water** from **Liquids** list and click on the **Add** button. Water will be added in the Project fluids.

- After selecting the fluid, click on the **Next** button. The **Wizard-Default Solid** dialog box will be displayed; refer to Figure-8.

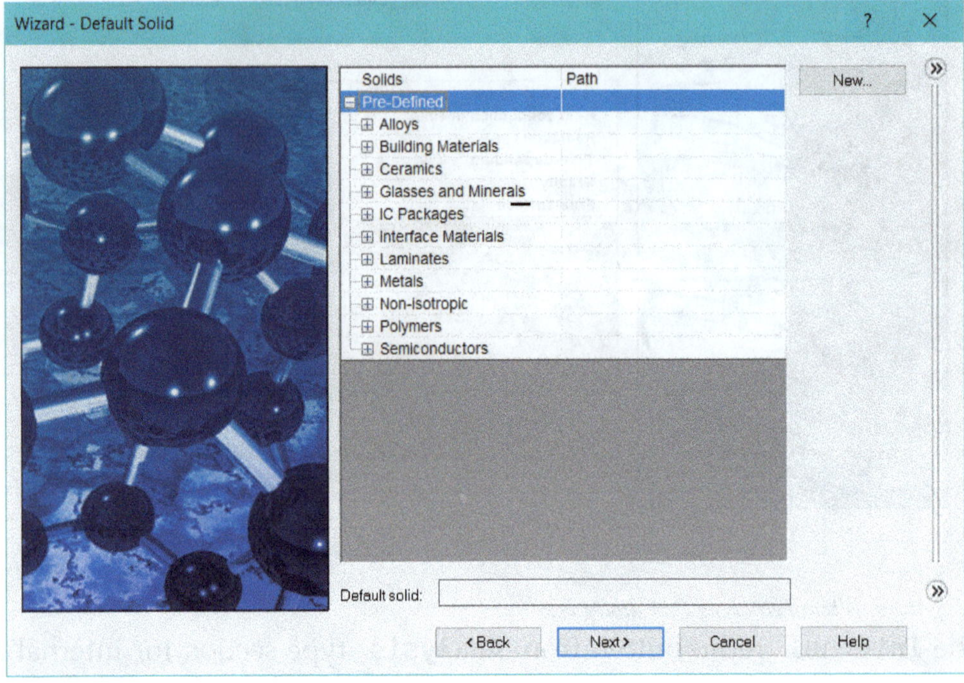

Figure-8. Wizard-Default Solid dialog box

- In this dialog box, we need to select the material of model in which the fluid flows.
- Select the **Aluminium** option from **Metals** list of **Solids** section. The **Aluminium** will be added in **Default Solid** option and will be selected as metal of solid.
- After selecting the metal, click on the **Next** button. The **Wizard- Wall Conditions** dialog box will be displayed; refer to Figure-9.

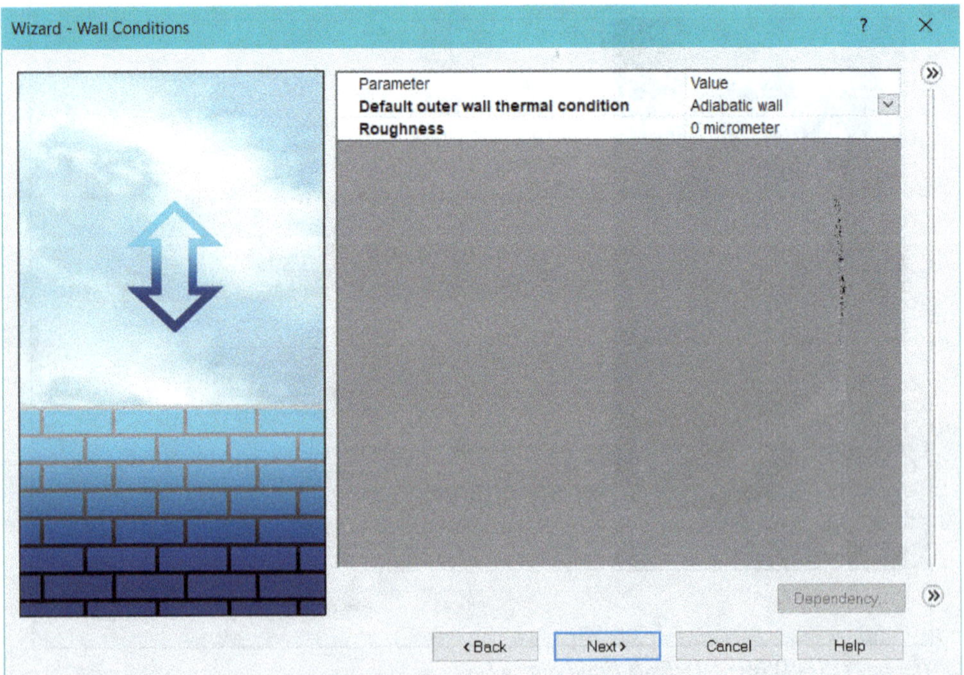

Figure-9. Wizard-Wall Conditions dialog box

- Click in the **Default outer wall thermal condition** drop-down and select on the **Adiabatic wall** option.

- After specifying the parameters, click on the **Next** button. The **Wizard-Initial Conditions** dialog box will be displayed; refer to Figure-10.

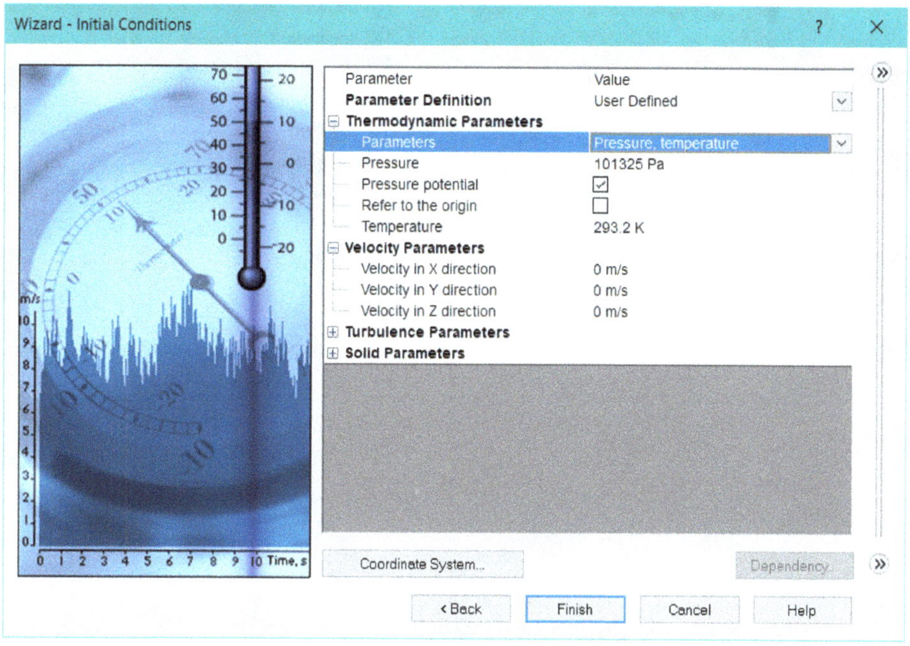

Figure-10. Wizard-Initial and Ambient Conditions dialog box

- Click on the **Finish** button from **Wizard-Initial Conditions** dialog box. A new project will be added. If the ends of your component are not closed as they are in our case then the **Flow Simulation** information box will be displayed; refer to Figure-11.

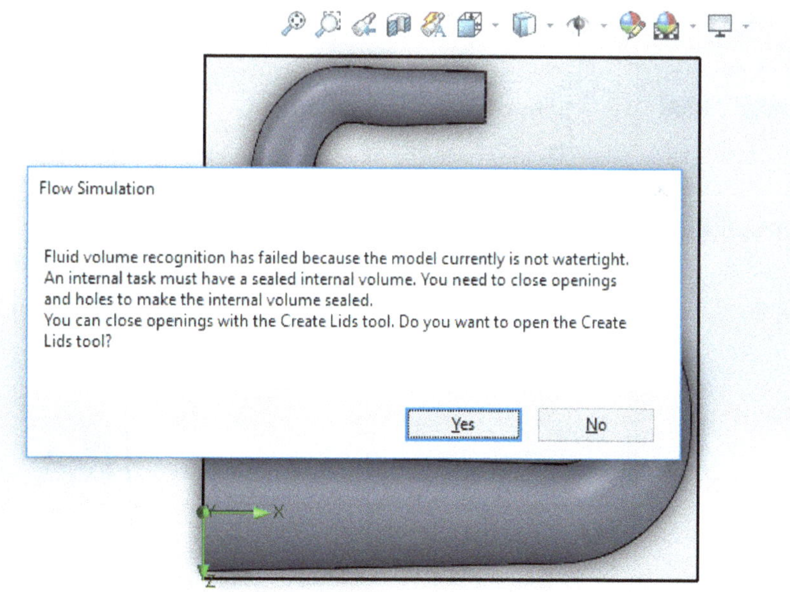

Figure-11. The Flow Simulation information box

- Click on the **Yes** button to create the lids on the model. The **Create Lids PropertyManager** will be displayed; refer to Figure-12.

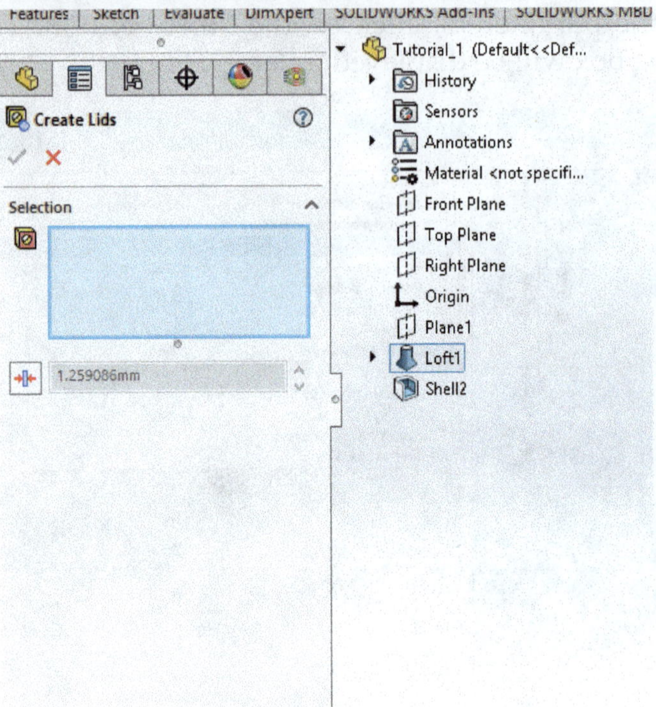

Figure-12. The Create Lids PropertyManager

- Click on the outer surface of fluid entrance and fluid exit to create a lid; refer to Figure-13.

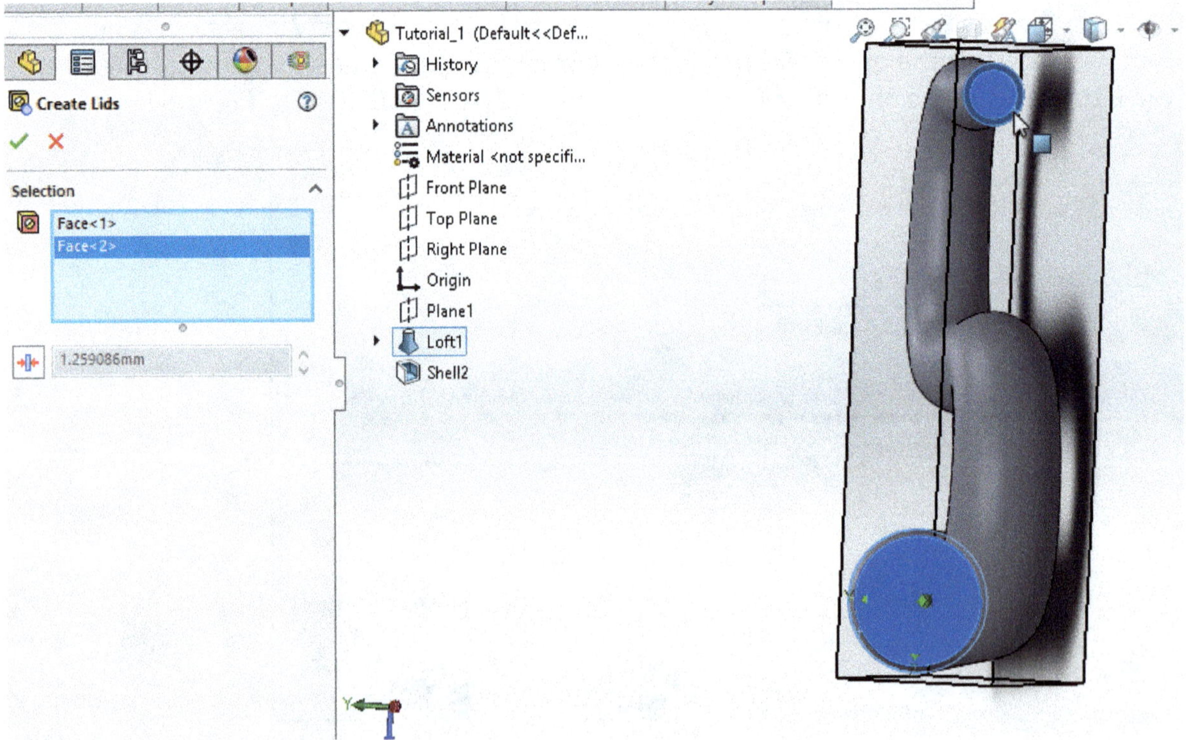

Figure-13. Selection of surface for creating lid

- The selected surface will be displayed in **Selection** box. Click on the **OK** button from **Create Lids PropertyManager** to create the lids. Another **Flow Simulation** information box will be displayed; refer to Figure-14.

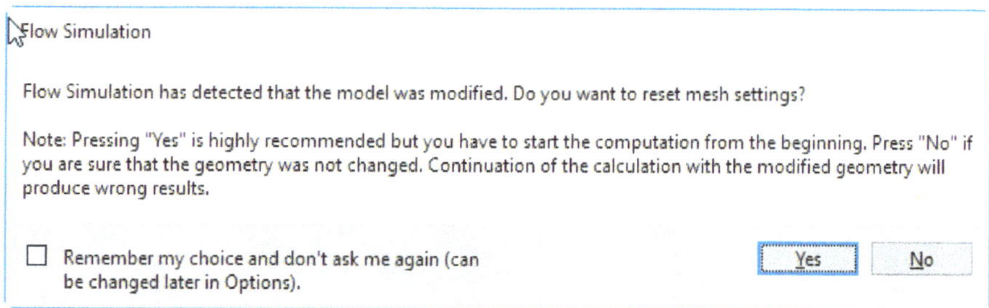

Flow Simulation
Flow Simulation has detected that the model was modified. Do you want to reset mesh settings?
Note: Pressing "Yes" is highly recommended but you have to start the computation from the beginning. Press "No" if you are sure that the geometry was not changed. Continuation of the calculation with the modified geometry will produce wrong results.
Remember my choice and don't ask me again (can be changed later in Options).
Yes
No

Figure-14. Resetting Mesh Settings box

- Click on the **Yes** button from the displayed box to reset the mesh setting of model. The setting will be re-adjusted as per model specifications.

Activating Section View

For displaying flow of fluid inside the tube, we need to activate section view by using **Section View** tool. The procedure to use this tool is discussed next.

- Click on the **Section View** tool from **Heads-up View Toolbar**; refer to Figure-15. The **Section View PropertyManager** will be displayed; refer to Figure-16.

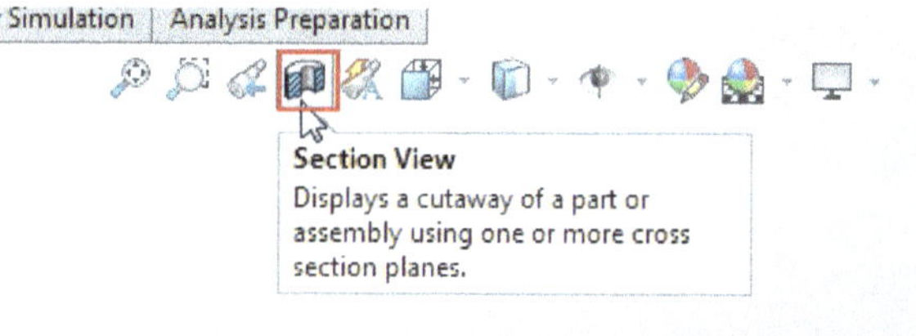

Displays a cutaway of a part or assembly using one or more cross section planes.

Figure-15. Section View button

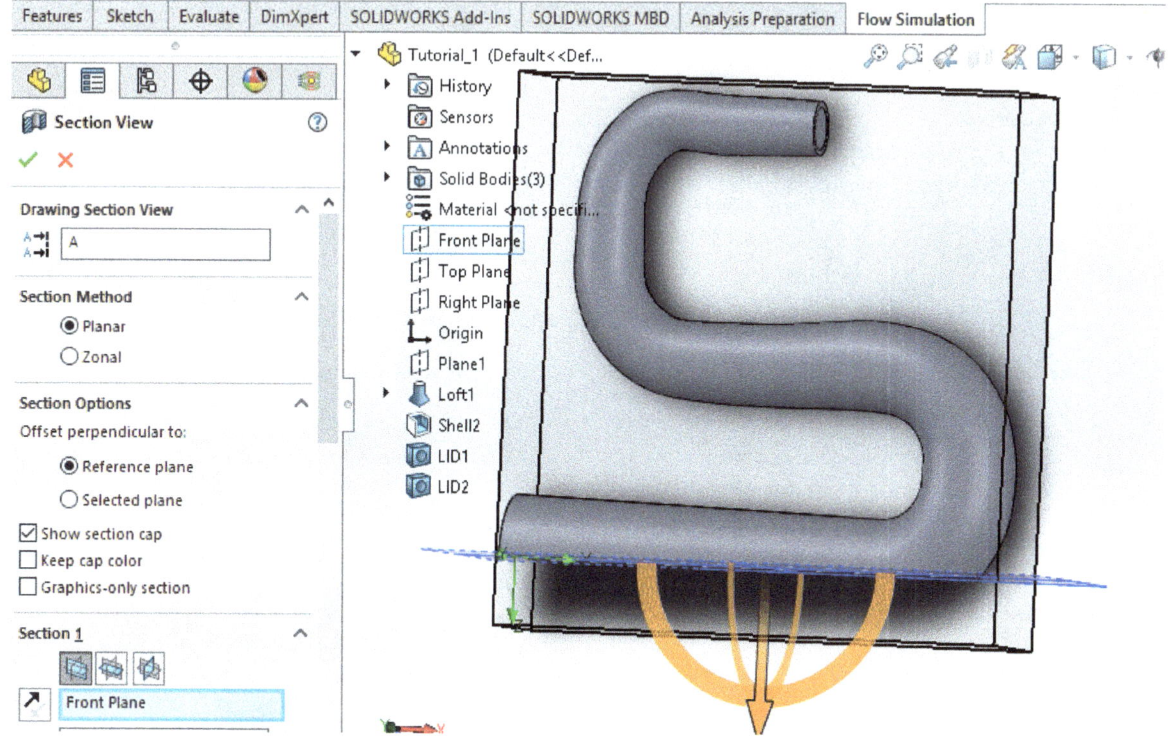

Section Options

Solid Bodies(3)

Figure-16. The Section View PropertyManager

- Click on the **Top Plane** from **Design Tree**. The section view from top plane will be displayed.
- After selecting the **Top Plane**, click on the **OK** button. The section view will be displayed; refer to Figure-17.

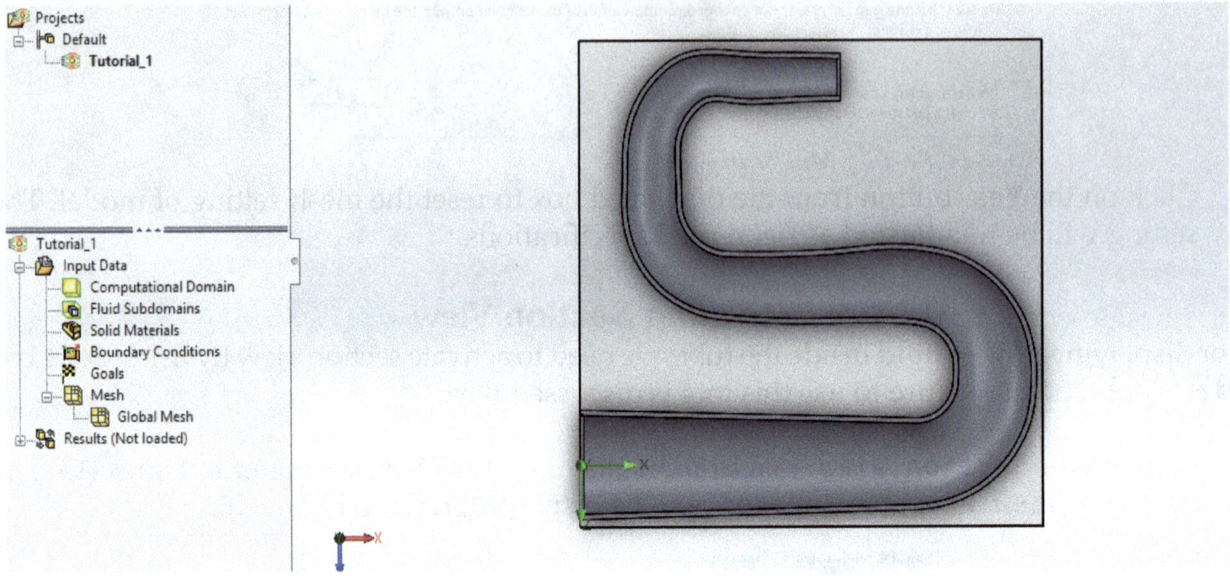

Figure-17. Section view of Tutorial 1

Specifying Boundary Conditions

- Right-click on the **Boundary Conditions** option from **Flow Simulation Analysis tree**. The right-click menu will be displayed.
- Click on the **Insert Boundary Condition** button from the displayed menu. The **Boundary Condition PropertyManager** will be displayed along with the model; refer to Figure-18.

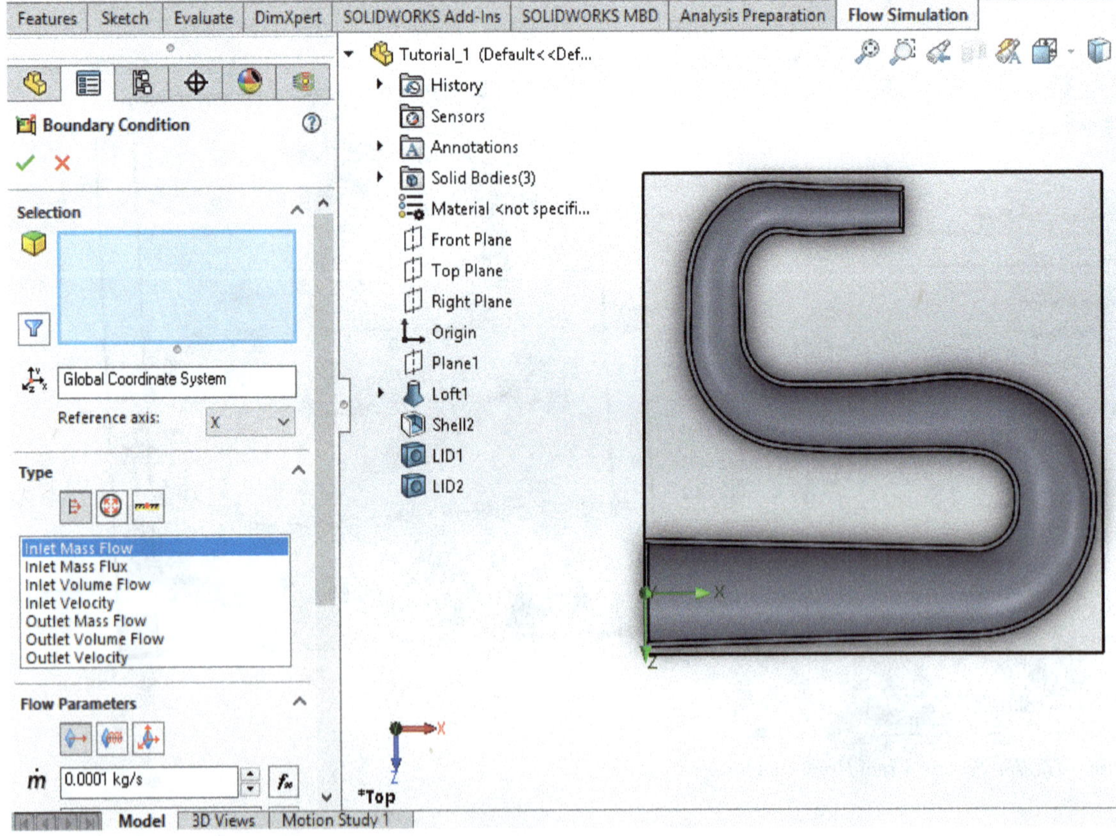

Figure-18. Boundary Condition Property Manager

- The **Selection** box is active by default. You need to select the internal face of fluid entrance; refer to Figure-19.

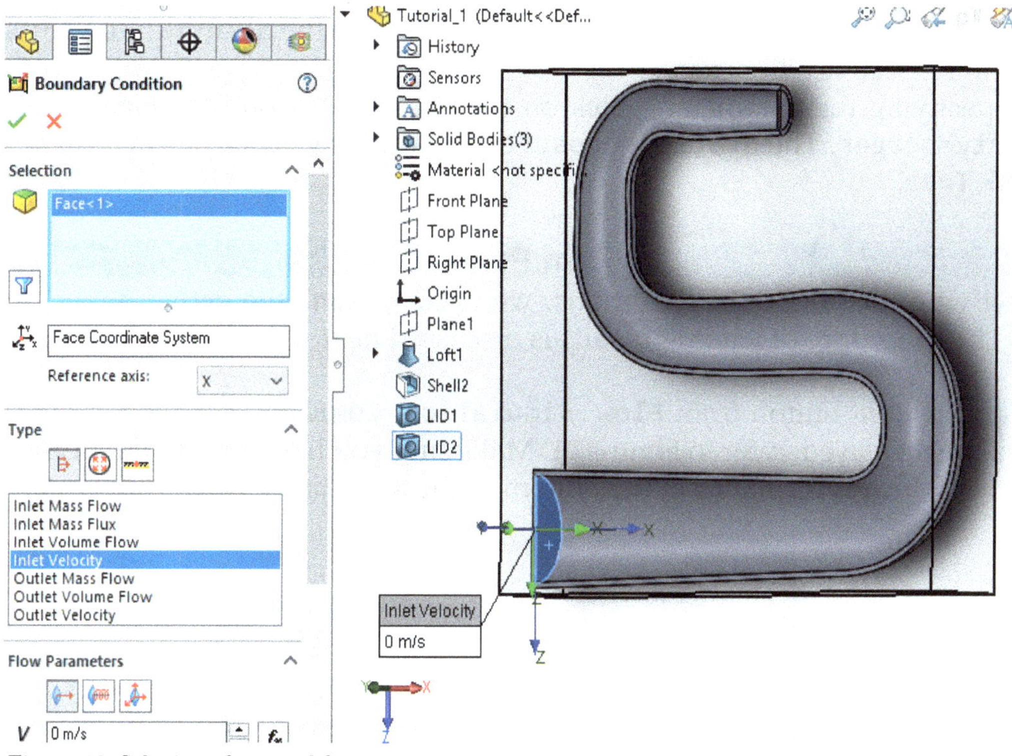

Figure-19. Selection of internal face

- The selected face will be displayed in **Selection** box. Select the **Inlet Velocity** option from **Type** section to specify the boundary condition.
- Click in the **Velocity Normal to Face** edit box from the **Flow Parameters** section and enter the value as **5** m/s.
- After specifying the parameters, click on the **OK** button from **Boundary Condition PropertyManager**. The **Inlet Velocity 1** boundary condition will be added in **Deign Tree**.
- Right-click again on the **Boundary Conditions** option from **Design Tree** and click on the **Insert Boundary Condition** button. The **Boundary Condition PropertyManager** will be displayed.
- The **Selection** box is active by default. You need to click on the internal face of pipe on the other side from where fluid depart; refer to Figure-20. The selected face will be displayed in selection box.

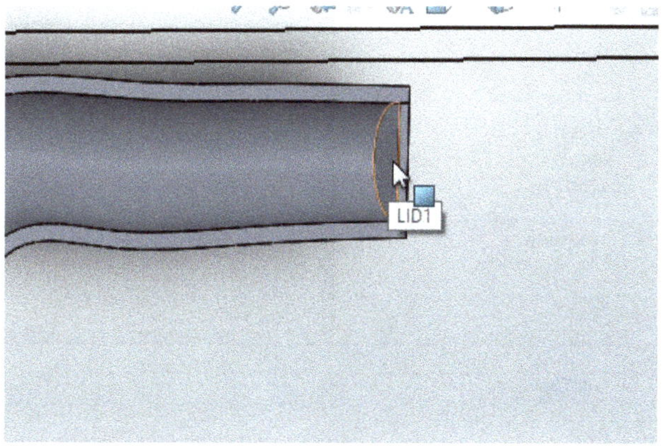

Figure-20. Selection of second face

- Click on the **Pressure Openings** button from **Type** section of **Boundary Condition PropertyManager**. The options related to **Pressure Openings** button will be displayed.
- Click on the **Static Pressure** button from the **Type of Boundary Condition** section. The static pressure will be displayed on the model. Keep the default values 101325 Pa and 293.2K in respective edit boxes.
- After specifying the parameters, click on the **OK** button from **Boundary Condition PropertyManager**. The **Static Pressure 1 Boundary Condition** will be added in **Design Tree**.

Run Project

After specifying the boundary conditions, we need to **Run** the active project for further analyses. The procedure of loading results is discussed next.

- Click on the **Run** button from **Flow Simulation Command Manager**. The **Run** dialog box will be displayed; refer to Figure-21. Make sure you have selected the **Load Results** check box which is displayed at the bottom of the **Run** dialog box.

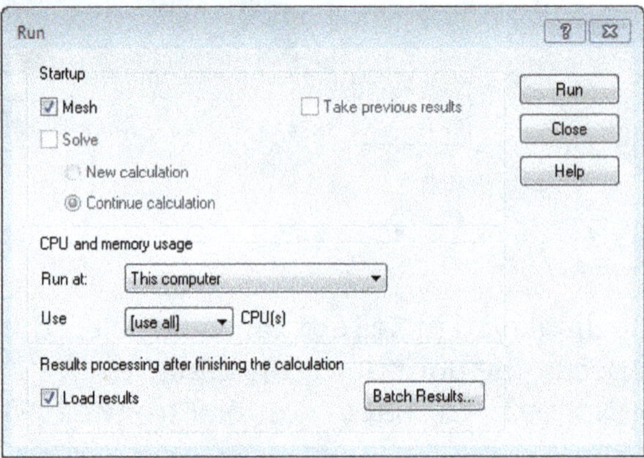

Figure-21. Run dialog box

- After specifying the parameters, click on the **Run** button. The **Solver** dialog box will be displayed while solving the current project; refer to Figure-22.

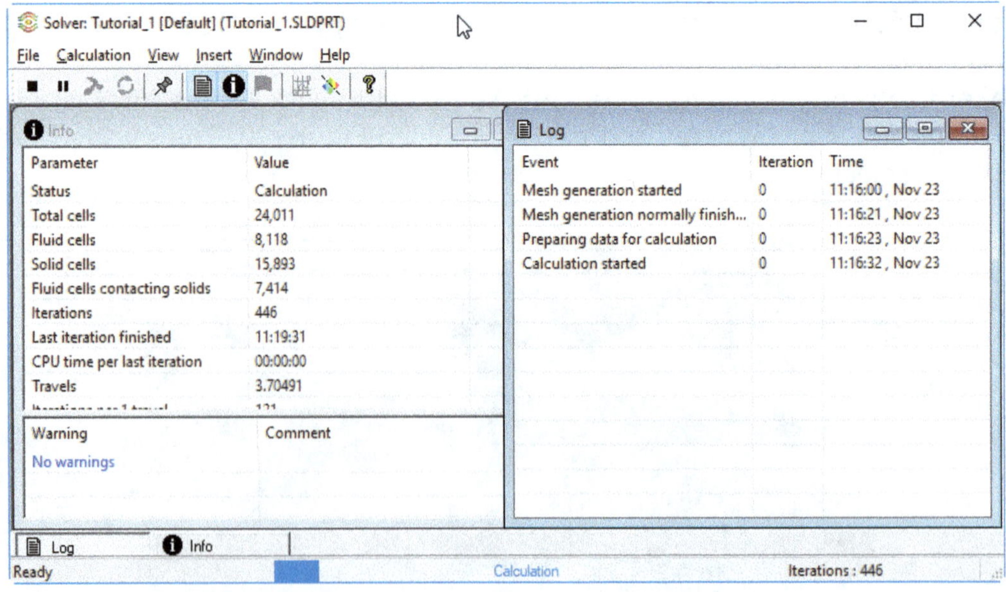

Figure-22. Solver dialog box

- When solver has finished the calculation process. The details of solved calculation will be displayed in **Info** and **Log** boxes of the dialog box. Make sure there is **No warnings** in the **Warning** section, which means we are doing right in the procedure. If there is any error displayed then rectify it as suggested by system.
- Click on the **Close** button from **Run** dialog box. The analysis results will be loaded and displayed in **Results Design Tree**; refer to Figure-23.

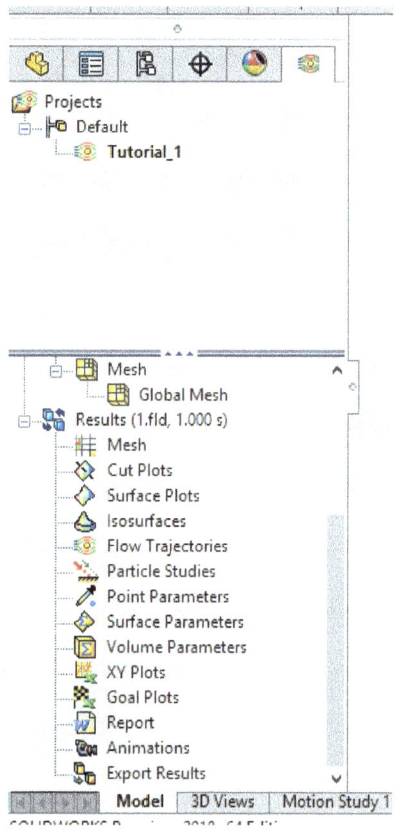

Figure-23. Loaded Results

Some of the analysis results like Mesh, Surface Plots, Flow Trajectories, and Particle Studies are discussed next.

Mesh Analysis

- Right-click on the **Mesh** feature from **Design Tree** and click on the **Insert** button from displayed menu. The **Mesh PropertyManager** will be displayed
- Select the **Use all Faces** check box from **Surfaces** section to use all faces of model. The **Section plane or Planer Surface** selection box is activated by default.
- Click on the **Top Plane** from **Feature Manager Design Tree**. The **Top Plane** will be selected and displayed in selection box.
- Click on the **Color By** drop-down and select the **Refinement level** button for the analysis.
- After specifying the parameters, click on the **OK** button from **Mesh PropertyManager**; refer to Figure-24. The mesh will be generated and displayed on the model; refer to Figure-25.

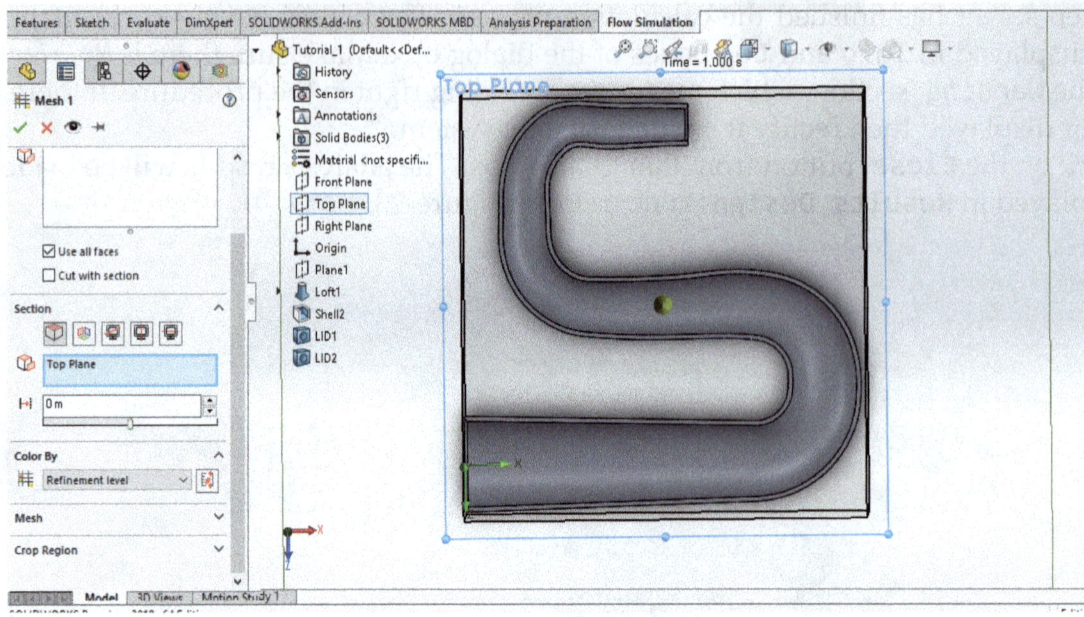

Figure-24. Selected Mesh parameters

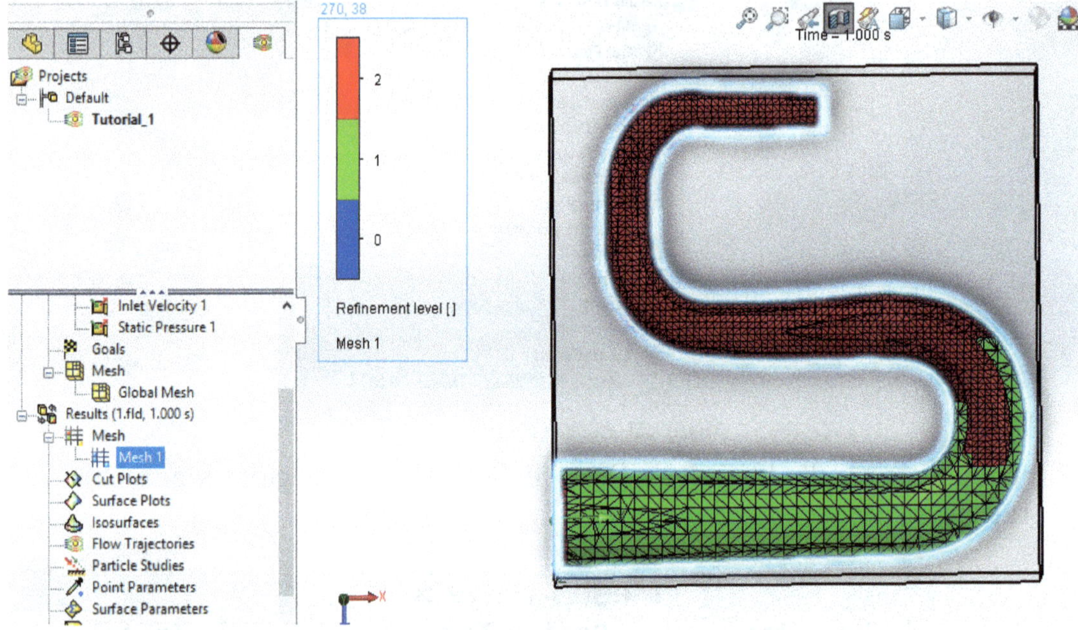

Figure-25. Generated Mesh for tutorial 1

- Right-click on the **Mesh 1** feature from **Design Tree** and click on the **Hide** button to hide the analysis of Mesh 1.
- Now, we will create an another mesh by displaying colored channels. For this, right-click on the **Mesh** feature and click on the **Insert** button. The **Mesh PropertyManager** will be displayed.
- Click in the **Display** drop-down from **Display** section and select the **Channels** button. The options related to the **Channels** button will be displayed.
- Click in the **Maximum** edit box from **Channels** section and the value as **0.04** m.
- Click in the **Minimum** edit box from **Channels** section and enter the value as **0.0006** m.
- After specifying the parameters, click on the **OK** button from **Mesh PropertyManager**. The **Mesh** feature will be added in **Mesh** option; refer to Figure-26.

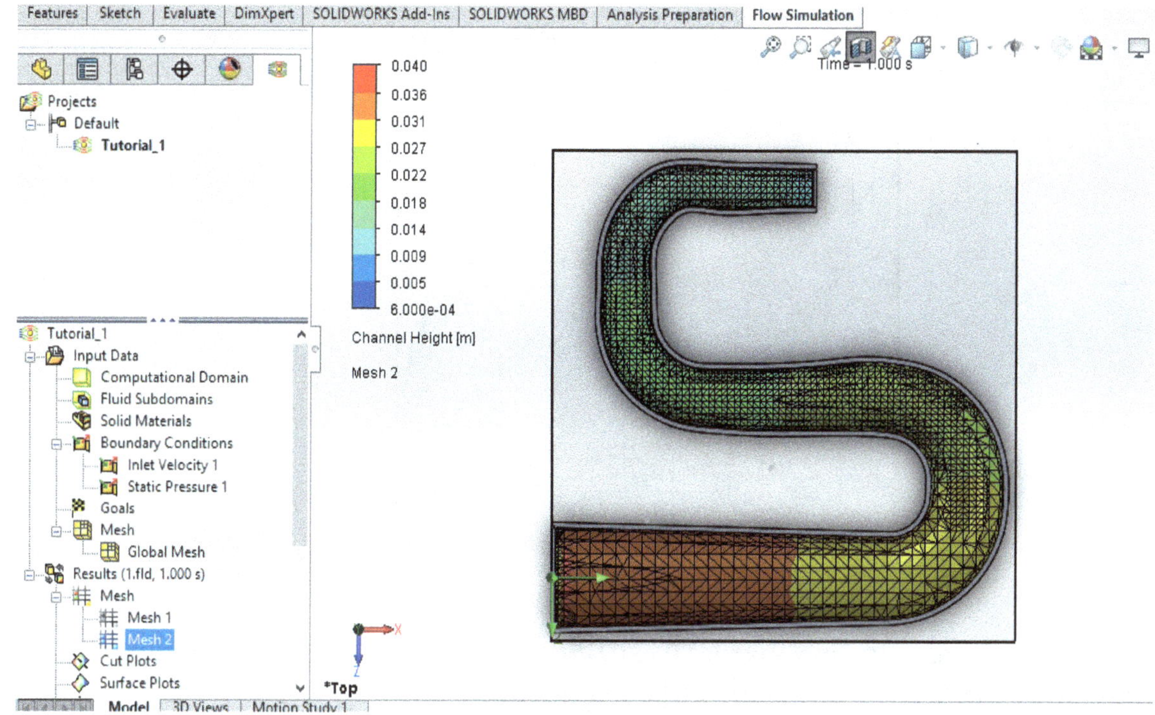

Figure-26. Mesh generation of tutorial 1

- Right-click on the **Mesh** feature from **Design Tree** and click on the **Hide** button to hide the analysis of Mesh.

Note that you need to check the broken mesh cells in the mesh plot.

Surface Plots

Now, we need to check the pressure and temperature distribution on the model faces. So, we are going to apply the **Surface Plots** feature on the current model. The procedure is discussed next.

- Right-click on the **Surface Plots** feature from **Design Tree** and click on the **Insert** button from the displayed menu. The **Surface Plot PropertyManager** will be displayed.
- The **Selection** box of **Surface Plot PropertyManager** is active by default. You need to click on the internal surface of model to select; refer to Figure-27. The selected surface will be displayed in **Selection** box.

Figure-27. Selecting surface for Surface Plot

- Click on the **Contours** button from **Display** section and select the **Pressure** parameter from **Parameter** drop-down of **Contours** section.
- Click in the **Number of Levels** edit box from **Contours** section and enter the value as **12**.
- After specifying the parameters, click on the **OK** button from **Surface Plot PropertyManager**; refer to Figure-28.

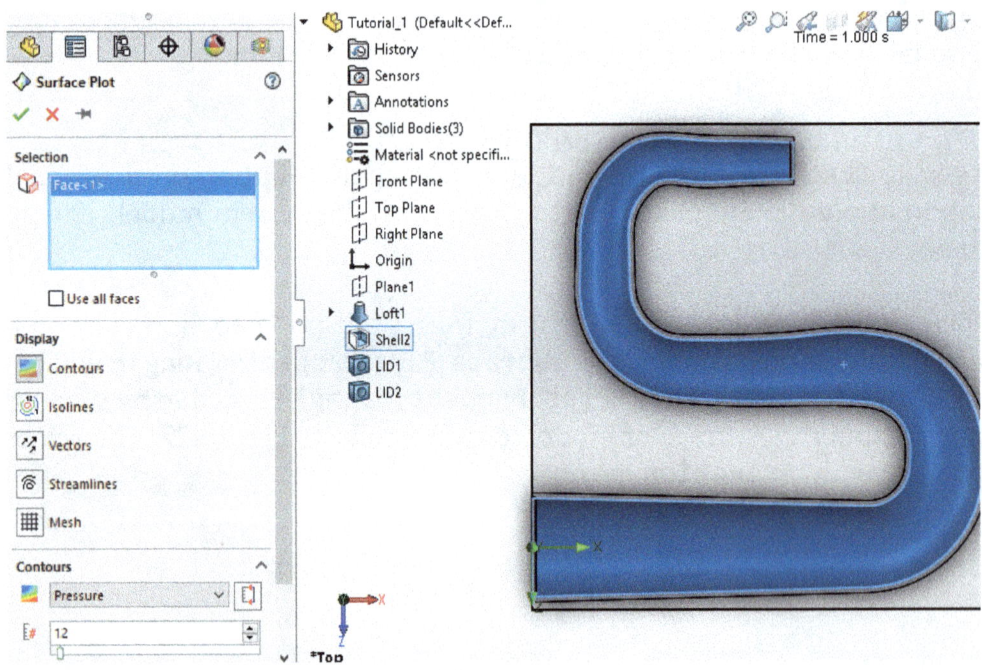

Figure-28. Specifying parametrs for Surface Plot

- When the result is calculated, the model will be displayed along with the contour of pressure distribution; refer to Figure-29.

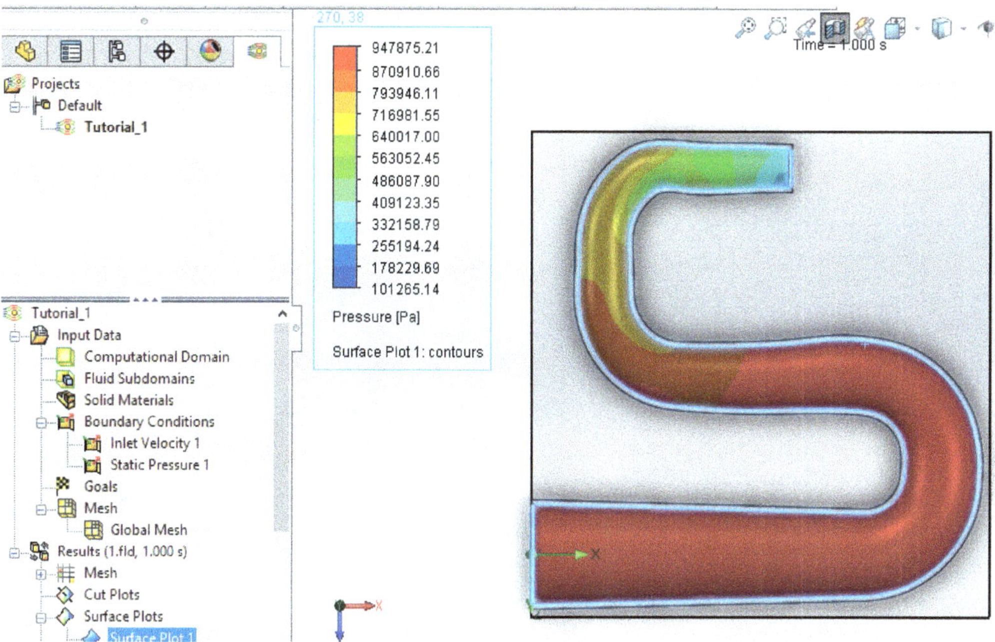

Figure-29. Surface Plot of Tutorial 1

- Right-click on the **Surface Plot 1** feature and click on the **Show Plot Minimum** and **Show Plot Maximum** button to view the maximum and minimum values of plot.
- Now, click on the **Hide** button from the right-click menu of **Surface Plot 1** feature to hide the result.
- Click on the **Insert** button from the right-click menu of **Surface Plots** feature, again. The **Surface Plots PropertyManager** will be displayed.
- Select the **Use All faces** check box from **Selection** section to select all the faces of model.
- Click on the **Isolines** and **Vectors** buttons from **Display** section for analysis and deselect **Contours** button.
- Click on the **Parameter** drop-down from **Isolines** section and select the **Temperature (Solid)** parameter.
- Click in the **Number of Levels** edit box from **Isolines** section and enter the value as **10**.
- Click in the **Width** edit box from **Isolines** section and enter the value as **1**.
- Click on the **Dynamic Vectors** button from the **Vectors** section.
- Click on the **Parameter** drop-down from **Vectors** section and select the **Heat Flux** button.
- Click in the **Spacing** and **Max Arrow Size** edit boxes from **Vectors** section and enter the value as **16** and **64** respectively.
- After specifying the parameters, click on the **OK** button from **Surface Plot PropertyManager**. The **Surface Plot 2** feature will be added in the **Design Tree**; refer to Figure-30.

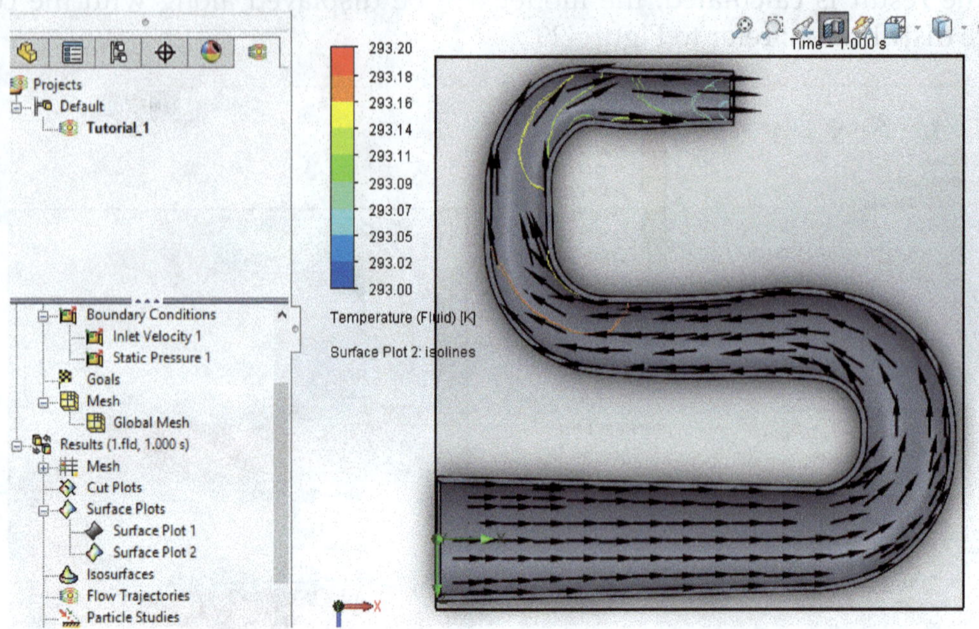

Figure-30. Surface Plot 2

- Now, click on the **Hide** button from the right-click menu of **Surface Plot 2** feature to hide the representation.

Flow Trajectories

Now, we will check the flow trajectories of the current model. The procedure is discussed next.

- Right-click on the **Flow Trajectories** feature and click on the **Insert** button. The **Flow Trajectories PropertyManager** will be displayed.
- The **Selection** box is activated by default. Click on the **Top Plane** from **Feature Manager Design Tree**. The **Top Plane** will be selected and displayed in **Selection** box.
- Click in the **Number of Points** edit box from **Starting Points** section and enter the value as **20**.
- Click on the **Draw Trajectories As** drop-down from **Appearance** section and select the **Pipe** button.
- Click on the **Color By** drop-down from **Appearance** section and select the **Velocity** option.
- After specifying the parameters, click on the **OK** button from **Flow Trajectories PropertyManager**. The calculation of flow trajectories will be started and display the representation on model when completed; refer to Figure-31.

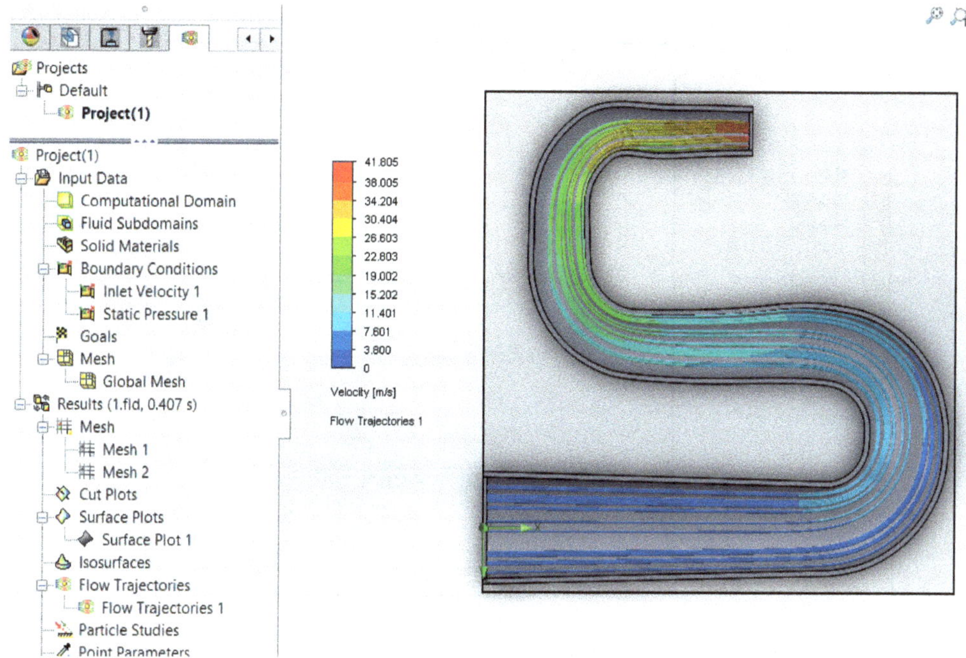

Figure-31. Flow Trajectories representation of Tutorial 1

- To view the visual representation of the current **Flow Trajectories**, right-click on the recently created **Flow Trajectories 1**. The right-click menu will be displayed; refer to Figure-32.

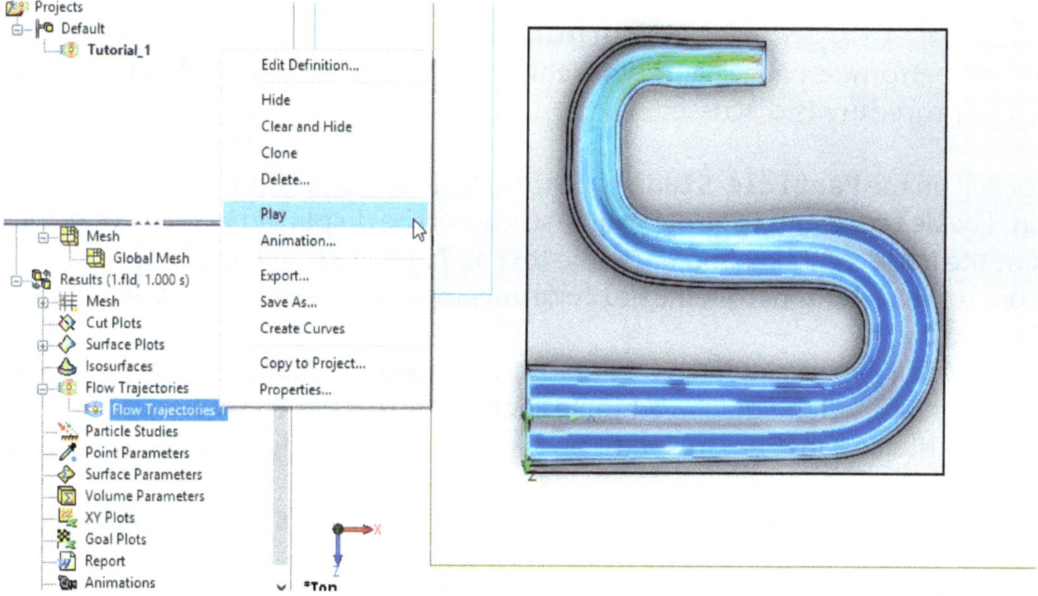

Figure-32. Right-click menu of Flow Trajectories 1

- Click on the **Play** button from the displayed menu. The animation of flow trajectories will be started on the model; refer to Figure-33.

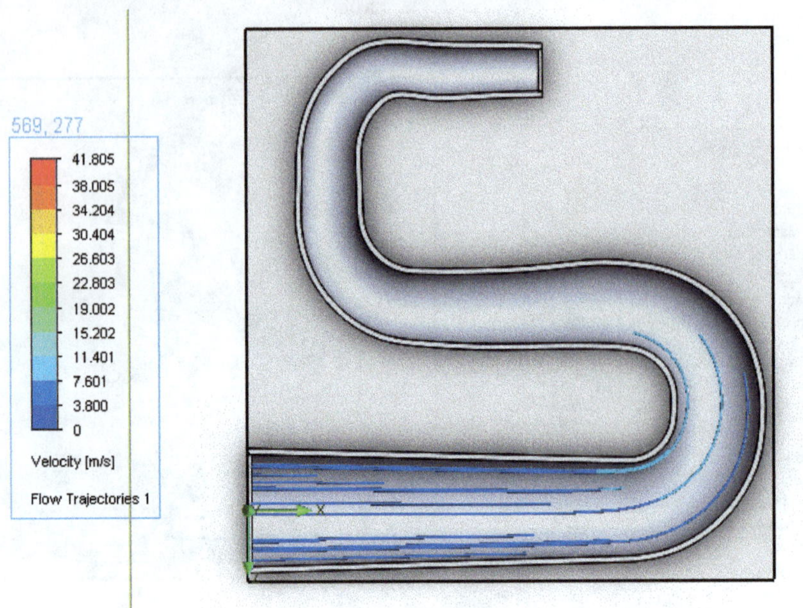

Figure-33. Animation on Flow Trajectories

- To stop the animation, click on the **Stop** button from the right-click menu of **Flow Trajectories** feature.
- Now, click on the **Hide** button from the right-click menu of **Flow Trajectories 1** feature to hide the representation.

Particle Study

Now, we will perform a particle study on the current model to see behavior of fluid inside the tube. The procedure is discussed next.

- Right-click on the **Particle Studies** feature from **Design Tree** and click on the **Wizard** button. The **Welcome** message of particle study will be displayed in the **PropertyManager**.
- Click in the **Name** edit box and enter the text as **Tutorial 1 Particle Study**.
- Click on the **Next** button from the **Welcome** message. The **Injection 1 PropertyManager** will be displayed.
- The **Selection** box is activated by default. You need to select the internal face of model. The selected face will be displayed in **Selection** box; refer to Figure-34.
- Click in the **Number of Points** edit box from the **Starting Points** section and enter the value as **20**.
- Click in the **Diameter** edit box from **Particle Properties** section and enter the value as **0.0001** m.
- Select the **Water** from **Liquid Parameter** selection list. The water will be displayed in **Selected Parameter** box.
- Click in the **Mass Flow Rate** edit box in **Mass Flow Rate** section and enter the value as **1** kg/s.
- Select the **Relative** radio button from **Initial Particle Velocity** section and **Initial Particle Temperature** section. Specify **Velocity Magnitude** as **5**.
- After specifying the parameters, click on the **Next** button from **Injection 1 PropertyManager**; refer to Figure-34. The **Physical Settings PropertyManager** will be displayed.

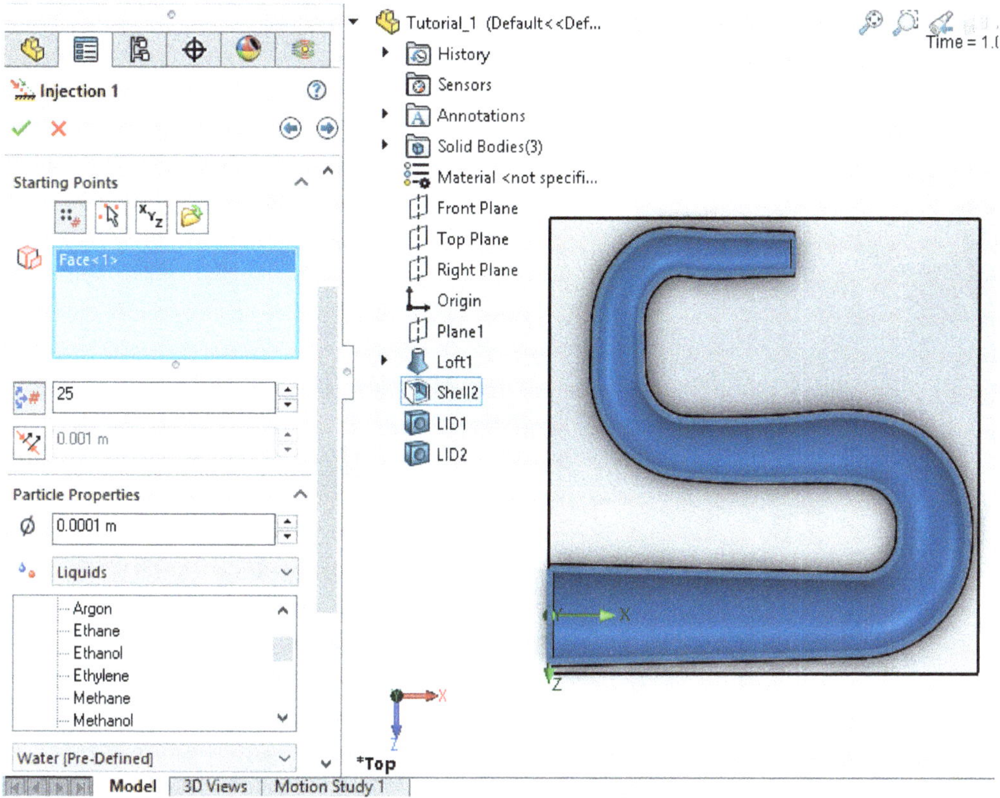

Figure-34. Injection 1 for Tutorial 1

- To calculate mass accumulation rates, select the **Accretion** check box from **Physical Settings PropertyManager** and click on the **Next** button. The **Default Wall Condition PropertyManager** will be displayed.
- Click on the **Ideal reflection** button from **Condition** section to select the wall condition and click on the **Next** button. The **Calculation Settings PropertyManager** will be displayed.
- Select the **Trajectories and statistics** radio button from **Results Saving** section to visualize particle study.
- Click on the **Draw Trajectories As** drop-down from **Default Appearance** section and select the **Spheres** button.
- Click in the **Width** edit box from **Default Appearance** section and enter the value as **0.0007** m.
- Click on the **Color By** drop-down from **Default Appearance** section and select the **Velocity** button.
- Click in the **Maximum Length** edit box from **Constraints** section and enter the value as **1** m.
- Make sure you have selected the **Use CAD geometry** check box from **Constraints** section.
- After specifying the parameters, click on the **Next** button from **Calculation Settings PropertyManager**; refer to Figure-35. The **Run PropertyManager** will be displayed.

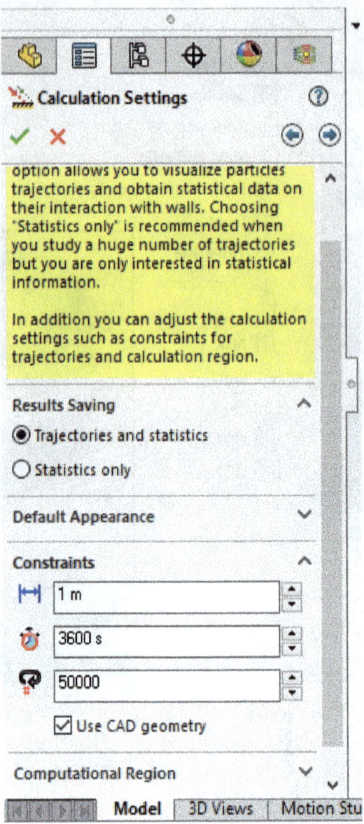

Figure-35. Calculation Settings PropertyManager for Tutorial 1

- Click on the **Run** button from **Run PropertyManager**. The **Progress** box will be displayed in which processes are running for the calculation of particle study. After the calculations, the **Wall Conditions** feature and **Injection 1** feature will be added in **Design Tree**; refer to Figure-36.

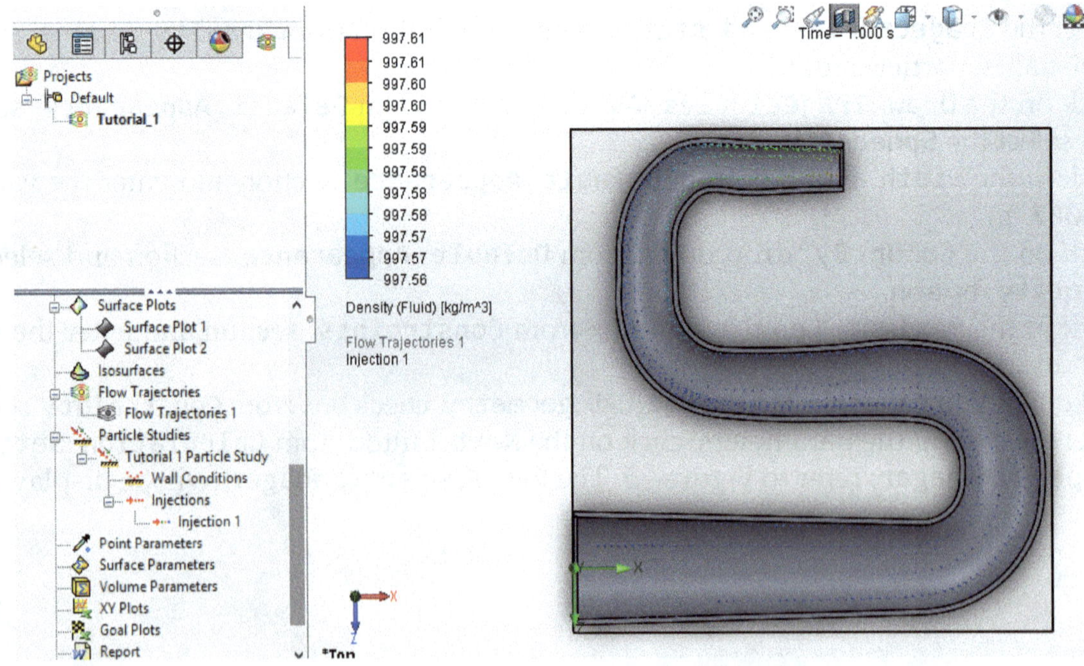

Figure-36. Injection 1 Tutorial 1

- To view the animation of particle study, you need to select the **Play** button from right-click shortcut menu of **Injection 1** feature. The animation of the particle study will be started on the model; refer to Figure-37

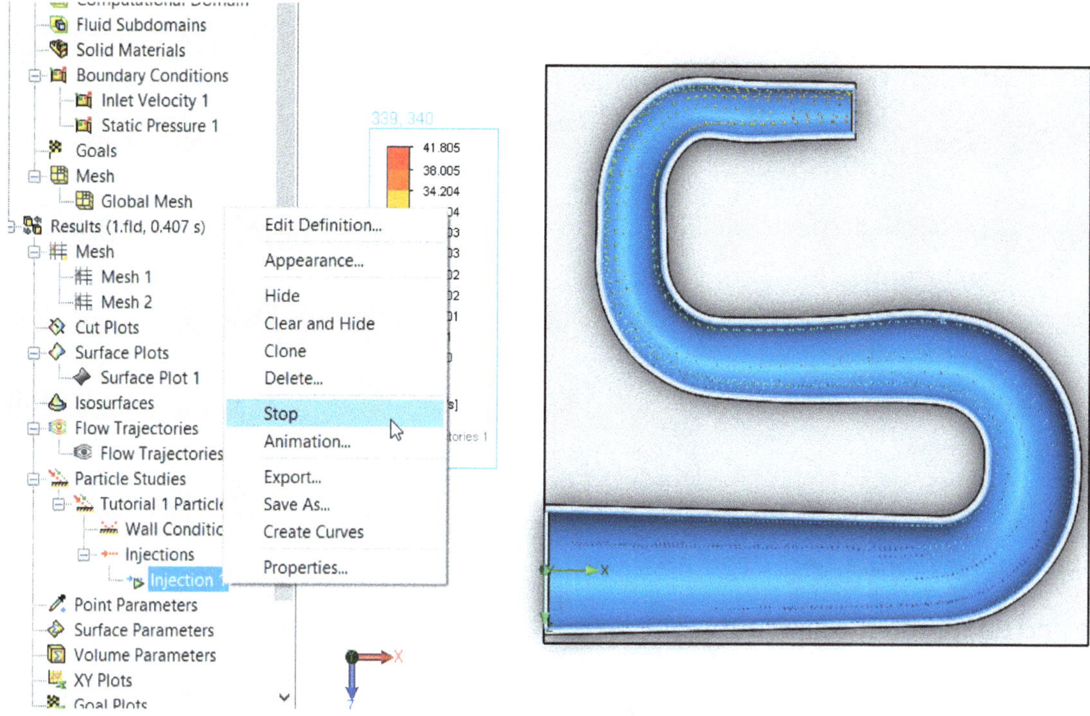

Figure-37. Animation of Injection

- Click on the **Stop** button from right-click menu of **Injection 1** to stop the animation.

Till now we have generated different results of analysis. Now, we will generate a report of analysis results.

Report

The procedure to generate a MS Word report of analysis results is discussed next.

- Click on the **Report** tool from **Flow Simulation Results** drop-down in the **Flow Simulation CommandManager** and click on the **Report** button from displayed menu. The **Report** dialog box will be displayed.
- Click on the **From Template** button from **Report** dialog box. The **Open** dialog box will be displayed.
- Select the **flow simulation report.dotx** template and click on the **Open** button. The generation of report will be started and report will be displayed in Microsoft Word upon completion of calculation; refer to Figure-38.

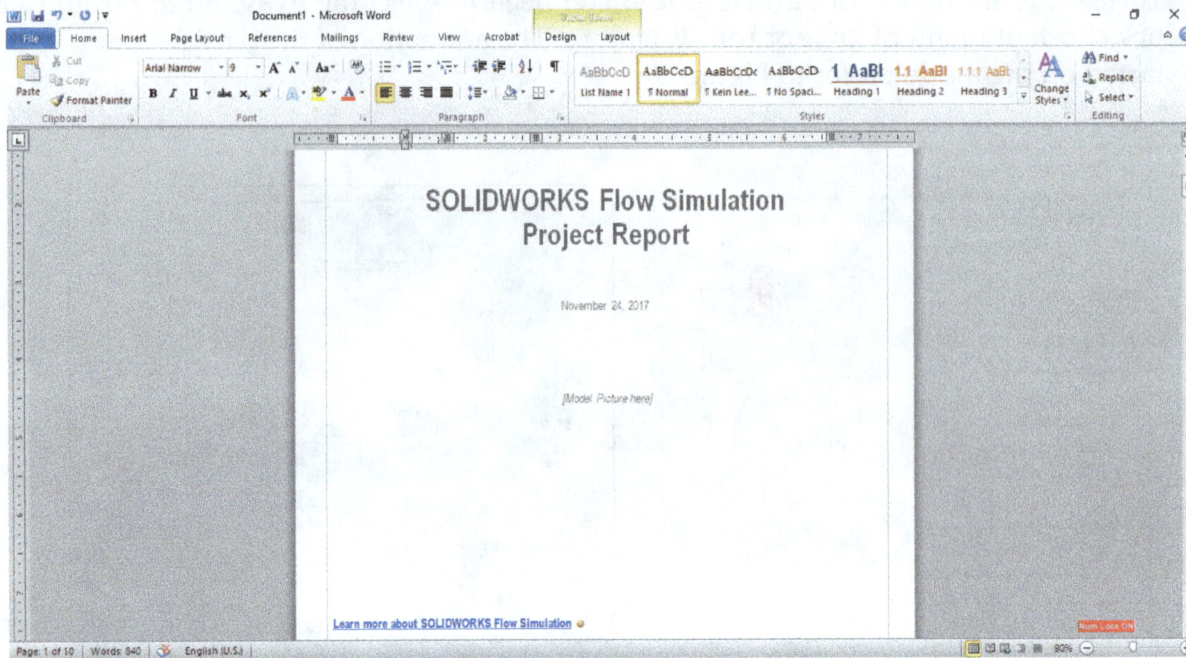

Figure-38. Report of Tutorial 1

- Save and share the file as required.

PRACTICAL 2

Flow Simulation is also used to study the fluid flow and heat transfer in a engineering equipment. In this tutorial, we will use flow simulation to determine the efficiency of counter flow heat exchanger and to observe the temperature and flow pattern inside it. With the help of flow simulation, the design engineer can gain insight into the physical process involved thus giving guidance for improvement to the design.

A convenient measure of heat exchanger performance is its "efficiency" in transferring a given amount of heat from one fluid at higher temperature to another fluid at lower temperature. The efficiency can be determined if the temperatures at all flow openings are known. In Flow Simulation, the temperatures at the fluid inlets are specified and the temperatures at the outlets can be easily determined. Heat exchanger efficiency is defined as follows:

$$\varepsilon = \frac{\text{actual heat transfer}}{\text{maximum possible heat transfer}}$$

Starting with model

- Double-click on the **Heat Exchanger-Tutorial** file from the resource folder. The welcome window of SOLIDWORKS software will be displayed along with the model; refer to Figure-39. Activate SolidWorks Flow Simulation Add-In if not done yet.

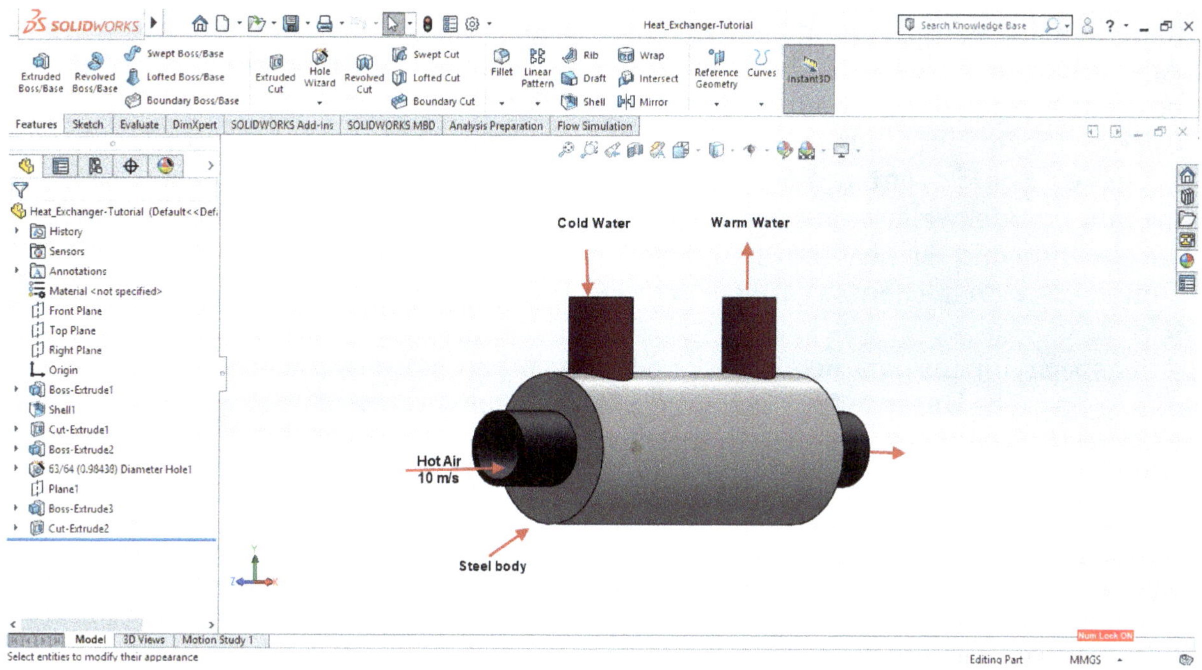

Figure-39. Heat exchanger model

Creating a Flow Simulation project

- Click on the **Flow Simulation CommandManager** from the **Ribbon**. The tools of flow simulation will be displayed.
- Click on the **Wizard** button from **Flow Simulation CommandManager**. The **Wizard-Project Name** will be displayed.
- Click in the **Project name** edit box from the dialog box and enter the name as **Heat Exchanger**.
- After specifying the parameters, click on the **Next** button from **Wizard-Project Name** dialog box. The **Wizard-Unit System** dialog box will be displayed.
- Click on the **SI** unit system from **Unit System** selection box and click on the **Next** button from **Wizard-Unit** System dialog box. The **Wizard-Analysis Type** dialog box will be displayed.
- Select the **Internal** radio button from **Analysis Type** section and select the **Exclude cavities without flow conditions** check box from **Consider closed cavities** section.
- Select the **Heat conduction in Solids** check box and **Gravity** check box from **Physical features** selection list. Click on the **Next** button. The **Wizard-Default Fluid** dialog box will be displayed.
- Click on the **Liquid Node** button from **Fluids** selection box and select the **Water** option. Click on **Add** button from **Wizard-Default Fluid** dialog box. The selected fluid will be added in **Project Fluids** section for the current project.
- Click on the **Gases Node** button from **Fluids** selection box and select the **Air** option. Click on **Add** button from **Wizard-Default Fluid** dialog box. The **Air** will be added in **Project Fluids** section for the current project.
- After specifying the parameters, click on the **Next** button from **Wizard - Default Fluid** dialog box. The **Wizard-Default Solid** dialog box will be displayed.
- In this dialog box, we are going to specify the default solid material applied to all solid component.

- Click on the **Alloys Node** and select the **Stainless Steel 302** option. The selected material will be displayed in **Default Solid:** box.
- After selecting the material, click on the **Next** button from **Wizard-Default Solid** dialog box. The **Wizard-Wall Conditions** dialog box will be displayed.
- Click on the **Default outer wall thermal condition** drop-down and select the **Heat Transfer coefficient** option. This condition allows you to define the heat transfer from the outer model walls to an external fluid (not modeled) by specifying the reference fluid temperature and the heat transfer coefficient value.
- Double-click on the **Heat transfer coefficient** edit box and specify the value as **16.3 W/m^2/K**.
- After specifying the parameters, click on the **Next** button from **Wizard-Wall Conditions** dialog box. The **Wizard-Initial Conditions** dialog box will be displayed.
- Click in the **Pressure** edit box from **Thermodynamic Parameters Node** and enter the value as **2 atm**. The **Flow Simulation** will automatically convert the entered value to the selected system of units.
- After specifying the parameters, click on the **Finish** button to finish the wizard from **Wizard-Initial Conditions** dialog box. The **Flow Simulation** information box will be displayed which was asking for creating the lids on the openings and closings of model.
- Create the lids of this model using **Create Lids** tool. The procedure of creating lids was discussed earlier.
- Now, click on the **Yes** button from **Flow Simulation** information box, The **Create Lids PropertyManager** will be displayed. Create the lids as required; refer to Figure-40. Click on **Yes** button from subsequent dialog boxes.

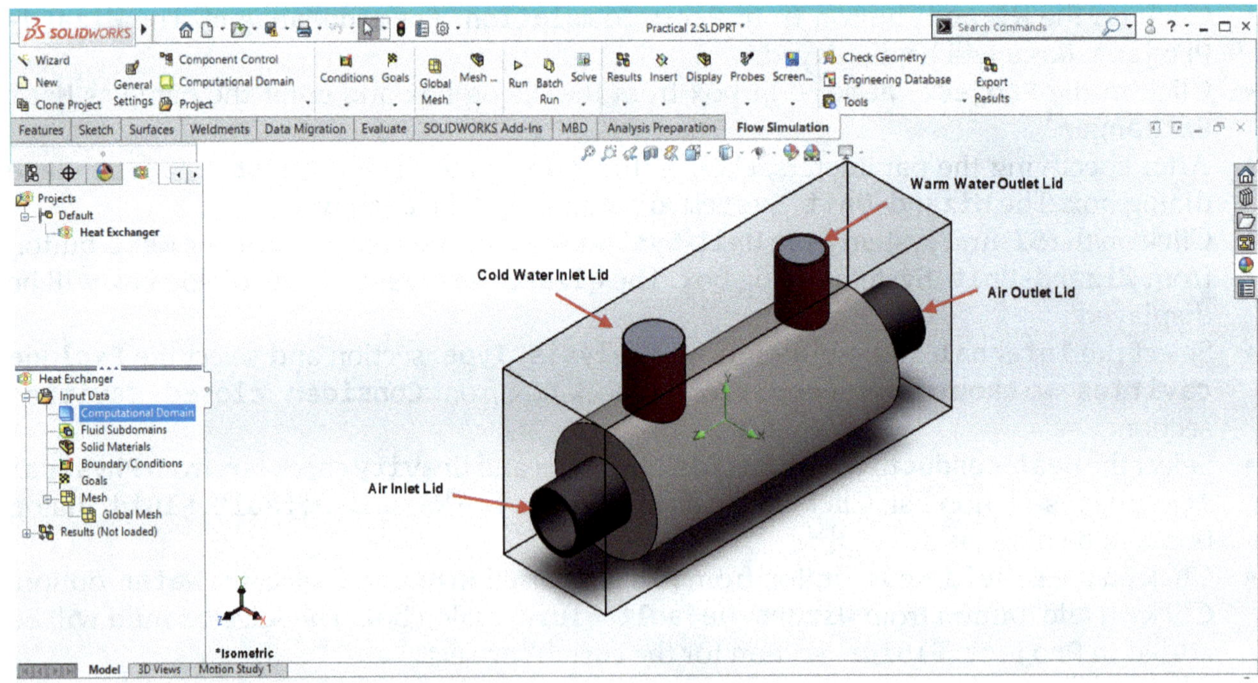

Figure-40. Heat Exchanger after creating lids

Specifying Symmetry Condition

- Right-click on the **Computational Domain** option from **Flow Simulation Analysis Design Tree** and click on the **Edit Definition** button. The **Computational Domain PropertyManager** will be displayed; refer to Figure-41.

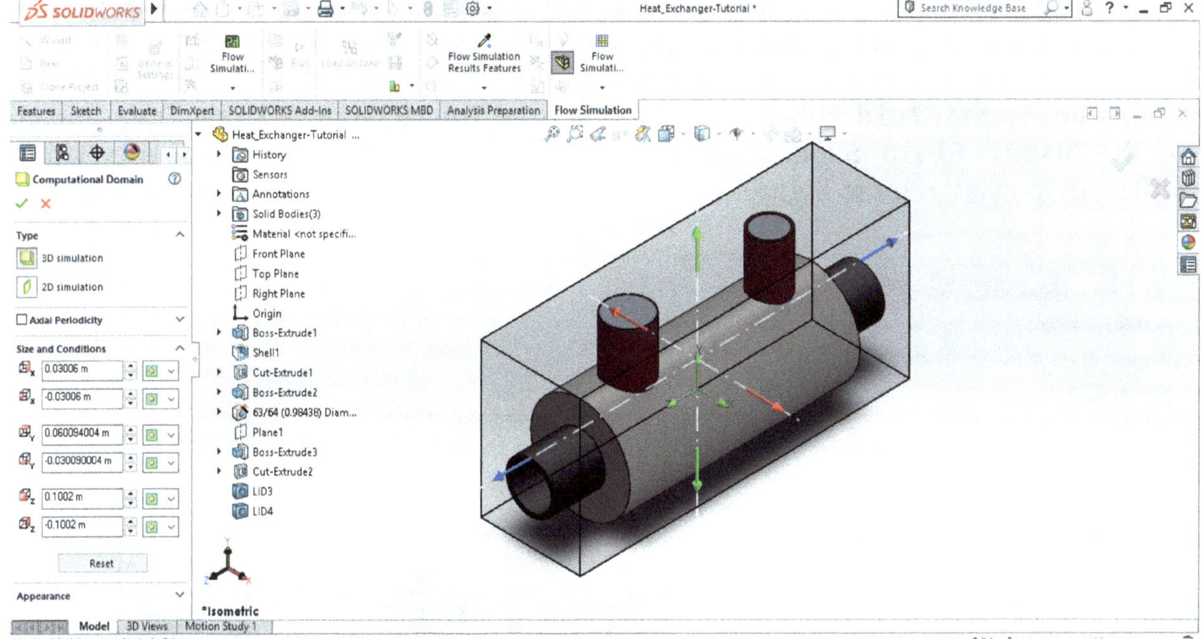

Figure-41. The Computational Domain Feature Manager of Heat Exchanger

- Click on the **Boundary Condition** drop-down from **X max** option and select the **Symmetry button**; refer to Figure-42.

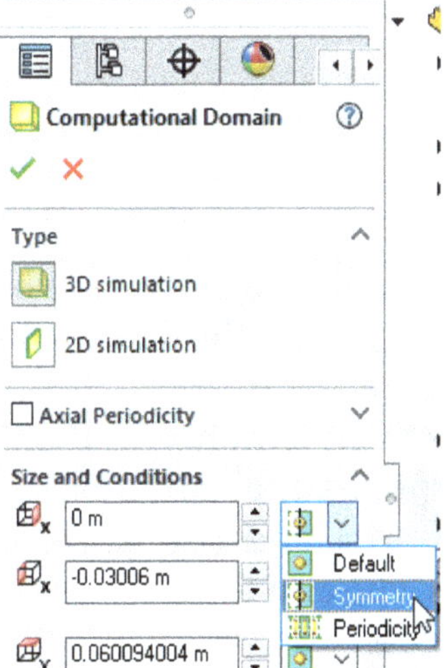

Figure-42. Boundary Condition drop-down

- Click in the **X max** edit box from **Size and Conditions** section and enter the value as **0 m**.
- After specifying the parameters, click on the **OK** button from **Computational Domain PropertyManager**. The parameters will be updated.

Specifying a Fluid Subdomain

Since, we have selected water as a default fluid in the **Wizard** dialog box, we need to specify another fluid type and select another fluid (air) for the fluid region inside the tube through which the hot air flows. We can do this by creating a Fluid Subdomain parameter. We will specify air as a fluid at a initial temperature of 600 K and flow velocity of -10 m/s as the initial conditions in the selected fluid region.

- Right-click on the **Fluid Subdomains** option from **Input Data Design Tree** and click on the **Insert Fluid Subdomain** button. The **Fluid Subdomain PropertyManager** will be displayed; refer to Figure-43.

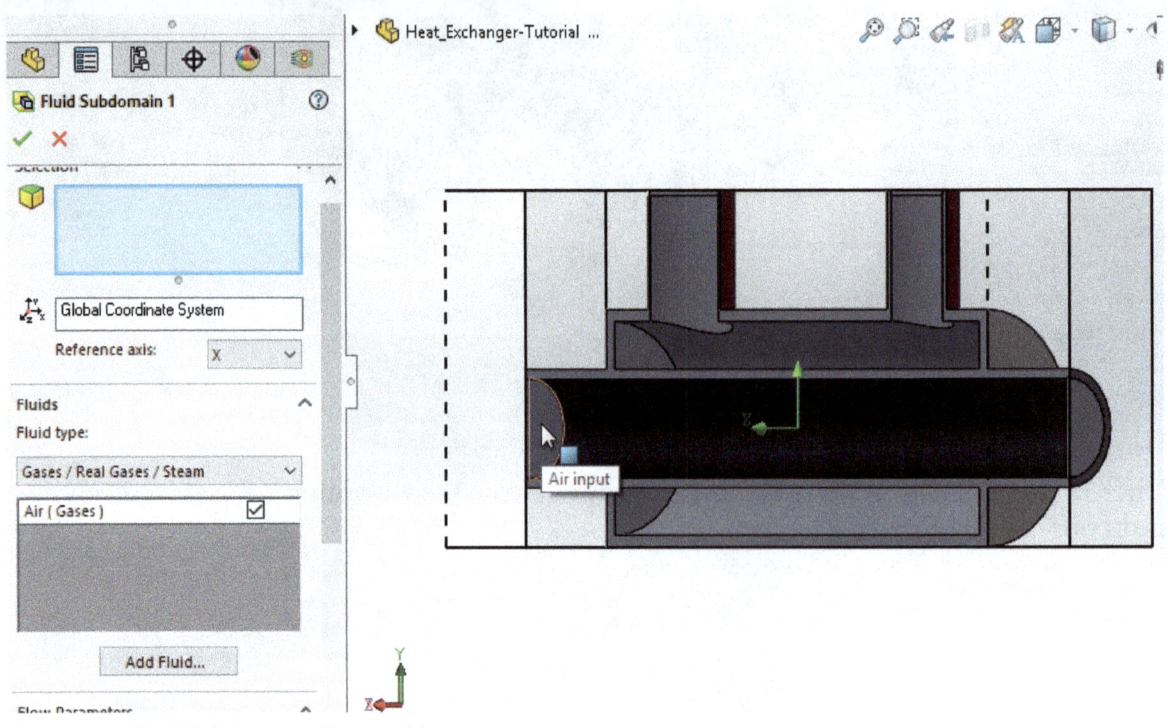

Figure-43. Fluid Subdomain 1 PropertyManager

- The selection box is active by default. Select the face as shown in above figure.
- From the **Fluid type** drop-down, select the **Gases/Real Gases/Steam** option and then select the **Air (Gases)** as a fluid.
- Click in **velocity in Z Direction** edit box and enter the value as **-10 m/s**.
- Click in the **Pressure** and **Temperature** edit boxes from **Thermodynamic Parameters** section and enter the value as **1 atm** and **600 K** respectively.
- After specifying the parameters, click on the **OK** button. The **Fluid Subdomain 1** feature will be added in the **Design Tree**.
- Similarly, we are going to create a second fluid domain taking water as a fluid.
- Right-click on the **Fluid Subdomains** option from **Input Data Design Tree** and click on the **Insert Fluid Subdomains** button. The **Fluid Subdomain PropertyManager** will be displayed.
- The selection box is active by default. You need to select internal face of **Cold Water Lid** to specify the subdomain; refer to Figure-44.

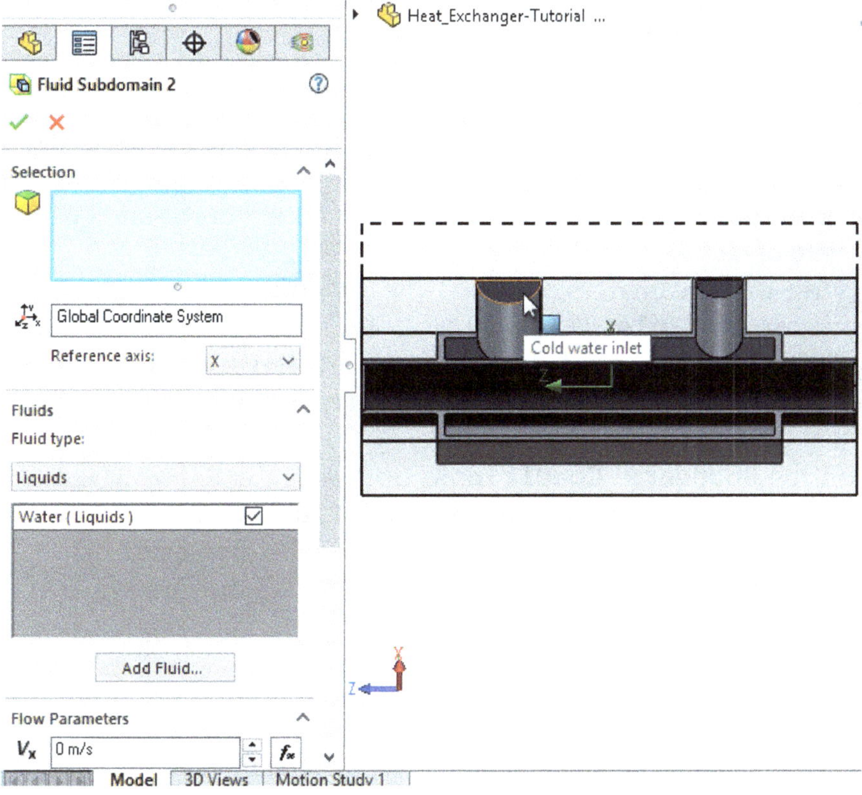

Figure-44. Fluid Subdomain 2

- From the **Fluid type** drop-down, select the **Water (Liquids)** as a fluid.
- After specifying the parameters, click on the **OK** button. The **Fluid Subdomain 2** feature will be added in **Input Data Design Tree**.

Specifying Boundary Condition

- Right-click on the **Boundary Conditions** option in the **Flow Simulation Analysis Tree** and click on the **Insert Boundary Condition** button from the displayed right-click menu. The **Boundary Condition PropertyManager** will be displayed.
- The selection box is active by default. You need to select the internal face of water inlet lid; refer to Figure-45.

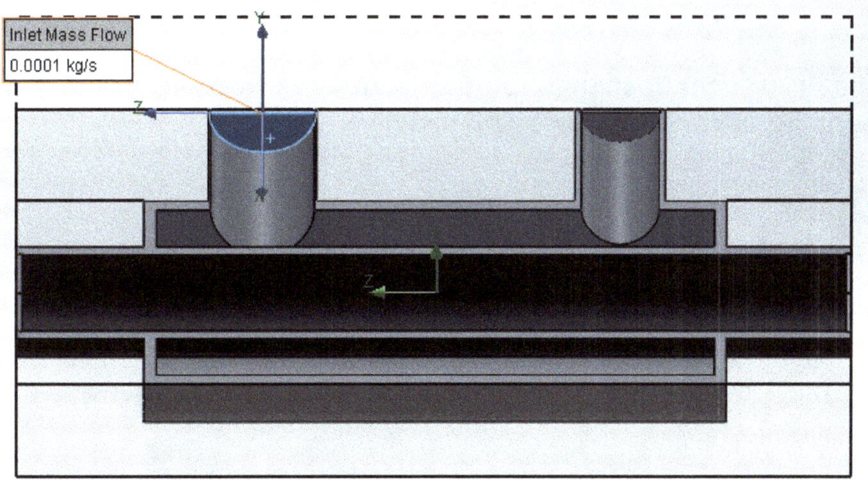

Figure-45. Selecting internal face of Cold Water Inlet Lid

- The selected face will be displayed in selection box.

- Click on the **Inlet Mass Flow** boundary condition from **Types of Boundary Condition** list.
- Click on the **Normal to Face** button from **Flow Parameter** section. The parameters related to **Normal to Face** will be displayed.
- Click in the **Mass Flow Rate** edit box and enter the value as **0.01** kg/s. Since the symmetry plane halves the opening, we need to specify a half of the actual mass flow rate. This boundary condition specifies that water enters in the shell of heat exchanger at a mass flow rate of 0.02 kg/s and temperature of 293.2 k.
- After specifying the parameters, click on the **OK** button from **Boundary Condition PropertyManager**. The **Inlet Mass Flow 1** feature will be added in **Analysis Tree**.
- Rename the **Inlet Mass Flow 1** feature to **Inlet Mass Flow - Cold Water**.
- Now, we are going to specify the water outlet **Environment Pressure**.

- Right-click on the **Boundary Conditions** option in the **Flow Simulation Analysis Tree** and click on the **Insert Boundary Condition** option from the displayed menu. The **Boundary Condition PropertyManager** will be displayed.
- The selection box is active by default. You need to select the inner face of water outlet lid; refer to Figure-46.

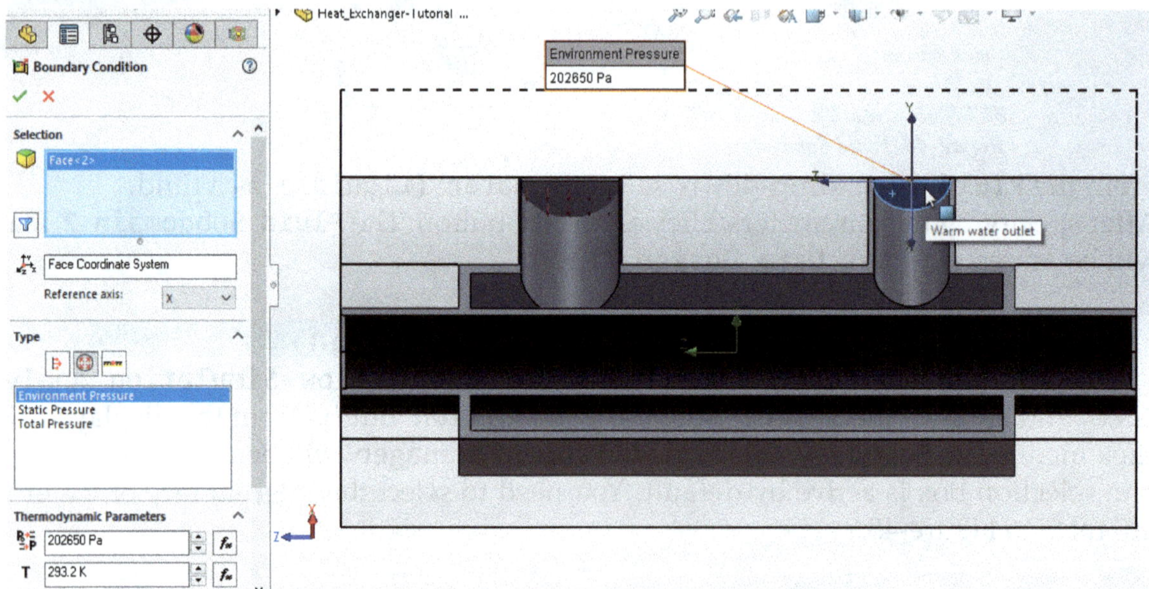

Figure-46. Selecting inner face of water outlet lid

- Click on the **Pressure Openings** button from **Type** section. The parameters related to **Environment Pressure** will be displayed.

- After selecting **Environment Pressure** option, click on the **OK** button from **Boundary Condition PropertyManager**. The **Environment Pressure 1** feature will be added in **Design Tree**.
- Rename the **Environment Pressure 1** feature to **Environment Pressure-Warm Water**.

Next, we will specify the boundary conditions for the hot air flow.

- Right-click on the **Boundary Conditions** option in the **Flow Simulation Analysis Tree** and click on the **Insert Boundary Condition** button from the displayed right-click menu. The **Boundary Condition PropertyManager** will be displayed.
- The selection box is active by default. You need to select the inner face of air inlet lid; refer to Figure-47.

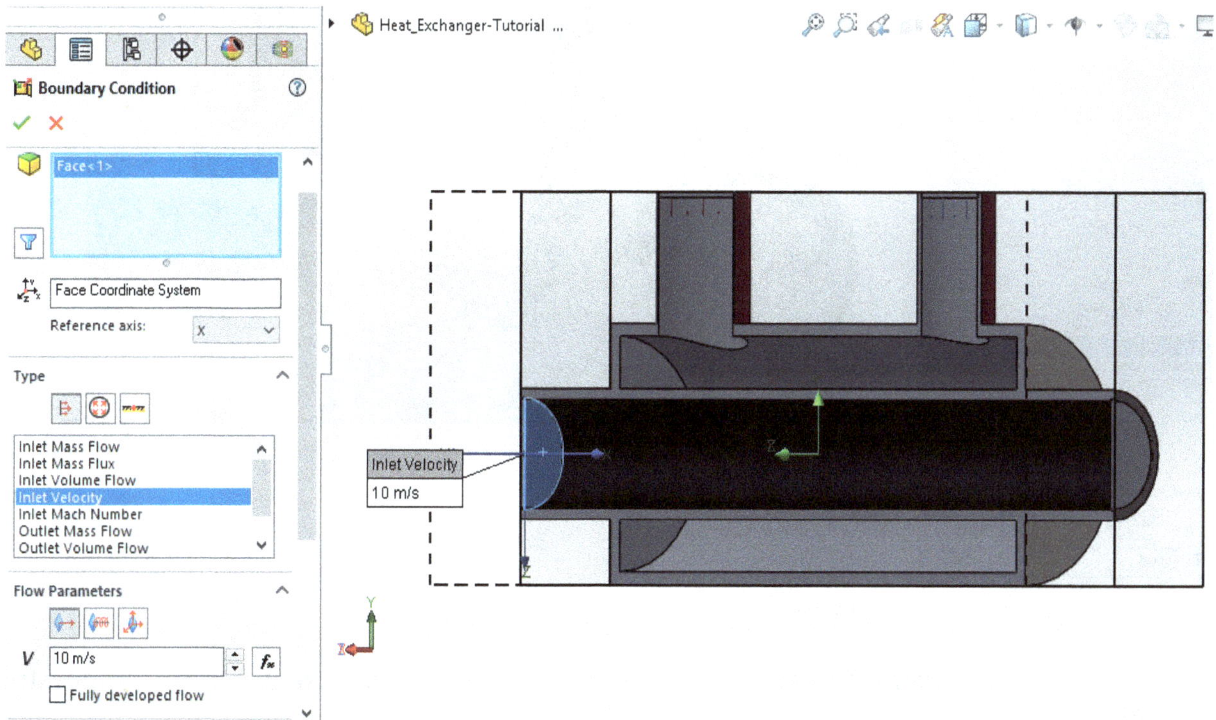

Figure-47. Selecting inner face for air inlet

- Select the **Inlet Velocity** option from the **Types of Boundary Condition** list.
- Click on the **Normal to Face** button from **Flow Parameters** section to select.
- Click in the **Velocity Normal to Face** edit box from **Flow Parameters** section and enter the value as **10 m/s**.
- The parameters of **Thermodynamic Parameters** section were stated in **Wizard-Fluid Subdomain** dialog box earlier.
- After specifying the parameters, click on the **OK** button from **Boundary Condition PropertyManager**. The **Inlet Velocity 1** feature will be added in **Analysis Tree**.
- Rename the **Inlet Velocity 1** feature as **Inlet Velocity-Hot air**.

- Now, we are going to create an another boundary condition whose procedure is discussed next.

- Right-click on the **Boundary Conditions** option in the **Flow Simulation Analysis Tree** and click on the **Insert Boundary Condition** button from the displayed right-click menu. The **Boundary Condition PropertyManager** will be displayed.
- The selection box is active by default. You need to select the inner face of air outlet lid; refer to Figure-48.

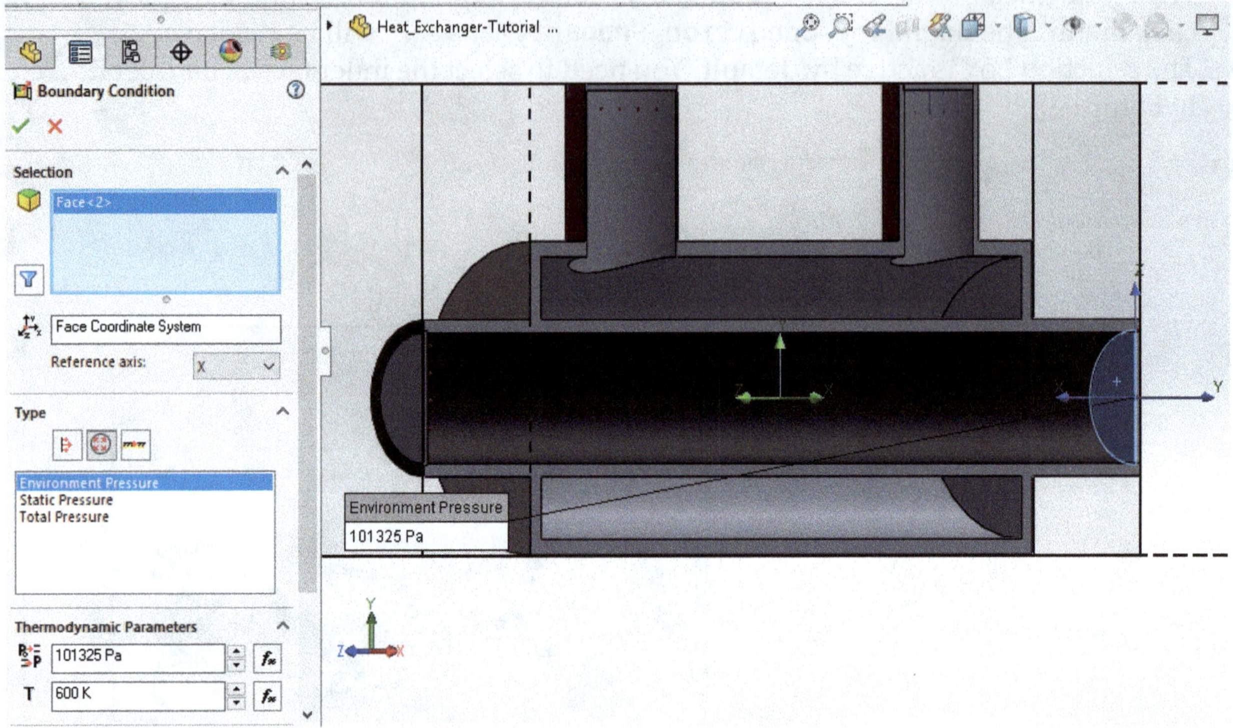

Figure-48. Selecting face of water outlet lid

- Select the **Environment Pressure** option from **Pressure Openings** list. Set temperature as 293 K.
- After specifying the parameters, click on the **OK** button from **Boundary Conditions PropertyManager**. The **Environment Pressure 1** feature will be added in **Analysis Tree**.
- Rename the **Environment Pressure 1** feature to **Environment Pressure - Air**.

Specifying Solid Material

Notice that the auxiliary lids on the openings and closings are solid. Since the material for the lids is the default stainless steel, they will have an influence on the heat transfer. You cannot suppress or disable them in the **Component Control** dialog box, because boundary conditions must be specified on solid surfaces in contact with the fluid region. However, you can exclude the lids from the heat conduction analysis by specifying the lids as insulators. The procedure is discussed next.

- Right-click on the **Solid Materials** option and click on the **Insert Solid Material** button from the displayed menu. The **Solid Material PropertyManager** will be displayed.

- The selection box is active by default. You need to select all lids of current model; refer to Figure-49.

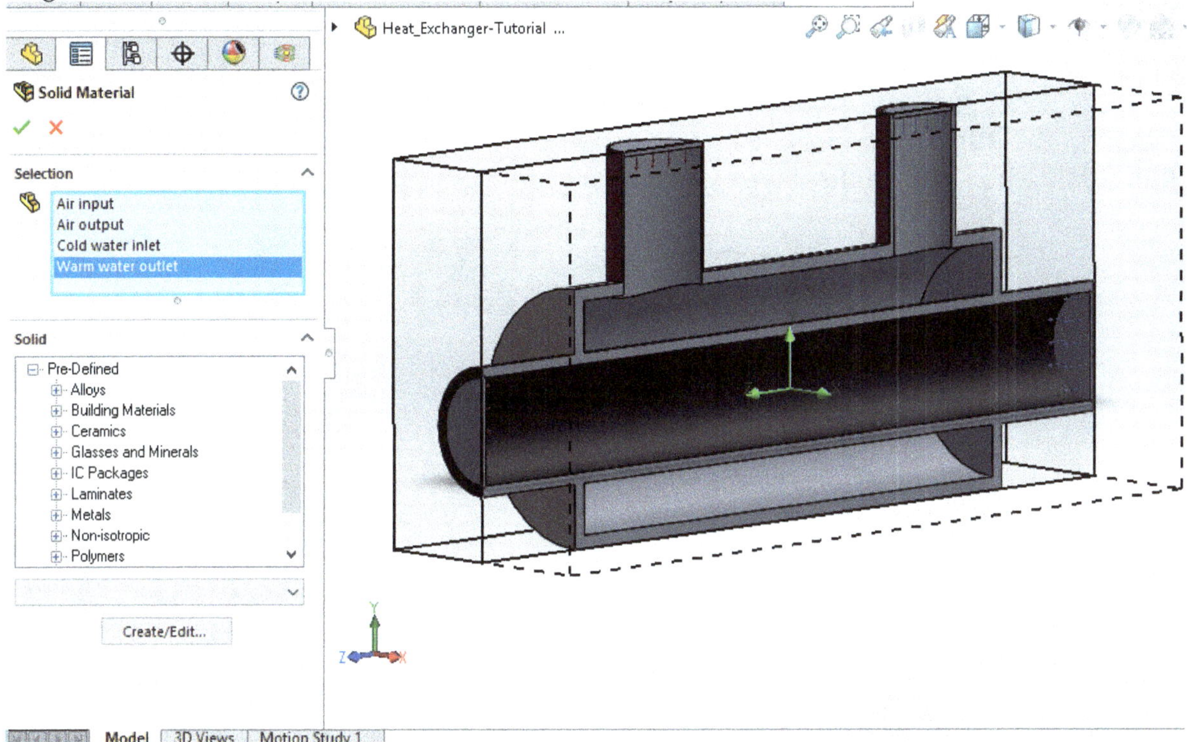

Figure-49. Selecting lids for Solid Material

- Click on the **Glasses and Minerals Node** from the **Solid** selection list and click on the **Insulator** material from the expanded list; refer to Figure-50.

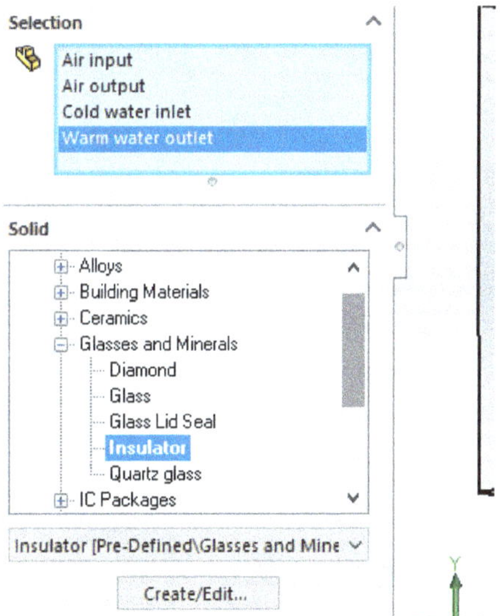

Figure-50. Selecting Insulator

- After selecting the required parameter, click on the **OK** button from **Solid Material PropertyManager**. The **Insulator Solid Material 1** feature will be added in **Analysis Tree**.

- The thermal conductivity of the insulator substance is zero. Hence, there is no heat transferred through an insulator.
- Rename the **Insulator Solid Material 1** feature to **Insulator**.

Running the Calculation

- Click on the **Run** button from **Flow Simulation CommandManager** of the **Ribbon**. The **Run** dialog box will be displayed; refer to Figure-51.

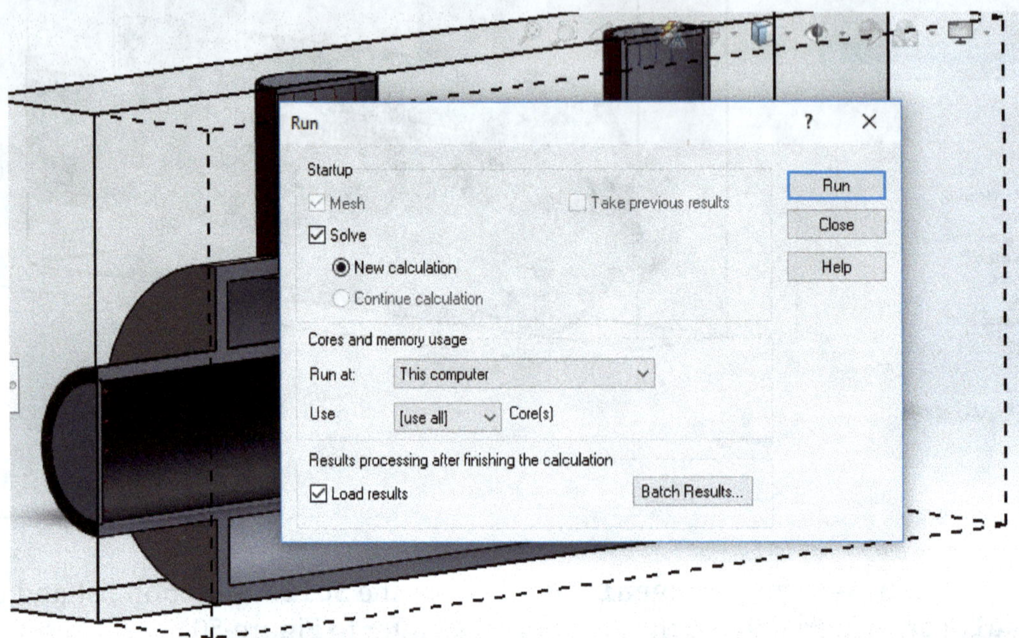

Figure-51. Run dialog box of Heat Exchanger

- Click on the **Run** button from **Run** dialog box. The **Solver: Heat Exchanger** dialog box will be displayed; refer to Figure-52.

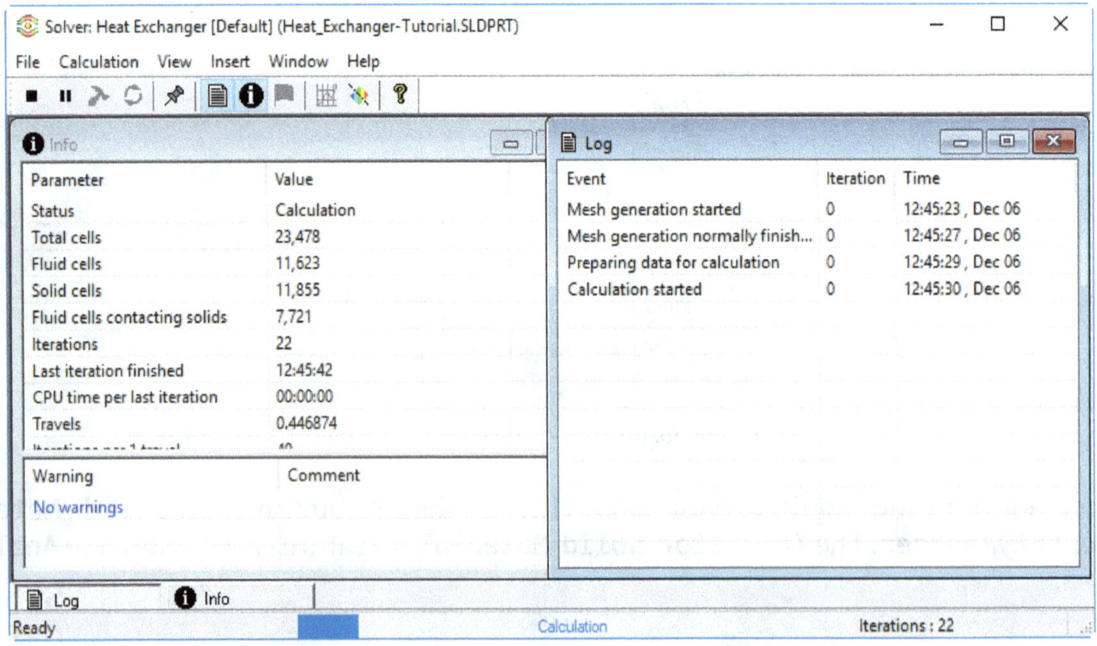

Figure-52. Solver Heat Exchanger

- After finishing of calculation, the solver show a notification "Solver is finished". Now, close the **Solver** dialog box. The results will be displayed in **Results** node; refer to Figure-53.

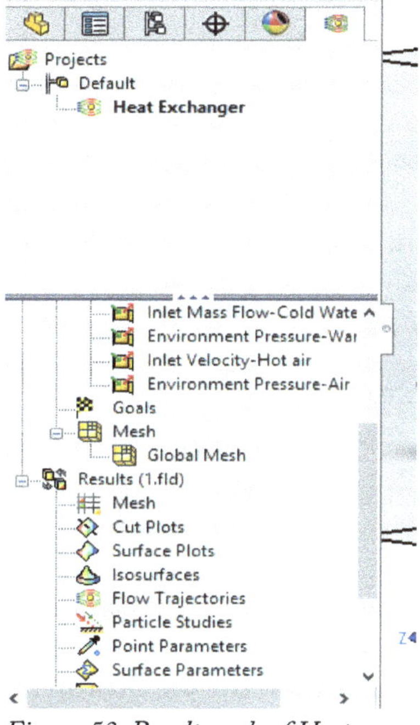

Figure-53. Results node of Heat Exchanger

Viewing Cut Plots

- Right-click on the **Cut Plots** option from **Results** node and click on the **Insert** button from the displayed menu. The **Cut Plot PropertyManager** will be displayed.
- The **Selection Plane** box is active by default. You need to click on the **Right Plane** from **Design Tree** to select.
- Click on the **Contours** and **Vectors** button from **Display** section for displaying the analysis.
- Click on the **Parameter** drop-down from **Contours** section and select the **Temperature** option from the displayed list.
- Click in the **Number of Levels** edit box from **Contours** section and enter the value as **255**.
- Click on the **Parameter** drop-down from **Vectors** section and select the **Velocity** from the displayed list.
- Click on the **Adjust Minimum and Maximum** button from **Parameter** section. The parameters related to Adjust Minimum and Maximum will be displayed.
- Click in the **Maximum** edit box from **Vectors** section and enter the value as **0.004** m/s; refer to Figure-54.

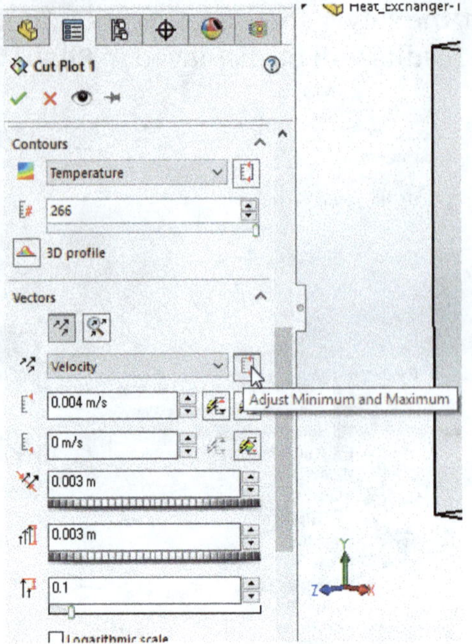

Figure-54. Specifying value of Maximum edit box

- After specifying the parameters, click on the **OK** button from **Cut Plot PropertyManager**. The **Cut Plot 1** feature will be added under **Cut Plots** option along with display on model; refer to Figure-55.

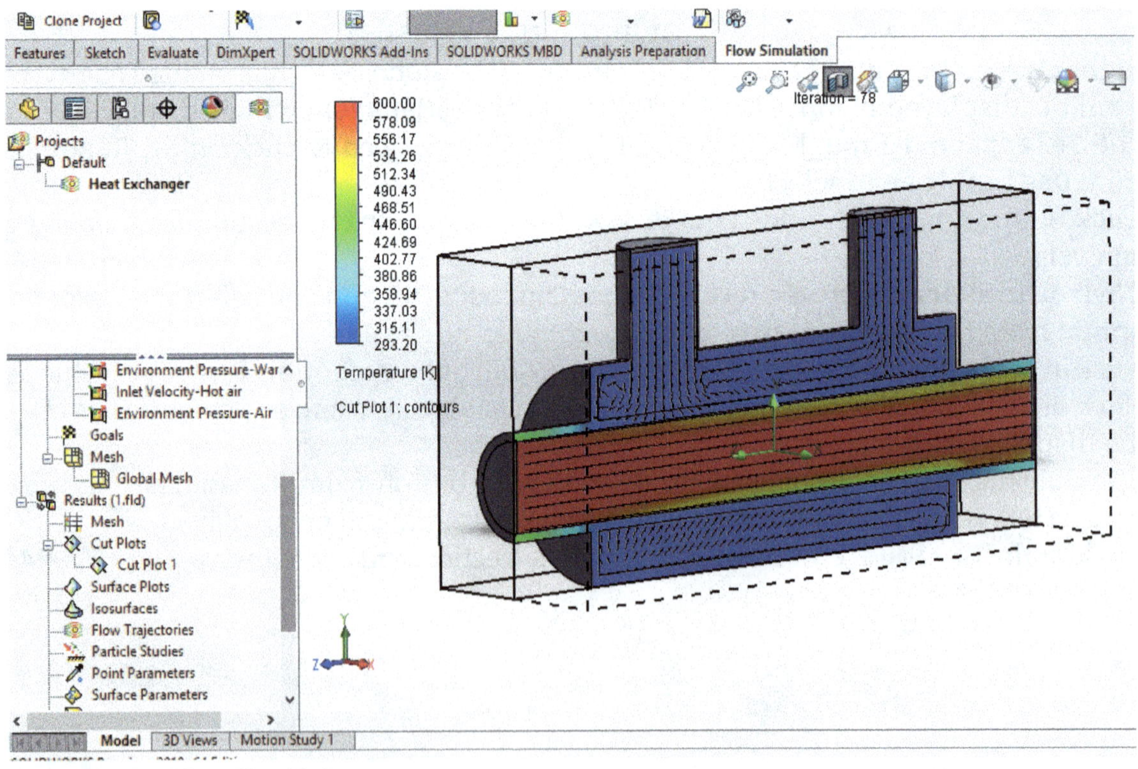

Figure-55. Cut Plot 1 Heat Exchanger

Displaying Flow Trajectories

- Right-click on the **Flow Trajectories** option from **Results Node** and click on the **Insert** button from the displayed menu. The **Flow Trajectories PropertyManager** will be displayed.
- The selection box is active by default. You need to click on the **Cold Water Inlet** lid from the **Design Tree** to select. The selected faces will be displayed in selection box.
- Click on the **Color By** drop-down from **Appearance** section and select the **Velocity** button from the displayed list of parameters.
- Click on the **Adjust Minimum/Maximum and Number of Levels** button from **Appearance** section. The parameters related this will be displayed; refer to Figure-56.

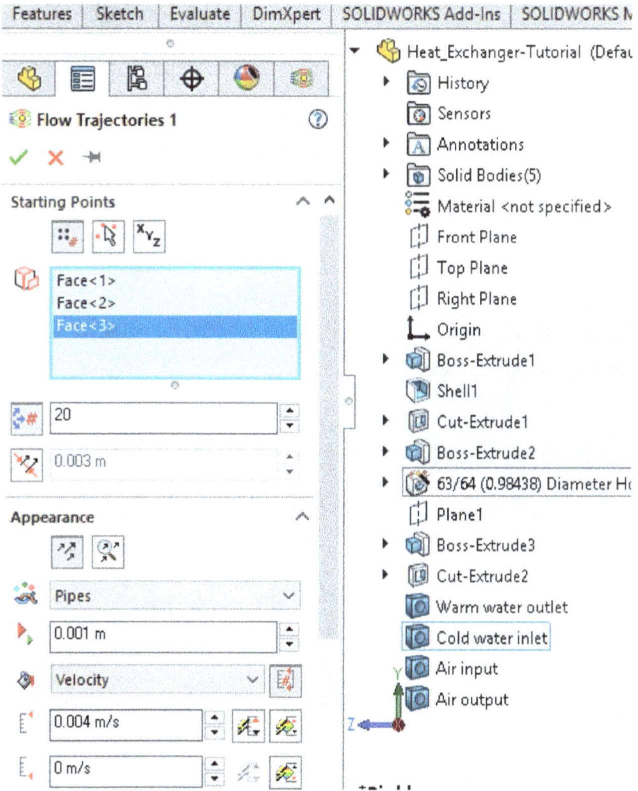

Figure-56. Selecting velocity parameter

- Click in the **Maximum** edit box from **Appearance** section and enter the value as **0.004 m/s**.
- After specifying the parameters, click on the **OK** button from **Flow Trajectories PropertyManager**. The **Flow Trajectories 1** feature will be added under **Flow Trajectories** node along with the displayed flow trajectories on the display screen; refer to Figure-57.

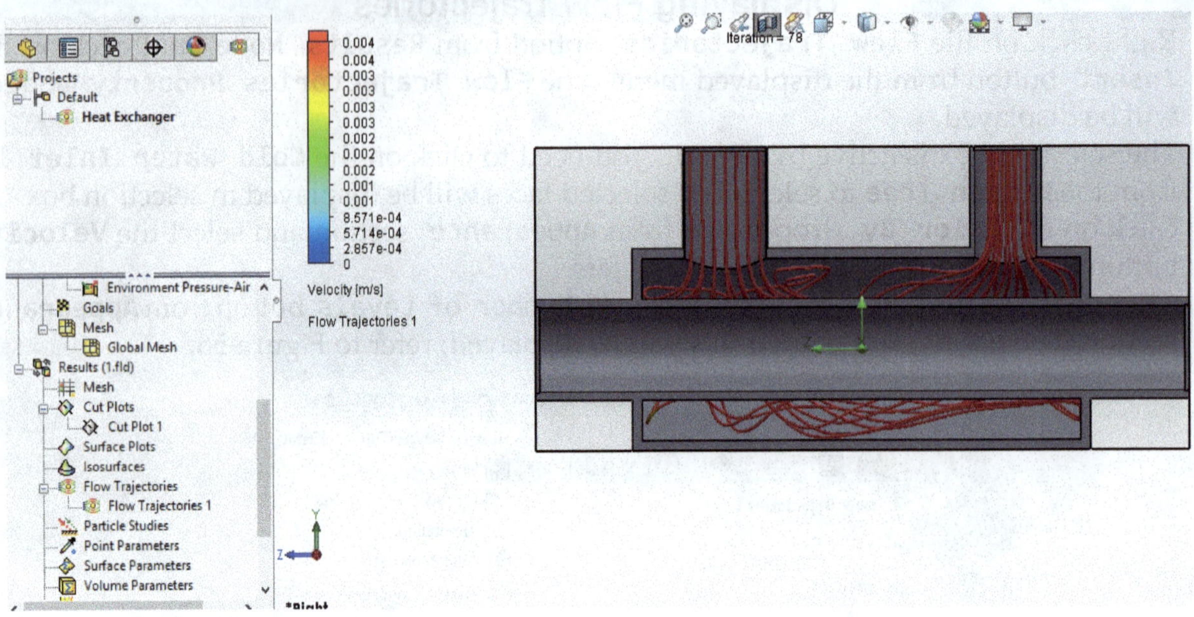

Figure-57. Displayed flow trajectories

- If you want to see the animation of current **Flow Trajectories** then right-click on the **Flow Trajectories 1** and click on the **Play** button. The animation will be started and displayed on the model; refer to Figure-58.

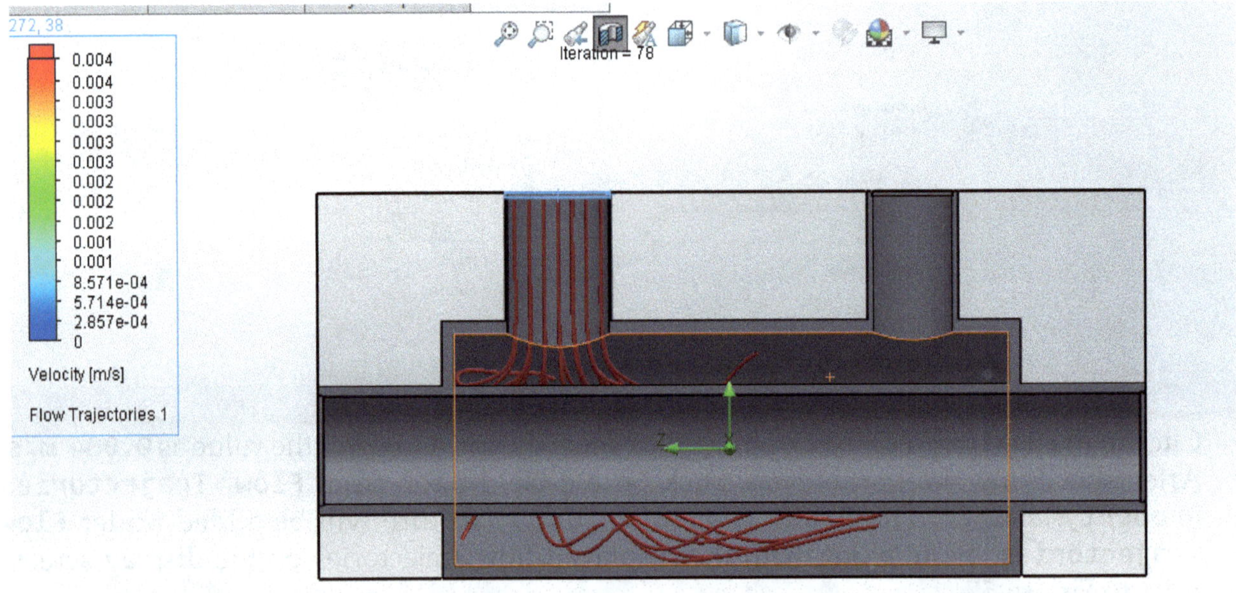

Figure-58. Animation of flow trajectories

Viewing the Surface Parameter

- Right-click on the **Surface Parameter** option from **Results Node** and click on the **Insert** button from the displayed menu. The **Surface Parameters PropertyManager** will be displayed.
- The **Selection** box of **Surface Parameter** is activated by default. You need to click on the lid for **Warm Water Outlet** from the **Design Tree**.

- Select the **Consider entire model** check box from **Selection** section for considering the entire model.
- Select the **All** check box from **Parameters** section to select all the parameters listed under **Parameters** section.
- Click on the **Show** button from **Options** section. The values of calculated parameter will be displayed at the bottom of screen.
- The local parameters were displayed at the left and integral parameters are displayed at the right; refer to Figure-59.

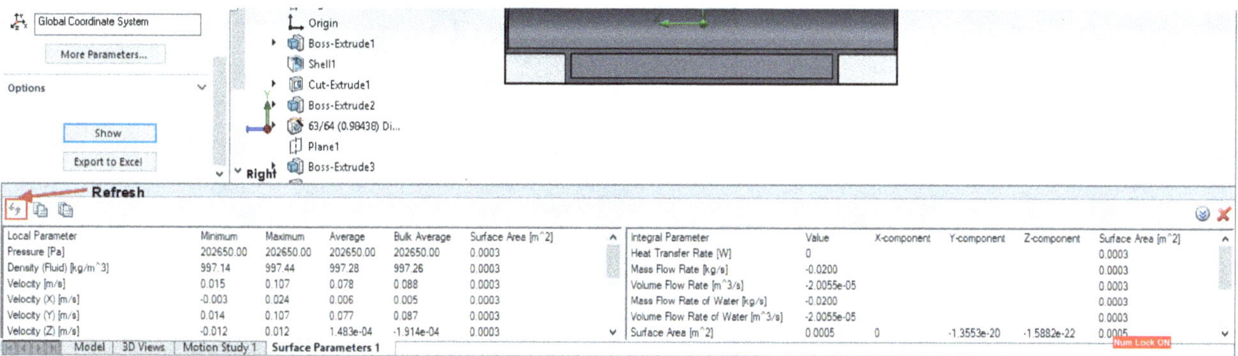

Local Parameter	Minimum	Maximum	Average	Bulk Average	Surface Area [m^2]
Pressure [Pa]	202650.00	202650.00	202650.00	202650.00	0.0003
Density (Fluid) [kg/m^3]	997.14	997.44	997.28	997.26	0.0003
Velocity [m/s]	0.015	0.107	0.078	0.088	0.0003
Velocity (X) [m/s]	-0.003	0.024	0.006	0.005	0.0003
Velocity (Y) [m/s]	0.014	0.107	0.077	0.087	0.0003
Velocity (Z) [m/s]	-0.012	0.012	1.483e-04	-1.914e-04	0.0003

Integral Parameter	Value	X-component	Y-component	Z-component	Surface Area [m^2]
Heat Transfer Rate [W]	0				0.0003
Mass Flow Rate [kg/s]	-0.0200				0.0003
Volume Flow Rate [m^3/s]	-2.0055e-05				0.0003
Mass Flow Rate of Water [kg/s]	-0.0200				0.0003
Volume Flow Rate of Water [m^3/s]	-2.0055e-05				0.0003
Surface Area [m^2]	0.0005	0	-1.3553e-20	-1.5882e-22	0.0005

Figure-59. Local and Integral parameters

- Now, click on the **Air Output** lid from **Design Tree** and click on the **Refresh** button from **Surface Parameters 1** tab.
- The local parameters will be updated. Scroll down to the **Temperature (Fluid)[K].** The average parameter of temperature is about **464**; refer to Figure-60.

Local Parameter	Minimum	Maximum	Average	Bulk Average	Surface Area [m^2]
Volume Fraction of Water [ ]	1.0000	1.0000	1.0000	1.0000	0.0003
Temperature (Fluid) [K]	293.69	599.67	464.55	326.62	0.0007
Relative Pressure [Pa]	0	1.64e-08	5.85e-09	1.44e-08	0.0007
Surface Heat Flux (Conductive) [W/m^2]	0	0	0		0.0008
Acoustic Power Level [dB]	0	17.89	1.05	0.16	0.0007
Acoustic Power [W/m^3]	1.194e-34	6.150e-11	1.077e-12	1.560e-13	0.0007

Figure-60. Local Parameter

- The values of integral parameters are displayed at the right side of the bottom pane. You can see that the mass flow rate of air is 0.022 kg/s. This value is calculated with the **Consider entire model** check box selected, i.e. taking into account the Symmetry condition.
- After checking the parameters, click on the **OK** button from **Surface Parameters PropertyManager**; refer to Figure-61. The **Surface Parameters 1** feature will be added under **Surface Parameter** option.

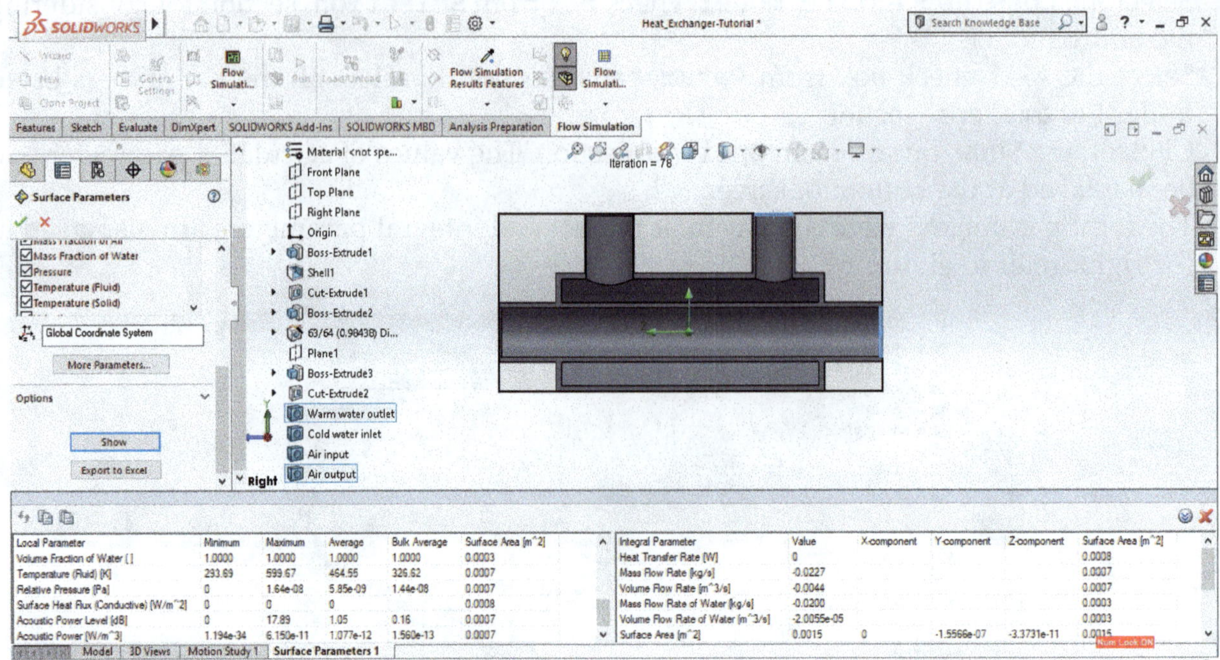

Figure-61. Checking parameters of Surface Parameters

Calculating the Heat Exchanger Efficiency

The heat exchanger efficiency can be easily calculated, but first we must determine the fluid with the minimum capacity rate (C=mc). In this example the water mass flow rate is 0.02 kg/s and the air mass flow rate is 0.022 kg/s. The specific heat of water at the temperature of 300 K is about five times greater than that of air at the temperature of 464 K. Thus, the air capacity rate is less than the water capacity rate. The heat exchanger efficiency is calculated as follows:

$$\varepsilon = \frac{T_{hot}^{inlet} - T_{hot}^{outlet}}{T_{hot}^{inlet} - T_{cold}^{inlet}},$$

Where T_{hot}^{inlet} is the temperature of the air at the inlet, T_{hot}^{outlet} is the temperature of the air at the outlet and T_{cold}^{inlet} 'is the temperature of the water at the inlet.

We already know the air temperature at the inlet (600 K) and the water temperature at the inlet (293.2 K), so using the obtained values of water and air temperatures at outlets, we can calculate the heat exchanger efficiency:

$$\varepsilon = \frac{T_{hot}^{inlet} - T_{hot}^{outlet}}{T_{hot}^{inlet} - T_{cold}^{inlet}}, = \frac{600 - 464}{600 - 293.2} = 0.44$$

As you can see, Flow Simulation is a powerful tool for heat-exchanger design calculations.

PRACTICE 1- FINDING HYDRAULIC LOSS

In engineering practice, the hydraulic loss of pressure head in any piping system is traditionally split into two components: the loss due to friction along straight pipe sections and the local loss due to local pipe features, such as bends, T-pipes, various cocks, valves, throttles, etc. Being determined, these losses are summed to form the total hydraulic loss. Generally, there are no problems in engineering practice to determine the friction loss in a piping system since relatively simple formulae based on theoretical and experimental investigations exist. The other matter is the local hydraulic loss (or so-called local drag). Here usually only experimental data are available, which are always restricted due to their nature, especially taking into account the wide variety of pipe shapes (not only existing, but also advanced) and devices, as well as the substantially complicated flow patterns in them. Flow Simulation presents an alternative approach to the traditional problems associated with determining this kind of local drag, allowing you to predict computationally almost any local drag in a piping system within good accuracy.

Model Description

This is a ball valve. Turning the handle closes or opens the valve; refer to Figure-62.

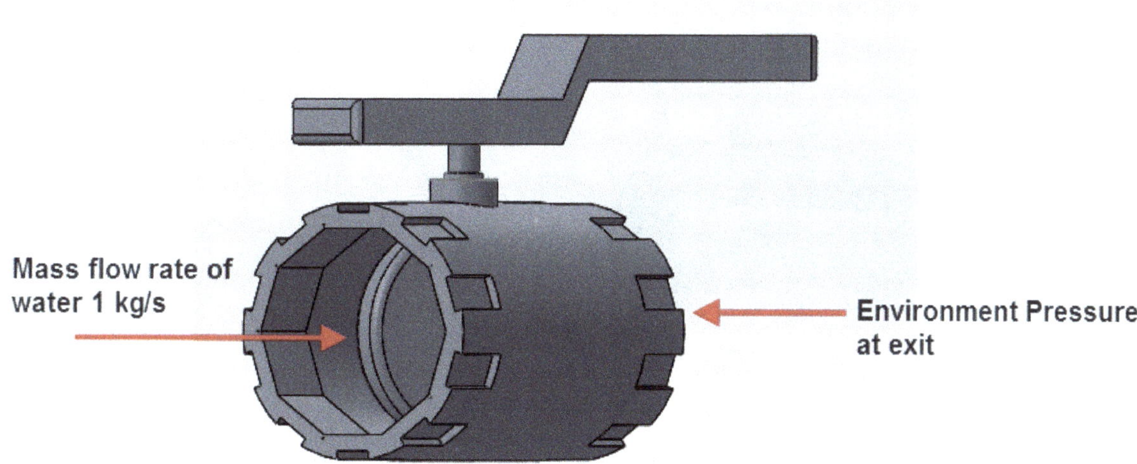

Figure-62. Ball Valve

The local hydraulic loss (or drag) produced by a ball valve installed in a piping system depends on the valve turning angle or on the minimum flow passage area governed by it. The latter depends also on a ball valve geometrical parameter, which is the ball-to-pipe diameter ratio governing the handle angle at which the valve becomes closed:

$$\theta = \arcsin\left[2\frac{D_{ball}}{D_{pipe}}\right]$$

The standard engineering convention for determining local drag is by calculating the difference between the fluid dynamic heads measured upstream of the local pipe feature (ball valve in our case) and far downstream of it, where the flow has become uniform (undisturbed) again. In order to extract the pure local drag the hydraulic friction loss in the straight pipe of the same length must be subtracted from the measured dynamic head loss.

In this practice, we will obtain pressure loss (local drag) in the ball valve whose handle is turned by an angle of 40 degree. The Valve analysis represents a typical Flow Simulation internal analysis.

To perform an internal analysis all the model openings must be closed with lids, which are needed to specify inlet and outlet flow boundary conditions on them. In any case, the internal model space filled with a fluid must be fully closed. You simply create lids as additional extrusions covering the openings. This displayed below; refer to Figure-63.

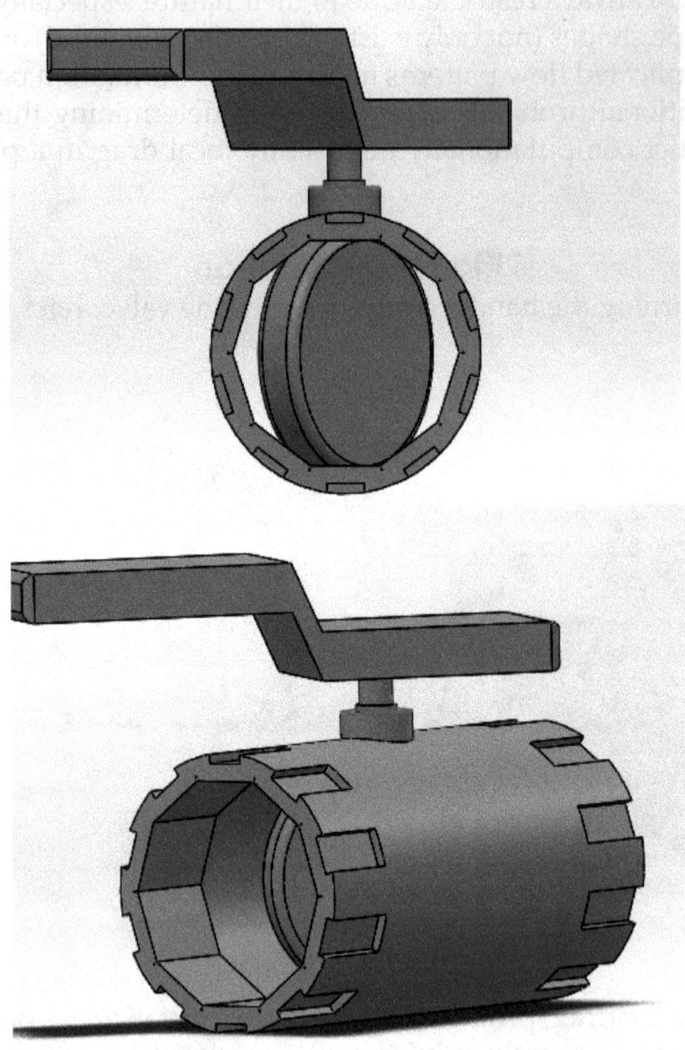

Figure-63. Ball Valve 1

PRACTICE 2 - CONJUGATE HEAT TRANSFER

This practice covers the basic steps required to set up a flow analysis problem including heat conduction in solids. This example is particularly pertinent to users interested in analyzing flow and heat conduction within electronics devices, although the basic principles are applicable to all thermal problems. It is assumed that you have already completed the Ball Valve Design tutorial since it teaches the basic principles of using Flow Simulation in greater detail.

In a typical assembly, there may be many features, parts or sub-assemblies that are not necessary for the analysis. Prior to creating a Flow Simulation project, it is a good practice to check the model to find components that can be removed from the analysis. Excluding these components reduces the computer resources and calculation time required for the analysis.

The assembly consists of the following components: enclosure, motherboard and two smaller PCBs, capacitors, power supply, heat sink, chips, fan, screws, fan housing, and lids. You can highlight these components by clicking them in the **FeatureManager Design Tree**. In this practice, the fan has a very complex geometry that may cause delays while rebuilding the model. Since, it is outside the enclosure, we can exclude it by suppressing it. The model is shown next; refer to Figure-64. You need to check the heat dissipation from electronic components by fan and conduction in solids. Note that CPU will work till maximum temperature of 50 °C.

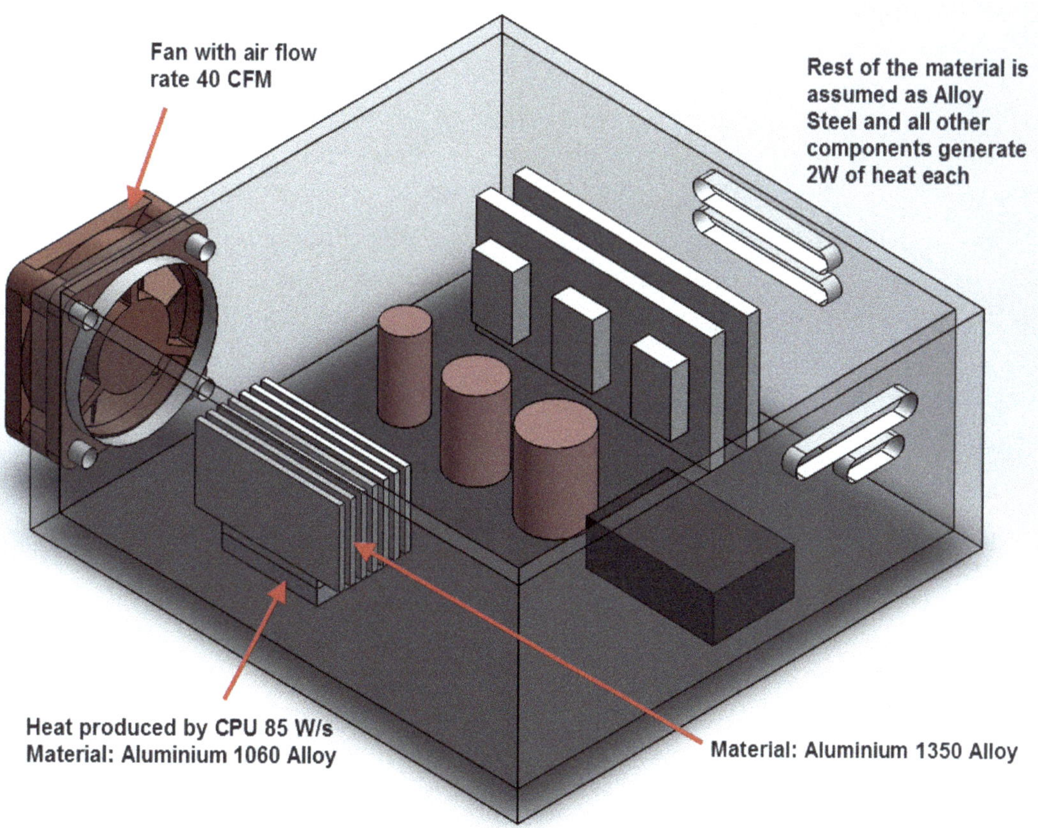

Figure-64. Practice 2 model

FOR STUDENT NOTES

Chapter 5

Advanced Boundary Conditions

The major topics covered in this chapter are:

- ***Porous Medium***
- ***Initial Condition***
- ***Defining Heat Sources***
- ***Creating Fan***
- ***Tracer Study***
- ***Transferred Boundary Conditions***
- ***Radiative Surface and Radiation Source***
- ***Heat Sink, Perforated Plate, Thermoelectric Cooler, Two Resistor, Heat Pipe***
- ***Thermal Joint and PCB***

Note that some of the tools discussed in this chapter are not available in standard SolidWorks Flow Simulation Application. These tools are part of Electronic Cooling module.

INTRODUCTION

In previous chapter, we have learnt about basic boundary conditions and we have performed analyses based on those conditions. In this chapter, we will learn about some advanced boundary conditions. The tools to create these boundaries conditions are discussed next.

POROUS MEDIUM

The porous medium is used when fluid can leak through the walls of tube or container. This type of medium is useful when you are performing analysis on assembly which has filters installed at inlets or outlets. In such cases, you will define filter as porous medium. The procedure to apply porous medium is given next.

- Click on the **Porous Medium** tool from the **Flow Simulation Features** drop-down in **Flow Simulation CommandManager** after starting the analysis. You can also click on the tool from menu bar as discussed earlier. The **Porous Medium PropertyManager** will be displayed; refer to Figure-1.

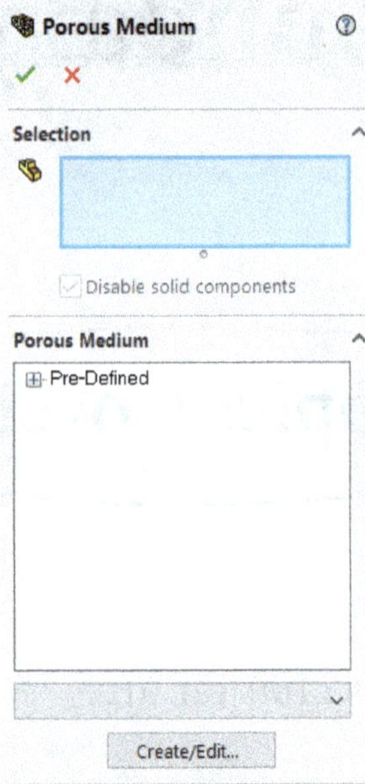

Figure-1. Porous Medium PropertyManager

- Select the part to which you want to apply porous medium properties. The options related to porous medium will be displayed in the list box; refer to Figure-2.

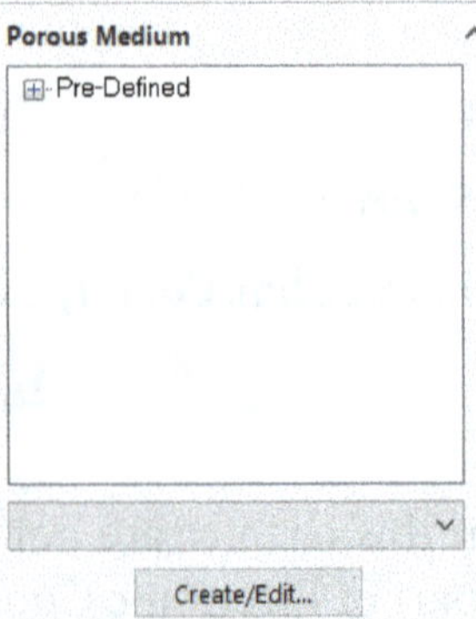

Figure-2. Porous Medium List box

- Select the **Isotropic** option if the material has same properties in all the directions. Select the **Screen Material** option if your porous object model is in the form of screen or derived to the form of screen. Select the **Unidirectional** option if your object is porous in one direction only. Expand the **UAF** category and selected the desired filter material from list of practically available materials.
- Click on the **OK** button from the **PropertyManager** to apply the porous medium; refer to Figure-3.

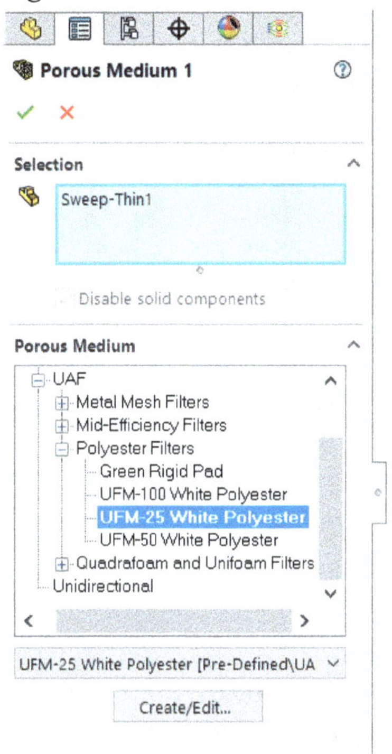

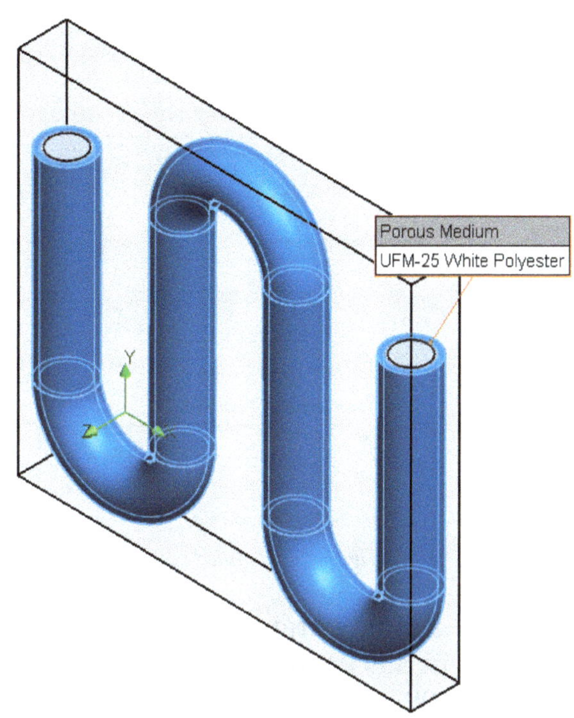

Figure-3. Pipe cover defined as porous material

INITIAL CONDITION

The **Initial Condition** tool in the **Flow Simulation Features** drop-down of **CommandManager** is used to specify the condition of fluid at the point of starting the analysis. For example, if the fluid was flowing at a speed of **1 m/s** in the tube before applying further mass flow then you can specify such conditions here. The procedure to use this tool is given next.

- Click on the **Initial Condition** tool from the **Tools > Flow Simulation >** menu of the Menu Bar. You can also activate the tool from **Ribbon**. The **Initial Condition PropertyManager** will be displayed; refer to Figure-4.

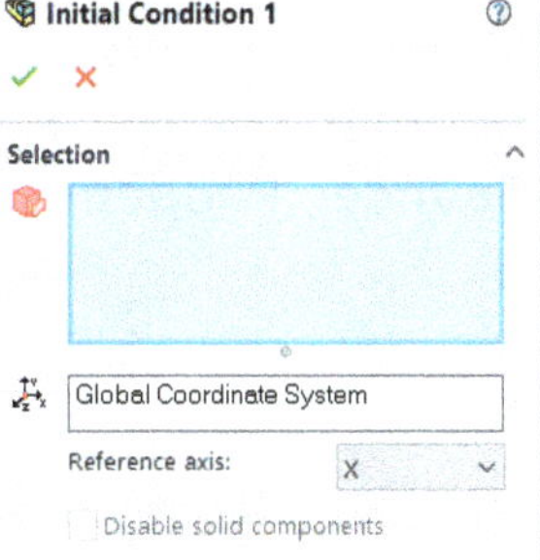

Figure-4. Initial Condition PropertyManager

- Select the face which is common to fluid and solid (face with porosity). The options in **PropertyManager** will be modified; refer to Figure-5.

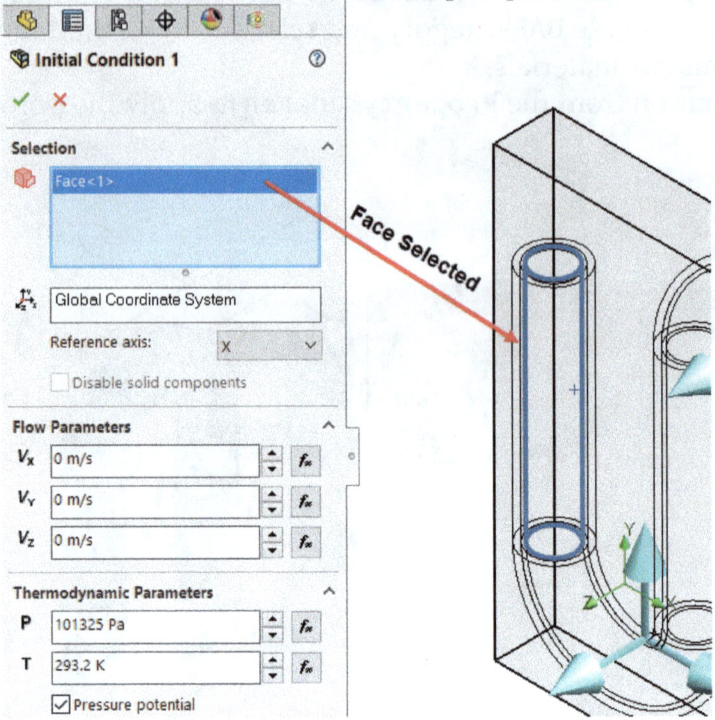

Figure-5. Modified options in Initial Condition PropertyManager

- Specify the desired parameters for initial condition of fluid at selected wall. Note that all the options in this **PropertyManager** have been discussed in previous chapters.
- Click on the **OK** button from the **PropertyManager** to apply the conditions.

DEFINING HEAT SOURCES

In SolidWorks Flow Simulation, you can define heat sources in analysis model by two ways:

1. Defining Surface Heat Source
2. Defining Volume Heat Source

The procedures for both the ways are discussed next.

Defining Surface Heat Source

Surface heat source is used when heat is generated from selected surface(s) only. The procedure to apply surface heat source is given next.

- Click on the **Surface Source** tool from the **Tools > Flow Simulation > Insert** menu; refer to Figure-6. The **Surface Source PropertyManager** will be displayed; refer to Figure-7.

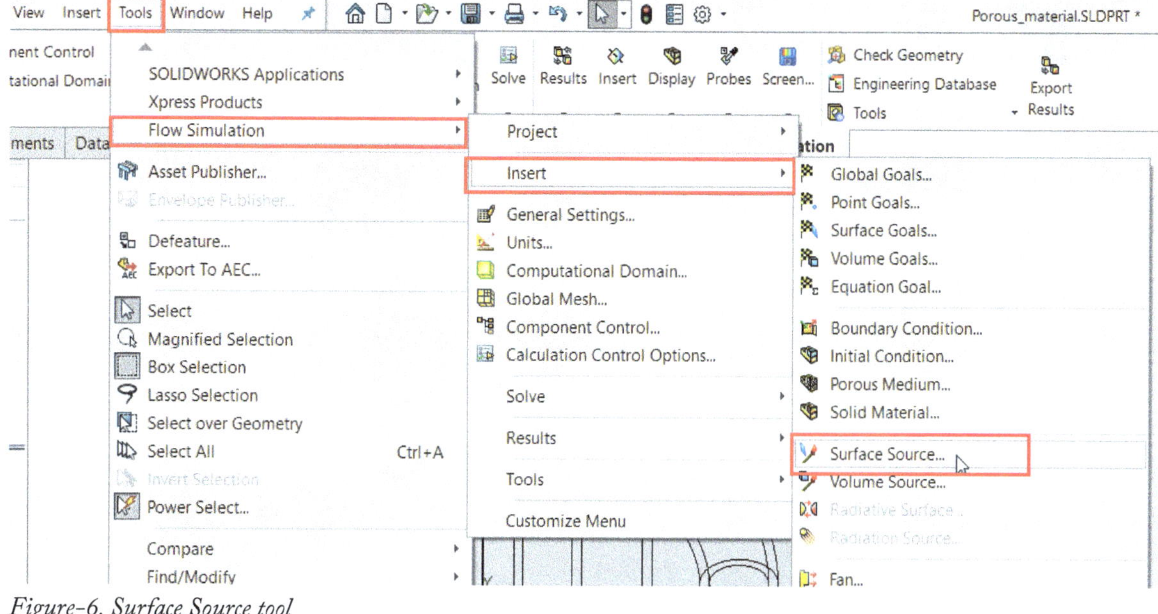

Figure-6. Surface Source tool

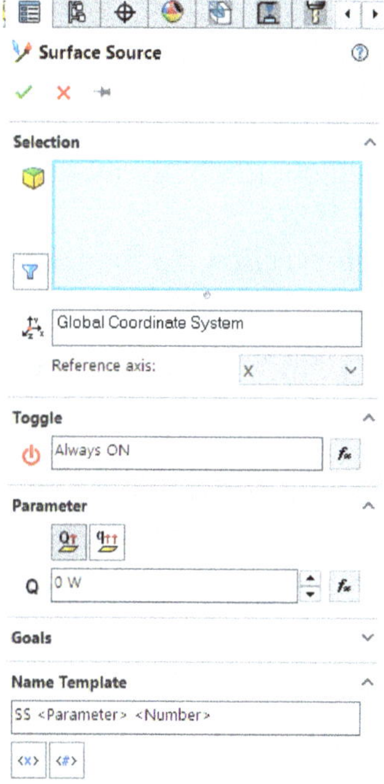

Figure-7. Surface Source PropertyManager

- Select the face(s)/surface(s) which you want to be used as heat generation source.
- There are two ways to define heat generation rate. You can set either total heat generation by selected surfaces or you can set the heat generation rate per area of surface. Select the **Heat Generation Rate** button from the **Parameter** section of **PropertyManager** if you want to specify the total heat generation by all selected surfaces. Select the **Surface Heat Generation Rate** button from the **Parameter** section of **PropertyManager** if you want to specify the total heat generation per square meter of surface.
- Specify the desired heat generation rate value in the edit box of the **Parameters** section; refer to Figure-8.

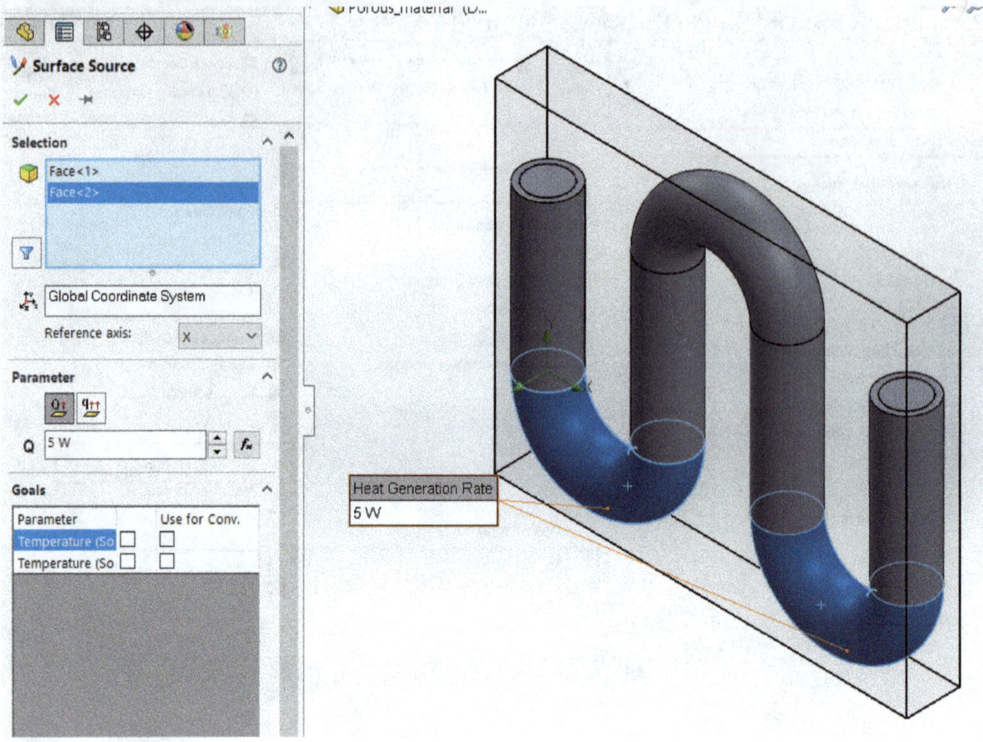

Figure-8. Heat generation

- Expand the **Goals** rollout and select check boxes for temperature parameters to be used as goal to convergence.
- After setting the desired parameters, click on the **OK** button.

Defining Volume Heat Source

Volume heat source is used when heat is generated in selected volume only. The procedure to apply volume heat source is given next.

- Click on the **Volume Source** tool from the **Tools > Flow Simulation > Insert** menu. The **Volume Source PropertyManager** will be displayed; refer to Figure-9.

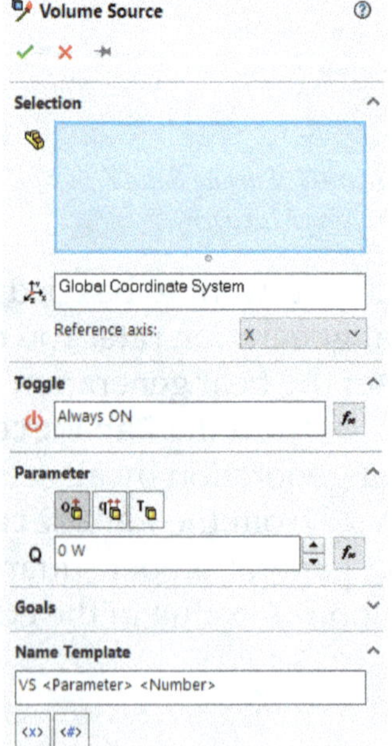

Figure-9. Volume Source PropertyManager

- Select the part/component to be used as heat source.
- Select the desired button from the **Parameter** section and specify the respective value of heat generation/temperature.
- Select the desired check boxes from the **Goals** rollout to create respective goals and use them for convergence.
- Click on the **OK** button from the **PropertyManager** to apply heat source.

You can use the Radiative Surface and Radiation Source tools of the **Tools > Flow Simulation > Insert >** menu in the same way. Note that these tools are active only when Radiation is applied to the study.

APPLYING RADIATIVE SURFACE PROPERTIES

The **Radiative Surface** tool is used to set up radiation related properties of the selected surfaces. The procedure to use this tool is given next.

- Click on the **Radiative Surface** tool from the **Tools > Flow Simulation > Insert** menu. The **Radiative Surface PropertyManager** will be displayed; refer to Figure-10.

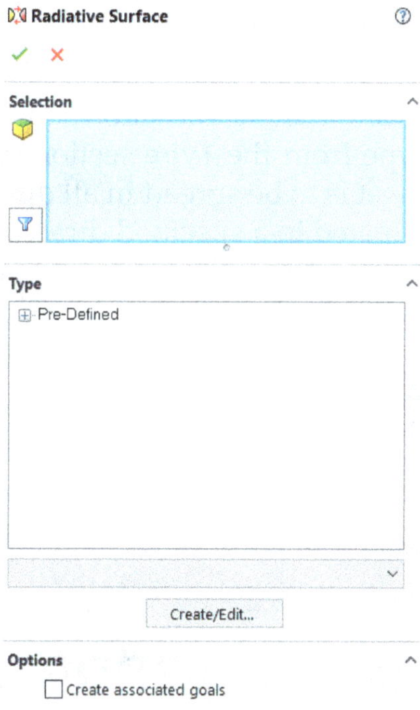

Figure-10. Radiative Surface PropertyManager

- Select the face(s) to which you want to apply radiative properties.
- Expand the **Pre-Defined** category from the **Type** selection box and select the desired radiative property.
- Click on the **OK** button from the **PropertyManager** after setting the properties.

APPLYING RADIATION SOURCE

The **Radiation Source** tool is used to create source of radiation for analysis. The procedure to create radiation source is given next.

- Click on the **Radiation Source** tool from the **Tools > Flow Simulation > Insert** menu. The **Radiation Source PropertyManager** will be displayed; refer to Figure-11.

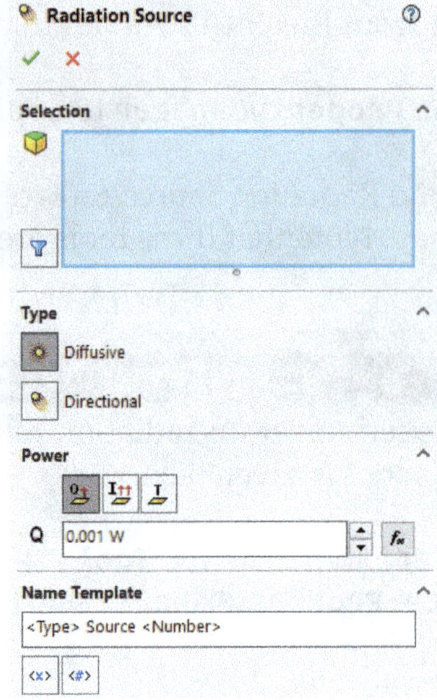

Figure-11. Radiation Source PropertyManager

- Select the face(s) which you want to make radiation source.
- Select the radiation source type from the **Type** section of the **PropertyManager**. Select the **Diffusive** button if the heat is to be spread in all directions. Select the **Directional** button if the heat is to be dissipated in a specified direction.
- On selecting the **Diffusive** button, the **Direction** section is displayed in the **PropertyManager**. Select the **Define As Normal to Plane** button from the **Direction** section if you want the radiation direction to be perpendicular to selected face/plane. Select the **Define As 3D Vector** button if you want to specify the components of heat radiation in each direction. Enter the desired values in respective edit boxes.
- Select the desired button from the **Power** section and specify the heat radiation power value in respective edit box.
- After specifying desired parameters, click on the **OK** button from the **PropertyManager** to create the radiation source.

APPLYING FAN PROPERTIES

The **Fan** tool in SolidWorks Flow Simulation is used to simulate the effect of fan in the analysis. The procedure to create fan is given next.

- Click on the **Fan** tool from the **Tools > Flow Simulation > Insert >** menu. The **Fan PropertyManager** will be displayed; refer to Figure-12.
- Select the **External Inlet Fan** option from the **Type** area if you want to insert an external fan at the inlet of fluid flow. Select the **External Outlet Fan** option if you want to insert an external fan at the outlet of the fluid flow. Select the **Internal Fan** option if the fan is placed inside the fluid flow tube. In our case, we are using the Internal Fan here.

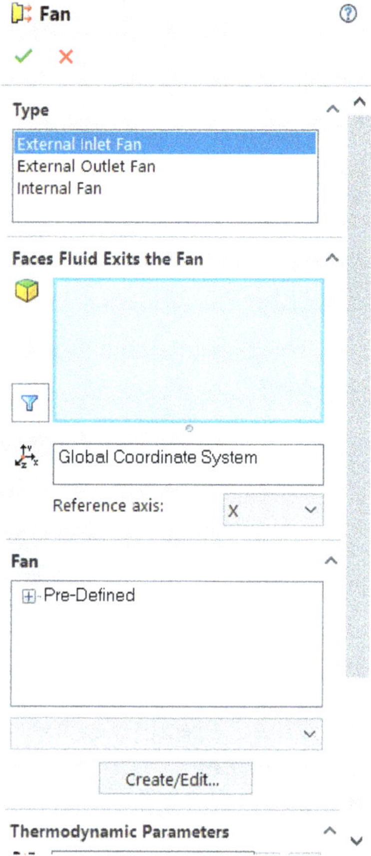

Figure-12. Fan PropertyManager

- Click in the **Faces Fluid Exits the Fan** selection box and select the face from where the fluid will come out after passing through the fan; refer to Figure-13.

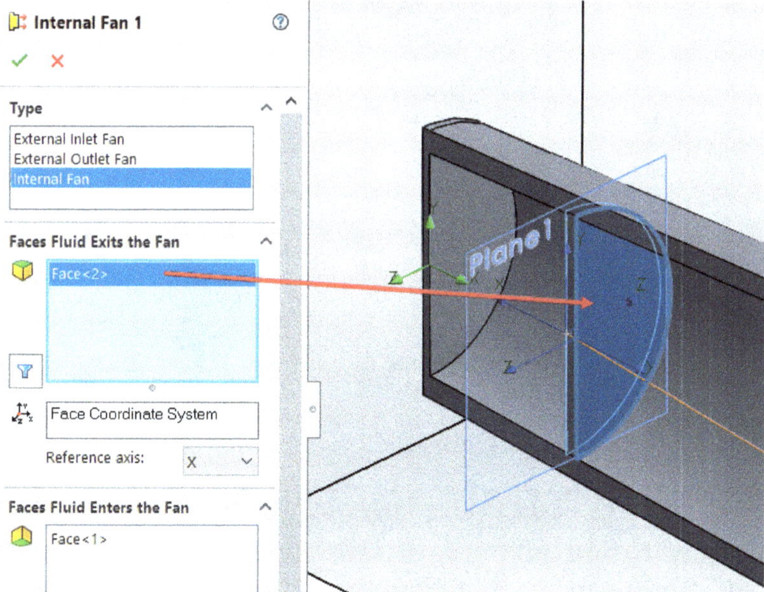

Figure-13. Face selected as exit for fan

- Click in the **Faces Fluid Enters the Fan** selection box and select the face from where fluid enters the fan; refer to Figure-14.

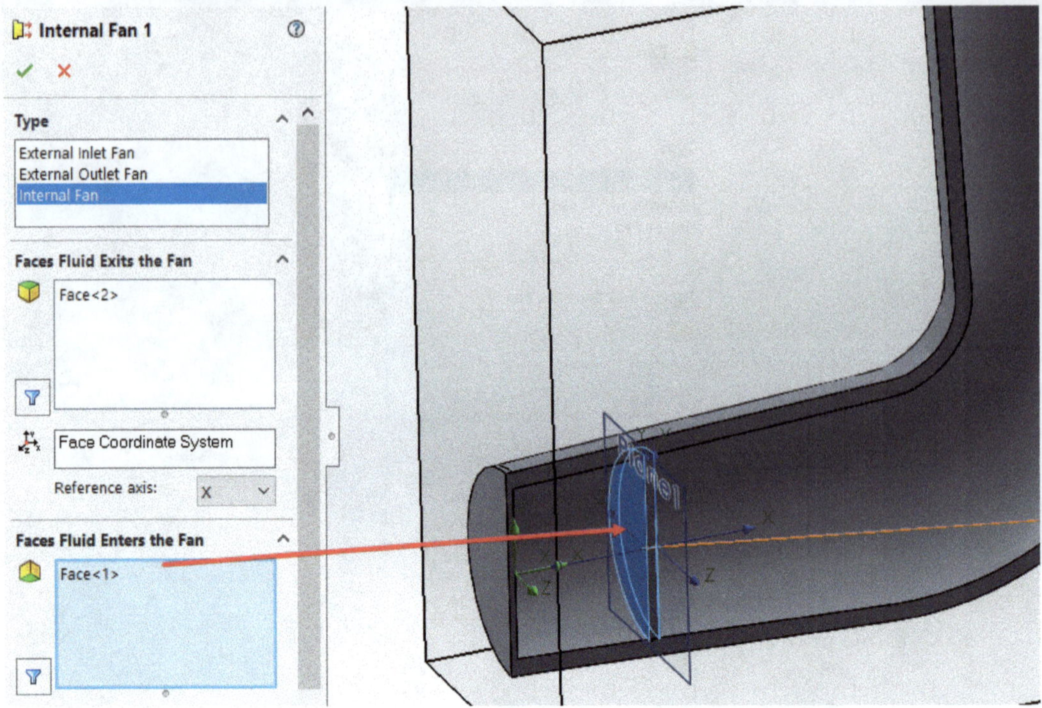

Figure-14. Face selected as entry for fan

- Set the directions of fan flow from the **Reference axis** drop-down in the **Faces Fluid Exits the Fan** and **Faces Fluid Enters the Fan** sections.
- Scroll-down in the **PropertyManager** and select the desired fan from the Fan list; refer to Figure-15.
- Set the other parameters as desired and click on the **OK** button from the **PropertyManager**.

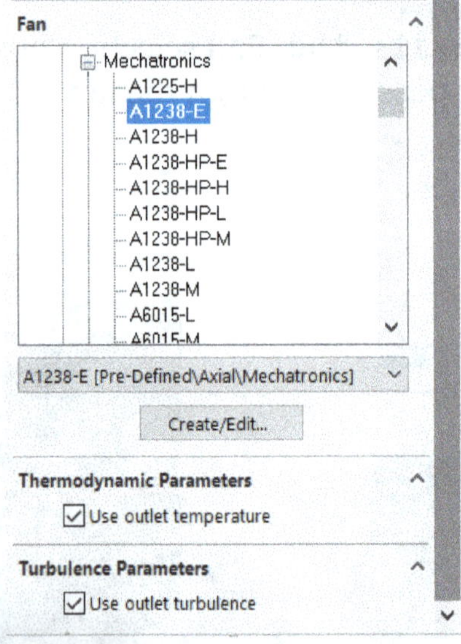

Figure-15. Fan selected from the list

TRACER STUDY

Tracer study is performed to find out traces of different gases in the flow of fluids. This feature is available for HVAC module users only. Sometimes, traces of dangerous gases are added in the fluid due to application requirement but their location during the flow of fluid is important to be taken care of. For example, when fire causes accumulation of Carbon Mono-oxide in a room then you need to study how CO will exit through the HVAC lines using the tracer study. Note that this is a post solve study. The procedure to create Tracer study is given next.

- Click on the **Tracer Study** tool from the **Tools > Flow Simulation > Insert >** menu. The **Tracer Study Settings PropertyManager** will be displayed; refer to Figure-16.

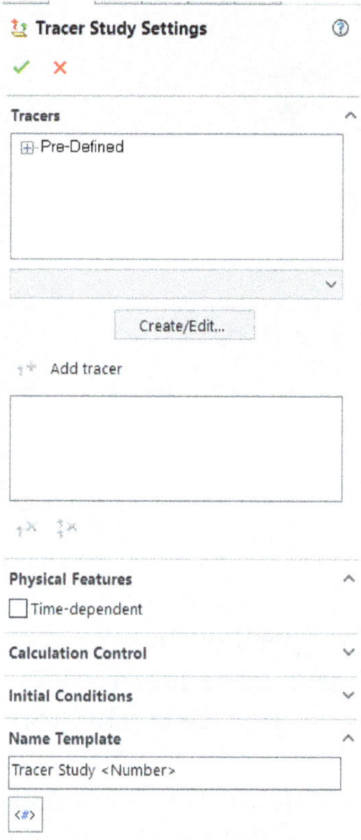

Figure-16. Tracer Study Settings PropertyManager

- Select the desired tracer from the **Tracers** selection box. Note that option selected here will define which gas(es) is to be traced.
- After selecting the desired tracer, click on the **Add tracer** button. The selected tracer will be added in the list. You can add more tracers by using the same procedure.
- Select the **Time-dependent** check box from the **Physical Features** section if you want to perform study for a specific time limit. After selecting this check box, specify total calculation time and time step values in their respective edit boxes of this section. The study will trace movement of CO during the flow of fluid.
- Expand the **Saving Result** section if you want to save the results at a specific time value or after specified time intervals. Note that this section is available only if you have selected the **Time-dependent** check box. The options of this section will be displayed as shown in Figure-17.

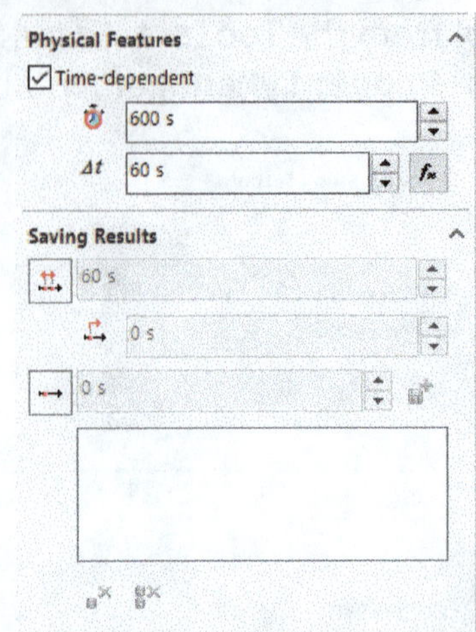

Figure-17. Saving Results section

- Click on the **Periodic Saving** button and specify the related parameters to perform periodic saving of results.
- Click on the **Save Moment** button to activate options for saving results at specified moment. Specify the time in seconds in the edit box and click on the **Add Save Moment** button to save results at the time specified in seconds.
- If you have not selected the **Time-dependent** check box in the **Physical Features** section of this **PropertyManager** then **Calculation Control** section is displayed in place of **Saving Results** section. Select the **Automatic** radio button from this section if the analysis is to converge automatically based on default goals. Select the **Maximum iterations** radio button to specify the number of iterations after which the system will stop solving the problem and display the results.
- Expand the **Initial Conditions** section to specify the initial condition of fluid mixture. The options will be displayed as shown in Figure-18.

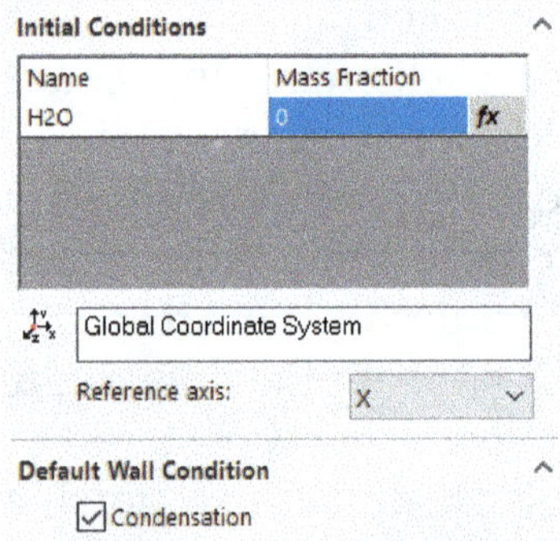

Figure-18. Initial Conditions section of PropertyManager

- Specify the mass fraction of the gas to be checked for traces at the beginning of analysis.
- Set the other parameters as desired and click on the **OK** button. A new tracer study will be added in the **Analysis Input Data Tree**; refer to Figure-19.

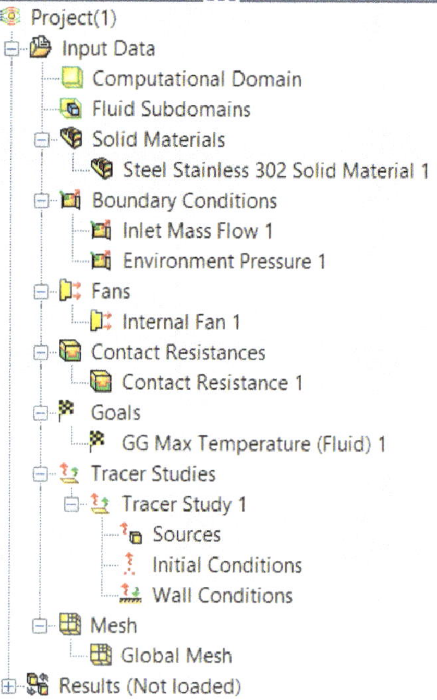

Figure-19. New tracer study added

Setting Source of Gas Traces

There are two ways to define the source of gas in fluid for trace study, surface source and volume source. Use the surface source option if gas is produced from a surface and select the volume source option if the gas is produced from a part in the model.

- Right-click on **Source** option from Tracer Study 1 created recently. The shortcut menu will be displayed as shown in Figure-20.

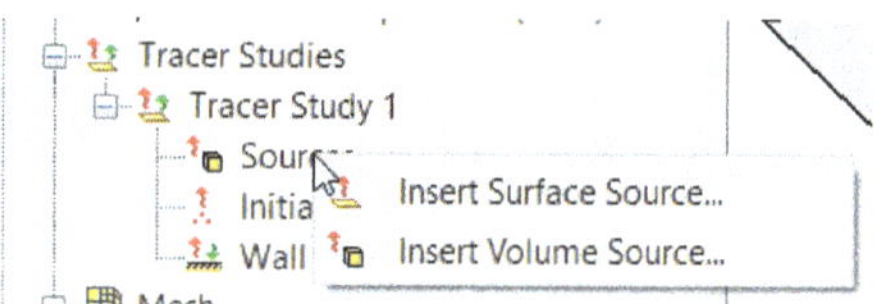

Figure-20. Shortcut menu for source options

- Select the desired option to define source of gases. In our case, we are using the **Insert Surface Source** option. On selecting this option, the **Surface Tracer Source PropertyManager** will be displayed; refer to Figure-21.

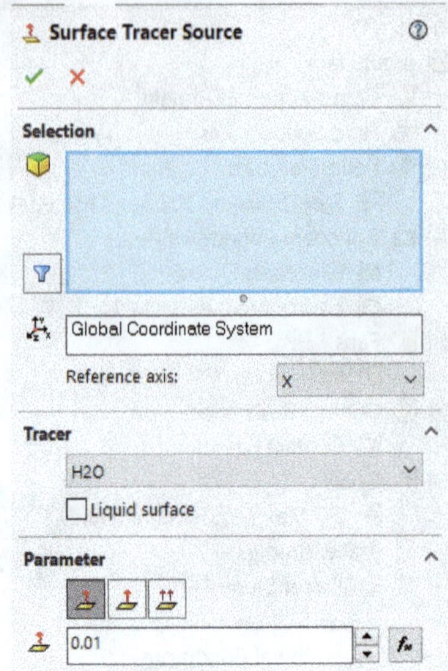

Figure-21. Surface Tracer Source PropertyManager

- Select the surface from where gas is being added in the fluid.
- Select the **Liquid surface** check box if traces of gas are over the surface of liquid or specify the desired mass flow rate in the **Parameter** section of the **PropertyManager**.

Similarly, specify the initial conditions and wall conditions for the tracer study. After specifying all the parameters, right-click on the study created viz. Tracer Study 1 from the Analysis Design Tree and select the **Run** option to perform the study; refer to Figure-22.

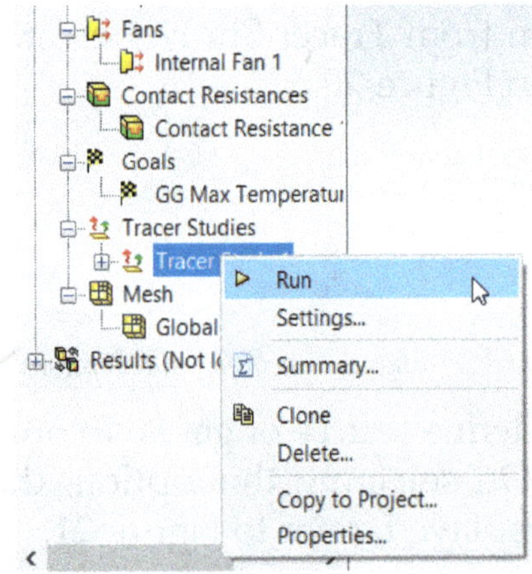

Figure-22. Running Tracer study

Once the analysis is complete, close the **Trace Study** tab and generate the results for traces of selected gas.

TRANSFERRED BOUNDARY CONDITION

The **Transferred Boundary Condition** tool is used to transfer the results of any previous analysis as boundary condition for your new analysis. The procedure to use this tool is given next.

- Click on the **Transferred Boundary Condition** tool from the **Tools > Flow Simulation > Insert >** menu. The **Selecting Boundaries** page of the dialog box will be displayed; refer to Figure-23.

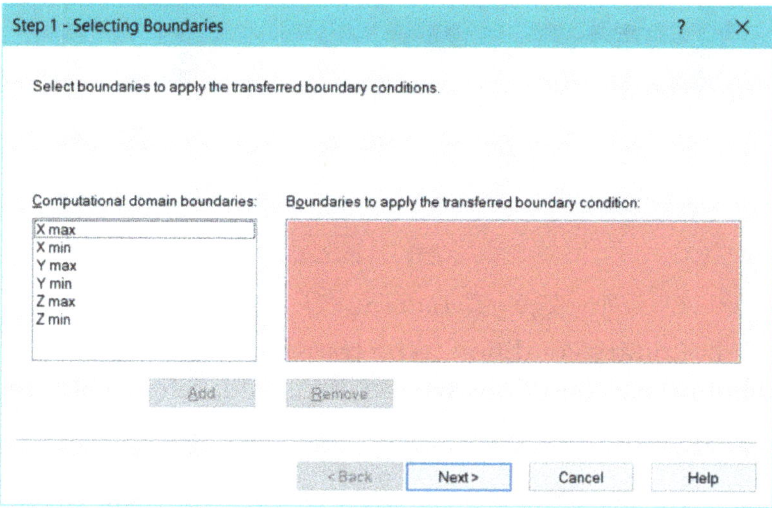

Figure-23. Selecting Boundaries page

- Select the desired option(s) from the domain list and click on the **Add** button. The boundary conditions of analysis results will be applied to the selected boundaries.
- Click on the **Next>** button. The **Selecting Results to Transfer** page of the dialog box will be displayed; refer to Figure-24.

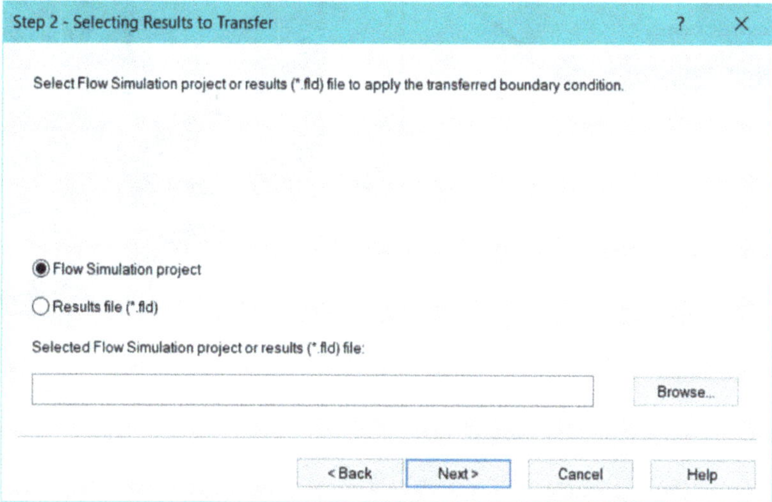

Figure-24. Selecting Results to Transfer page

- Select the **Flow Simulation project** radio button if you want to select a flow simulation project for transferring boundary condition from. Select the **Results file** radio button if you want to select result file of an analysis for transferring boundary condition from it.
- Click on the **Browse** button and select the desired project/result file.
- Click on the **Next>** button. The **Specifying Type of Condition** page of the dialog box will be displayed; refer to Figure-25.

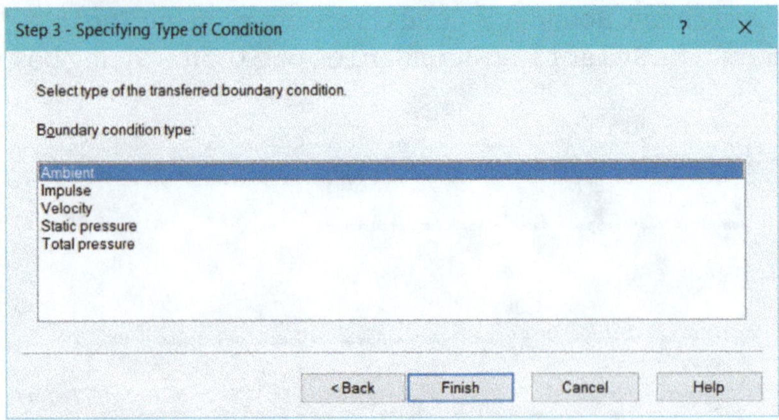

Figure-25. Specifying Type of Condition page

- Select the type of boundary condition to be transferred and click on the **Finish** button. The transferred boundary conditions will be displayed; refer to Figure-26.

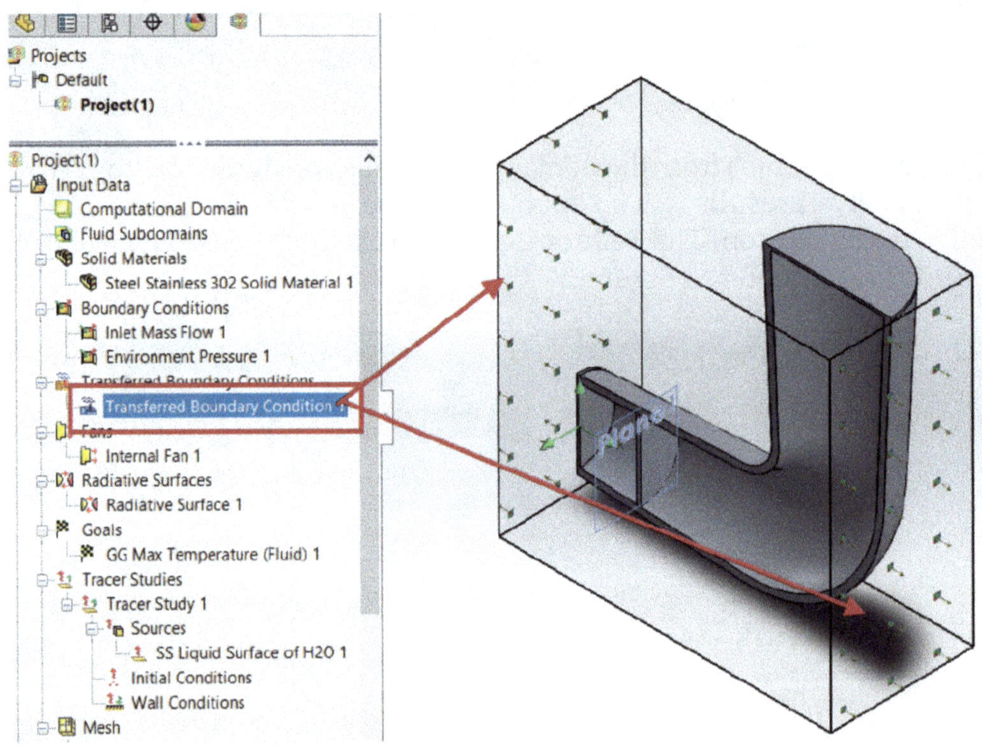

Figure-26. Transferred boundary conditions

APPLYING CONTACT RESISTANCE

The **Contact Resistance** tool is used to specify thermal resistance at solid-fluid or solid-solid boundaries. The procedure to use this tool is given next.

- Click on the **Contact Resistance** tool from the **Tools > Flow Simulation > Insert >** menu. The **Contact Resistance PropertyManager** will be displayed; refer to Figure-27.

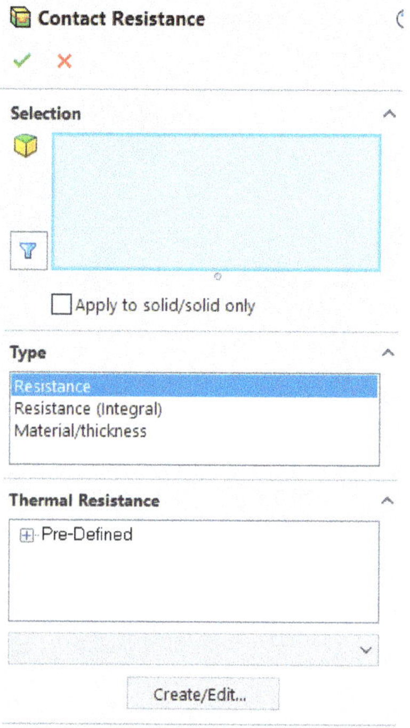

Figure-27. Contact Resistance PropertyManager

- Select the faces to which you want to apply thermal contact resistance.
- From the **Type** section, select the desired option. If you select the **Resistance** option then you will be asked to select a material from engineering database to specify the thermal resistance value. If you select the **Material/thickness** option then you need to specify material as well as thickness of the resistance layer in the **PropertyManager**.
- Specify the desired parameters and click on the **OK** button from the **PropertyManager** to apply contact resistance.

HEAT SINK SIMULATION

The **Heat Sink Simulation** tool is used to simulate the effect of heat sink on selected component. The procedure to use this tool is given next.

- Click on the **Heat Sink Simulation** tool from the **Tools > Flow Simulation > Insert >** menu. The **Heat Sink Simulation PropertyManager** will be displayed; refer to Figure-28.
- Select the component to which you want to apply heat sink.
- Click in the **Faces Fluid Enters the Heat Sink** selection box and select the faces from where the fluid (air or gas) enters; refer to Figure-29.
- Click in the **Faces Fluid Exits the Heat Sink** selection box and select the faces from where the fluid (air or gas) exits; refer to Figure-30.

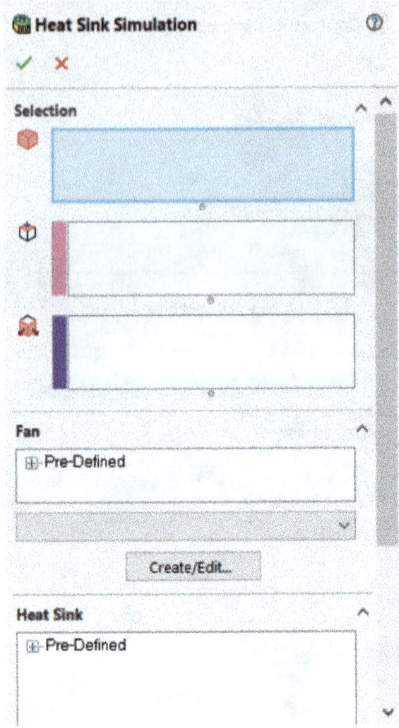

Figure-28. Heat Sink Simulation PropertyManager

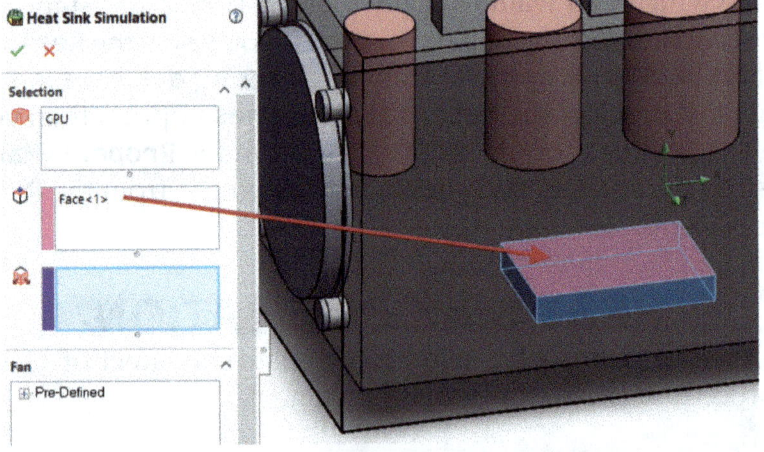

Figure-29. Face selected for entry of fluid

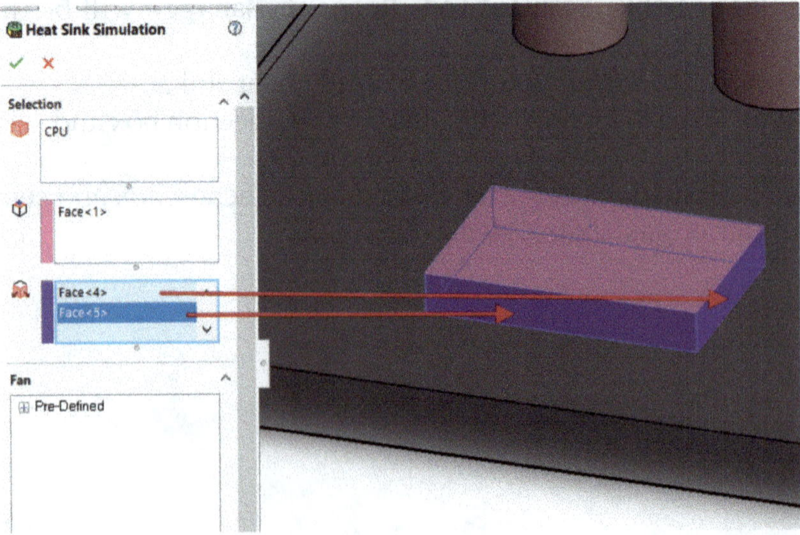

Figure-30. Faces selected for fluid exit

- Expand the **Fan** node and select the desired fan from the list.
- Select the desired heat sink from the **Heat Sink** section of the **PropertyManager**.
- Specify the heat generated from the source in the **Heat Generation Rate** edit box of **Source** section in the **PropertyManager**.
- Click on the **OK** button to apply heat sink parameters.

THERMOELECTRIC COOLER

The **Thermoelectric Cooler** tool is used to simulate thermoelectric cooler in the analysis. The procedure to use this tool is given next.

- Click on the **Thermoelectric Cooler** tool from the **Tools > Flow Simulation > Insert >** menu. The **Thermoelectric Cooler PropertyManager** will be displayed; refer to Figure-31.

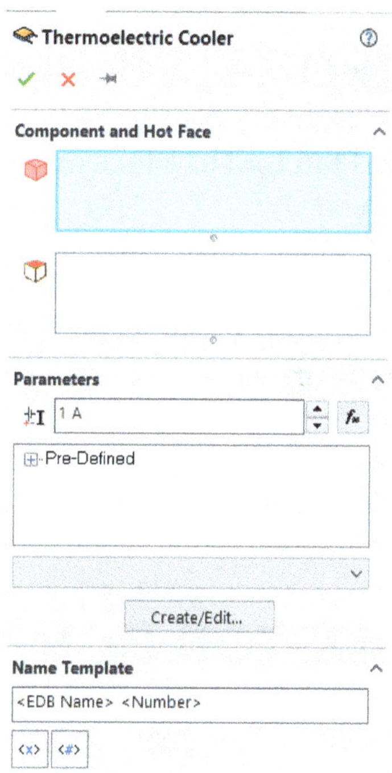

Figure-31. Thermoelectric Cooler PropertyManager

- Select the component to be used as thermoelectric cooler.
- Click in the **Hot Face** selection box and select the hot face to be cooled (the face in contact heat CPU or chip in case of computers).
- Specify the desired value of current in the **Parameters** edit box.
- Select the desired thermoelectric cooler from the list box; refer to Figure-32 and click on the **OK** button.

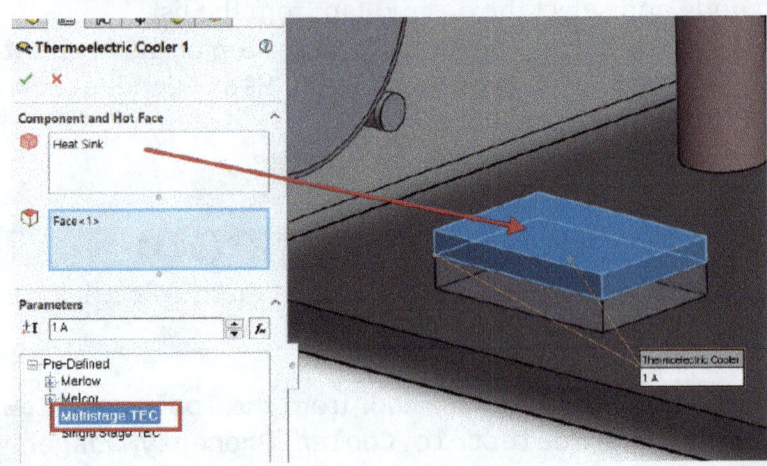

Figure-32. Thermoelectric Cooler created

PERFORATED PLATE

The **Perforated Plate** tool is used to apply perforated plate properties to selected face. This tool is useful when you have applied fan tool to a face and want the face to be perforated. The procedure to use this tool is given next.

- Click on the **Perforated Plate** tool from the **Tools > Flow Simulation > Insert >** menu. The **Perforated Plate PropertyManager** will be displayed asking you to select face on which fan or environment pressure has been specified.
- Select the face on which fan or environment pressure was applied earlier. The **PropertyManager** will be displayed as shown in Figure-33.
- Select the desired perforated plate from the **Perforated Plane** section and click on the **OK** button.

TWO RESISTOR COMPONENT

The **Two Resistor Component** tool is used to simulate the heat conduction by various electronic components in contact over a PCB or heat conduction between electronic component and PCB. The procedure to use this tool is given next.

- Click on the **Two-Resistor Component** tool from the **Flow Simulation Features** drop-down in the **Ribbon**. The **Two-Resistor Component PropertyManager** will be displayed; refer to Figure-34.

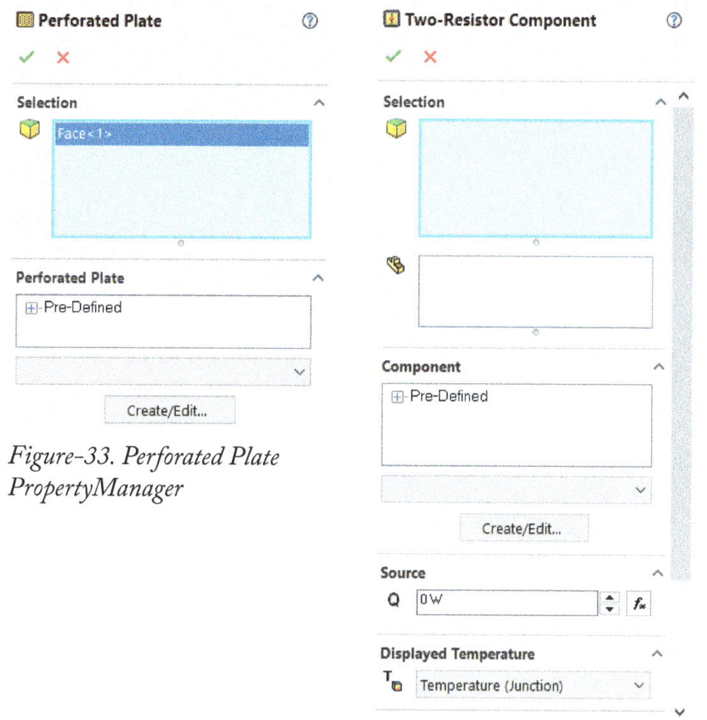

Figure-33. Perforated Plate PropertyManager

Figure-34. Two-Resistor Component PropertyManager

- Select the face on which components are embedded.
- Click in the **Components to Apply the Two-Resistor** selection box and select the components of circuit which generate heat.
- Select the desired component from the list in the **Component** section of the **PropertyManager**.
- Click in the edit box of **Source** section and specify the heat generate rate value for electronic components.
- Set the other parameters as desired and click on the **OK** button.

ELECTRICAL CONDITION

The **Electrical Condition** tool is used to set the electrical properties of the selected faces like setting current, voltage, and so on. Based on specified electrical conditions, joule heat will be generated as per engineering database. The procedure to apply electrical condition is given next.

- Click on the **Electrical Condition** tool from the **Flow Simulation Features** drop-down in the **Ribbon**. The **Electrical Conditions PropertyManager** will be displayed; refer to Figure-35.

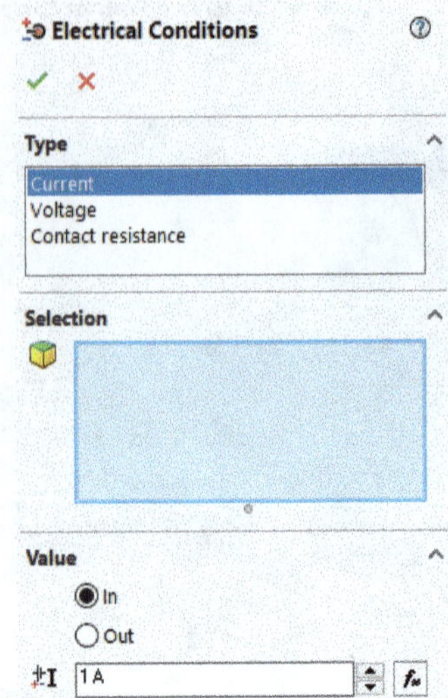

Figure-35. Electrical Conditions PropertyManager

- Select the desired type of electrical property from the **Type** section.
- Click in the selection box and select the faces on which you want to apply the electrical properties.
- Specify the desired parameters in the **Value** section of the **PropertyManager** and click on the **OK** button.

HEAT PIPE

The **Heat Pipe** tool is used to simulate heat transfer device in the analysis. The procedure to use this tool is given next.

- Click on the **Heat Pipe** tool from the **Tools > Flow Simulation > Insert >** menu. The **Heat Pipe PropertyManager** will be displayed; refer to Figure-36 and you will be asked to select the component to be used as heat pipe.
- Select the component to be used as heat pipe.
- Click in the **Heat In Faces** selection box and select the hot face of the pipe.
- Click in the **Heat Out Faces** selection box and select the cold face of the pipe. Note that heat will flow from hot face to cold face during the simulation.
- Specify the value of thermal resistance in the **Effective Thermal Resistance** edit box and click on the **OK** button to apply the properties.

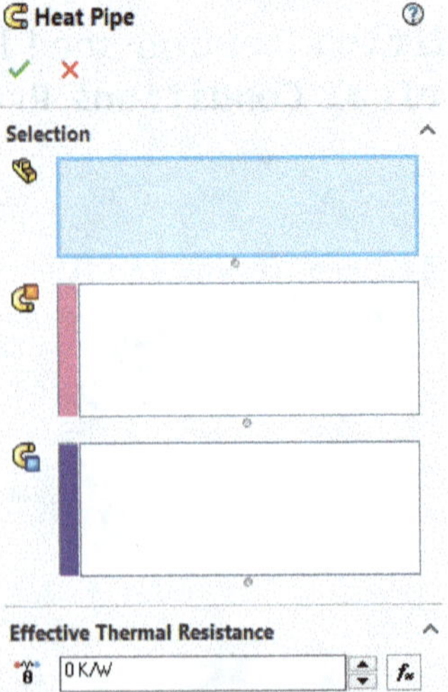

Figure-36. Heat Pipe PropertyManager

THERMAL JOINT

The **Thermal Joint** tool is used to create a joint between selected faces where heat can transfer from one object to another. A thermal joint allows to specify heat transfer rate between two faces. The procedure to use this tool is given next.

- Click on the **Thermal Joint** tool from the **Tools > Flow Simulation > Insert >** menu. The **Thermal Joint PropertyManager** will be displayed; refer to Figure-37.

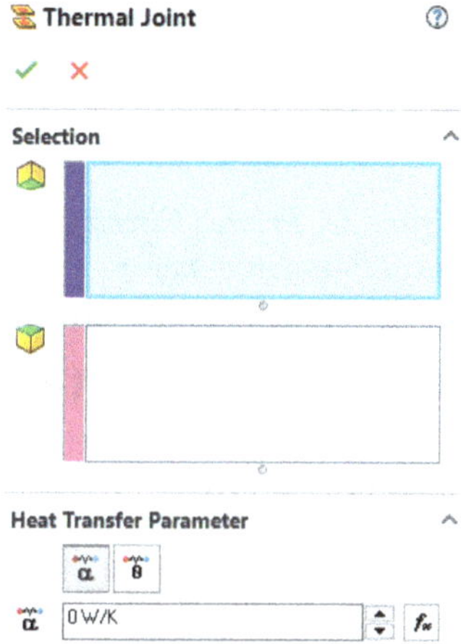

Figure-37. Thermal Joint PropertyManager

- Select the faces of first group and click in the **Second Group of Faces to Joint** selection box. You will be asked to select faces for second group.
- Select the faces for second group. Note that faces of two groups can or cannot be in direct contact.
- Click on the **Heat Transfer Coefficient (Integral)** button from the **Heat Transfer Parameter** section of the **PropertyManager** if you want to specify value of heat transfer coefficient. Click on the **Thermal Resistance (Integral)** button from the **Heat Transfer Parameter** section of the **PropertyManager** if you want to specify the thermal resistance value.
- Specify the desired value in **Heat Transfer Parameter** rollout and click on the **OK** button from the **PropertyManager** to apply the joint.

PRINTED CIRCUIT BOARD

The **Printed Circuit Board** tool is used to assign properties of PCB to selected body. The procedure is given next.

- Click on the **Printed Circuit Board** tool from the **Tools > Flow Simulation > Insert >** menu. The **Printed Circuit Board PropertyManager** will be displayed; refer to Figure-38.

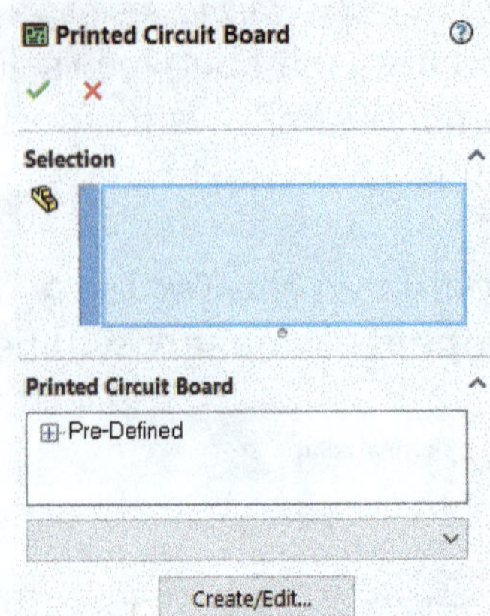

Figure-38. Printed Circuit Board PropertyManager

- Select the component to which PCB properties will be applied.
- Select the desired PCB template from the **Printed Circuit Board** section of the **PropertyManager**.
- Click on the **OK** button to apply the properties.

SELF-ASSESSMENT

Q1. Which of the following tools is used to represent filters for fluid flow in a CFD problem?

(a) Porous Medium
(b) Perforated Plate
(c) Heat Pipe
(d) Contact Resistance

Q2. Which of the following tools is used to define heat generation rate of a component?
(a) Surface Source
(b) Volume Source
(c) Radiation Source
(d) Thermoelectric Cooler

Q3. Which of the following tools is used to represent face with holes that allow fluid transfer with restrain after fan?

(a) Porous Medium
(b) Perforated Plate
(c) Heat Pipe
(d) Contact Resistance

Q4. Which of the following tools is used to define forced flow of fluid using table data?

(a) Boundary Condition
(b) Fan
(c) Initial Condition
(d) Thermoelectric Cooler

Q5. Which of the following tool is used to track flow of Nitrogen Di-oxide in an HVAC system?

(a) Heat Sink Simulation
(b) Tracer Study
(c) Contact Resistance
(d) Global Goals

Q6. Which of the following tool is used to define thermal resistance between fluid and solid boundary at contact?

(a) Two-Resistor Component
(b) Heat Sink Simulation
(c) Contact Resistance
(d) Heat Pipe

FOR STUDENT NOTES

FOR STUDENT NOTES

Chapter 6

Basics of CFD

Topics Covered

The major topics covered in this chapter are:

- ***Introduction***
- ***Conservation of Mass***
- ***Conservation of Momentum***
- ***Conservation of Energy***
- ***Variations of Navier-Strokes Equation***
- ***Steps of Computational Fluid Dynamics***
- ***Finite Difference Method***

INTRODUCTION

In Computational fluid dynamics, we are concerned about some basic properties of fluid at given instance of time like velocity, pressure, temperature, density, and viscosity. The Navier-Stokes equations are the broadly applied mathematical model to examine changes on those properties during dynamic and/or thermal interactions. The story of CFD starts with Navier-Strokes equation. This equation is based on conservation laws of mass, momentum, and energy. These law's can be defined as:

CONSERVATION OF MASS

The conservation of mass law states that the mass in the control volume can be neither created nor destroyed in accordance with physical laws. The conservation of mass, also expressed as Continuity Equation, states that the mass flow difference throughout system between inlet- and outlet-section is zero. In equation terms we can write as:

$$\frac{D_\rho}{D_t} + \rho\left(\nabla \cdot \vec{V}\right) = 0$$

Where, ρ is density, V is velocity and gradient operator ∇

$$\vec{\nabla} = \vec{i}\frac{\partial}{\partial x} + \vec{j}\frac{\partial}{\partial y} + \vec{k}\frac{\partial}{\partial z}$$

When density of fluid is constant, the flow is assumed as incompressible and then this equation represents a steady state process:

$$\frac{D_\rho}{D_t} = 0 \quad \text{so,} \quad \nabla \cdot \vec{V} = \frac{\partial u}{\partial x} + \frac{\partial v}{\partial y} + \frac{\partial w}{\partial z} = 0$$

here, u, v, and w are components of velocity at point (x,y,z) at time t.

Note that incompressible fluid is also called Newtonian fluid when stress/strain curve is linear.

CONSERVATION OF MOMENTUM

The momentum in a control volume is kept constant, which implies conservation of momentum that we call 'The Navier-Stokes Equations'. The description is set up in accordance with the expression of Newton's Second Law of Motion:

F=m.a

where, F is the net force applied to any particle, a is the acceleration, and m is the mass. In case the particle is a fluid, it is convenient to divide the equation to volume of particle to generate a derivation in terms of density as follows:

$$\rho\frac{DV}{D_t} = f = f_{body} + f_{surface}$$

in which f is the force exerted on the fluid particle per unit volume, and f_{body} is the applied force on the whole mass of fluid particles as below:

$$f_{body} = \rho \cdot g$$

Where, ρ is density, g is gravitational acceleration. External forces which are deployed through the surface of fluid particles, $f_{surface}$ is expressed by pressure and viscous forces as shown below:

$$f_{surface} = \nabla \cdot \tau_{ij} = \frac{\partial \tau_{ij}}{\partial x_i} = f_{pressure} + f_{viscous}$$

where τ_{ij} is expressed as stress tensor. According to the general deformation law of Newtonian viscous fluid given by Stokes, τ_{ij} is expressed as

$$\tau_{ij} = -p\delta_{ij} + \mu\left(\frac{\partial u_i}{\partial x_j} + \frac{\partial u_j}{\partial x_i}\right) + \delta_{ij}\lambda\nabla \cdot V$$

Hence, Newton's equation of motion can be specified in the form as follows:

$$\rho\frac{DV}{D_t} = \rho \cdot g + \nabla \cdot \tau_{ij}$$

Navier-Stokes equations of Newtonian viscous fluid in one equation gives:

$$\underbrace{\rho\frac{DV}{D_t}}_{I} = \underbrace{\rho \cdot g}_{II} - \underbrace{\nabla p}_{III} + \underbrace{\frac{\partial}{\partial x_i}\left[\mu\left(\frac{\partial v_i}{\partial x_j} + \frac{\partial v_j}{\partial x_i}\right) + \delta_{ij}\lambda\nabla \cdot V\right]}_{IV}$$

I : Momentum convection

II: Mass force

III: Surface force

IV: Viscous force

where, static pressure ρ and gravitational force ρ.g. The equation is convenient for fluid and flow fields both transient and compressible. D/D_t indicates the substantial derivative as follows:

$$\frac{D()}{D_t} = \frac{\partial()}{\partial t} + u\frac{\partial()}{\partial x} + v\frac{\partial()}{\partial y} + w\frac{\partial()}{\partial z} = \frac{\partial()}{\partial t} + V \cdot \nabla()$$

If the density of fluid is accepted to be constant, the equations are greatly simplified in which the viscosity coefficient μ is assumed constant and $\nabla \cdot V=0$ in equation. Thus, the Navier-Stokes equations for an incompressible three-dimensional flow can be expressed as follows:

$$\rho\frac{DV}{Dt} = \rho g - \nabla p + \mu\nabla^2 V$$

For each dimension when the velocity is V(u,v,w):

$$\rho\left(\frac{\partial u}{\partial t}+u\frac{\partial u}{\partial x}+v\frac{\partial u}{\partial y}+w\frac{\partial u}{\partial z}\right)=\rho g_x-\frac{\partial p}{\partial x}+\mu\left(\frac{\partial^2 u}{\partial x^2}+\frac{\partial^2 u}{\partial y^2}+\frac{\partial^2 u}{\partial z^2}\right)$$

$$\rho\left(\frac{\partial v}{\partial t}+u\frac{\partial v}{\partial x}+v\frac{\partial v}{\partial y}+w\frac{\partial v}{\partial z}\right)=\rho g_y-\frac{\partial p}{\partial y}+\mu\left(\frac{\partial^2 v}{\partial x^2}+\frac{\partial^2 v}{\partial y^2}+\frac{\partial^2 v}{\partial z^2}\right)$$

$$\rho\left(\frac{\partial w}{\partial t}+u\frac{\partial w}{\partial x}+v\frac{\partial w}{\partial y}+w\frac{\partial w}{\partial z}\right)=\rho g_z-\frac{\partial p}{\partial z}+\mu\left(\frac{\partial^2 w}{\partial x^2}+\frac{\partial^2 w}{\partial y^2}+\frac{\partial^2 w}{\partial z^2}\right)$$

p , u, v and w are unknowns where a solution is sought by application of both continuity equation and boundary conditions. Besides, the energy equation has to be considered if any thermal interaction is available in the problem.

CONSERVATION OF ENERGY

The Conservation of Energy is the first law of thermodynamics which states that the sum of the work and heat added to the system will result in the increase of energy of the system:

$$dE_t = dQ + dW$$

where dQ is the heat added to the system, dW is the work done on the system, and dE_t is the increment in the total energy of the system. One of the common types of energy equation is :

$$\rho\left[\underbrace{\frac{\partial h}{\partial t}}_{I}+\underbrace{\nabla\cdot(hV)}_{II}\right]=\underbrace{-\frac{\partial p}{\partial t}}_{III}+\underbrace{\nabla\cdot(k\nabla T)}_{IV}+\underbrace{\phi}_{V}$$

I : Local change with time

II: Convective term

III: Pressure work

IV: Heat flux

V: Heat dissipation term

The Navier-Stokes equations have a non-linear structure and various complexities so it is hardly possible to conduct an exact solution of those equations. Thus, with regard to the physical domain, both approaches and assumptions are partially applied to simplify the equations. Some assumptions also need to be applied to provide a reliable model in which the equation is carried out to further step in terms of complexity such as turbulence.

VARIATIONS OF NAVIER-STROKES EQUATION

The solution of the Navier-Stokes equations can be realized with either analytical or numerical methods. The analytical method is the process that only compensates solutions in which non-linear and complex structures of the Navier-Stokes equations are ignored within several assumptions. It is only valid for simple / fundamental cases such as Couette flow, Poisellie flow, etc. Almost every case in fluid dynamics comprises non-linear and complex structures in the mathematical model which cannot be ignored to sustain reliability. Hence, the solution of the Navier-Stokes equations are carried out with several numerical methods. Various parameters based on which Navier-Strokes equation can vary are given next.

Time Domain

The analysis of fluid flow can be conducted in either steady (time-independent) or unsteady (time-dependent) condition depending on the physical incident. In case the fluid flow is steady, it means the motion of fluid and parameters do not rely on change in time, the term $\partial()/\partial\tau=0$ where the continuity and momentum equations are re-derived as follows:

Continuity equation:

$$\frac{\partial(\rho u)}{\partial x}+\frac{\partial(\rho v)}{\partial y}+\frac{\partial(\rho w)}{\partial z}=0$$

The Navier-Stokes equation in x direction:

$$\rho\left(u\frac{\partial u}{\partial x}+v\frac{\partial u}{\partial y}+w\frac{\partial u}{\partial z}\right)=\rho g_x-\frac{\partial p}{\partial x}+\mu\left(\frac{\partial^2 u}{\partial x^2}+\frac{\partial^2 u}{\partial y^2}+\frac{\partial^2 u}{\partial z^2}\right)$$

While the steady flow assumption negates the effect of some non-linear terms and provides a convenient solution, variation of density is a hurdle that keeps the equation in a complex formation.

Compressibility

Due to the flexible structure of fluids, the compressibility of particles is a significant issue. Despite the fact that all types of fluid flow are compressible in a various range regarding molecular structure, most of them can be assumed to be incompressible in which the density changes are negligible. Thus, the term $\partial\rho/\partial t = 0$ is thrown away regardless of whether the flow is steady or not, as below:

Continuity equation:

$$\frac{\partial u}{\partial x}+\frac{\partial v}{\partial y}+\frac{\partial w}{\partial z}=0$$

The Navier-Stokes equation in x direction:

$$\rho\left(\frac{\partial u}{\partial t}+u\frac{\partial u}{\partial x}+v\frac{\partial u}{\partial y}+w\frac{\partial u}{\partial z}\right)=\rho g_x-\frac{\partial p}{\partial x}+\mu\left(\frac{\partial^2 u}{\partial x^2}+\frac{\partial^2 u}{\partial y^2}+\frac{\partial^2 u}{\partial z^2}\right)$$

As incompressible flow assumption provides reasonable equations, the application of steady flow assumption concurrently enables us to ignore non-linear terms where $\partial()/\partial t=0$. Moreover, the density of fluid in high speed cannot be accepted as incompressible in which the density changes are important. "The Mach Number" is a dimensionless number that is convenient to investigate fluid flow, whether incompressible or compressible.

$$Ma = \frac{V}{a} \leq 0.3$$

where, Ma is the Mach number, V is the velocity of flow, and a is the speed of sound at 340.29 m/s at sea level.

As in above equation, when the Mach number is lower than 0.3, the assumption of incompressibility is acceptable. On the contrary, the change in density cannot be negligible in which density should be considered as a significant parameter. For instance, if the velocity of a car is higher than 100 m/s, the suitable approach to conduct credible numerical analysis is the compressible flow. Apart from velocity, the effect of thermal properties on the density changes has to be considered in geophysical flows.

Low and High Reynolds Numbers

The Reynolds number, the ratio of inertial and viscous effects, is also effective on Navier-Stokes equations to truncate the mathematical model. While Re -> ∞, the viscous effects are presumed negligible and viscous terms in Navier-Stokes equations are removed. The simplified form of Navier-Stokes equation, described as Euler equation, can be specified as follows.

The Navier-Stokes equation in x direction:

$$\rho\left(\frac{\partial u}{\partial t} + u\frac{\partial u}{\partial x} + v\frac{\partial u}{\partial y} + w\frac{\partial u}{\partial z}\right) = \rho g_x - \frac{\partial p}{\partial x}$$

Even though viscous effects are relatively important for fluids, the inviscid flow model partially provides a reliable mathematical model as to predict real process for some specific cases. For instance, high-speed external flow over bodies is a broadly used approximation where inviscid approach reasonably fits. While Re<<1, the inertial effects are assumed negligible where related terms in Navier-Stokes equations drop out. The simplified form of Navier-Stokes equations is called either creeping flow or Stokes flow.

The Navier-Stokes equation in x direction:

$$\rho g_x - \frac{\partial p}{\partial x} + \mu\left(\frac{\partial^2 u}{\partial x^2} + \frac{\partial^2 u}{\partial y^2} + \frac{\partial^2 u}{\partial z^2} +\right) = 0$$

Having tangible viscous effects, creeping flow is a suitable approach to investigate the flow of lava, swimming of microorganisms, flow of polymers, lubrication, etc.

Turbulence

The behavior of the fluid under dynamic conditions is a challenging issue that is compartmentalized as laminar and turbulent. The laminar flow is orderly at which motion of fluid can be predicted precisely. Except that, the turbulent flow has various hindrances, therefore it is hard to predict the fluid flow which shows a chaotic behavior. The Reynolds number, the ratio of inertial forces to viscous forces, predicts the behavior of fluid flow whether laminar or turbulent regarding several properties such as velocity, length, viscosity, and also type of flow. Whilst the flow is turbulent, a proper mathematical model is selected to carry out numerical solutions. Various turbulent models are available in literature and each of them has a slightly different structure to examine chaotic fluid flow.

Turbulent flow can be applied to the Navier-Stokes equations in order to conduct solutions to chaotic behavior of fluid flow. Apart from the laminar, transport quantities of the turbulent flow, it is driven by instantaneous values. Direct numerical simulation (DNS) is the approach to solving the Navier-Stokes equation with instantaneous values. Having district fluctuations varies in a broad range, DNS needs enormous effort and expensive computational facilities. To avoid those hurdles, the instantaneous quantities are reinstated by the sum of their mean and fluctuating parts as follows:

$$instantaneous\ value = \overline{mean\ value} + fluctuating\ value'$$

$$u = \overline{u} + u'$$

$$v = \overline{v} + v'$$

$$w = \overline{w} + w'$$

$$T = \overline{T} + T'$$

where u, v, and w are velocity components and T is temperature.

Instead of instantaneous values which cause non-linearity, carrying out a numerical solution with mean values provides an appropriate mathematical model which is named "The Reynolds-averaged Navier-Stokes (RANS) equation". The fluctuations can be negligible for most engineering cases which cause a complex mathematical model. Thus, RANS turbulence model is a procedure to close the system of mean flow equations. The general form of The Reynolds-averaged Navier-Stokes (RANS) equation can be specified as follows:

Continuity equation:

$$\frac{\partial \overline{u}}{\partial x} + \frac{\partial \overline{v}}{\partial y} + \frac{\partial \overline{w}}{\partial z} = 0$$

The Navier-Stokes equation in x direction:

$$\rho\left(\frac{\partial \overline{u}}{\partial t} + \overline{u}\frac{\partial \overline{u}}{\partial x} + \overline{v}\frac{\partial \overline{u}}{\partial y} + \overline{w}\frac{\partial \overline{u}}{\partial z}\right) = \rho g_x - \frac{\partial \overline{p}}{\partial x} + \mu\left(\frac{\partial^2 u}{\partial x^2} + \frac{\partial^2 u}{\partial y^2} + \frac{\partial^2 u}{\partial z^2}\right)$$

The turbulence model of RANS can also vary regarding methods such as k-omega, k-epsilon, k-omega-SST, and Spalart-Allmaras which have been used to seek a solution for different types of turbulent flow.

Likewise, large eddy simulation (LES) is another mathematical method for turbulent flow which is also comprehensively applied for several cases. Tough LES ensures more accurate results than RANS, it requires much more time and computer memory. As in DNS, LES considers to solve the instantaneous Navier-Stokes equations in time and three-dimensional space.

Favre Time Averaging

SolidWorks Flow Simulation uses Favre-averaged Navier-Stokes equation for turbulent flow. In Favre-averaging, the time averaged equations can be simplified significantly by using the density weighted averaging procedure suggested by Favre. If Reynolds time averaging is applied to the compressible form of the Navier-Stokes equations, some difficulties arise. In particular, the original form of the equations is significantly altered. To see this, consider Reynolds averaging applied to the continuity equation for compressible flow.

Favre time averaging can be defined as follows. The instantaneous solution variable, ϕ , is decomposed into a mean quantity, $\tilde{\phi}$, and fluctuating component, ϕ'', as follows:

$$\phi = \tilde{\phi} + \phi''$$

The Favre time-averaging is then

$$\overline{\rho\phi(x_i, t)} = \frac{1}{T}\int_{t-T/2}^{t+T/2} \rho(x_i, t')\phi(x_i, t')\,\mathrm{d}t' = \overline{\rho\tilde{\phi}} + \overline{\rho\phi''} = \bar{\rho}\tilde{\phi}$$

Where,

$$\tilde{\phi}(x_i, t) \equiv \frac{1}{\bar{\rho}T}\int_{t-T/2}^{t+T/2} \rho(x_i, t')\phi(x_i, t')\,\mathrm{d}t', \quad \overline{\rho\phi''} \equiv 0$$

Favre-Averaged Navier-Stokes (FANS) Equations can be given as:

Continuity Equation:

$$\frac{\partial}{\partial t}(\bar{\rho}) + \frac{\partial}{\partial x_i}(\bar{\rho}\tilde{u}_i) = 0$$

Momentum Equation:

$$\frac{\partial}{\partial t}(\bar{\rho}\tilde{u}_i) + \frac{\partial}{\partial x_j}(\bar{\rho}\tilde{u}_i\tilde{u}_j + \bar{p}\delta_{ij}) = \frac{\partial}{\partial x_j}\left(\bar{\tau}_{ij} - \overline{\rho u_i'' u_j''}\right)$$

Favre-Averaged Reynolds Stress Tensor:

$$\lambda = -\overline{\rho u_i'' u_j''}$$

Turbulent Kinetic Energy:

$$\frac{1}{2}\overline{\rho u_i'' u_i''} = -\frac{1}{2}\lambda_{ii} = \bar{\rho}\tilde{k}$$

STEPS OF COMPUTATIONAL FLUID DYNAMICS

The process of solving Computational Fluid Dynamics can be defined in 5 stages which are: Creating Mathematical Model, Discretization of model, Analyzing with Numerical Scheme, Getting solution, and Post Processing (Visualization). These steps are discussed next.

Creating Mathematical Model

In this stage, various equations are defined based on physical properties, boundary conditions, and other real-world parameters of the problem. The mathematical model generally consists of partial differential equations, integral equations, and combinations of both. In this case, it will be Navier-Stokes equation with other mathematical equations to define boundary conditions and thermodynamic properties. Note that at this step there will be some assumptions which converting real problem into mathematical model like material might be assumed isotropic, heat conditions might be considered adiabatic and so on. These assumptions should generate permissible level of error in the solution and as a designer you should be aware of the quantum of those errors.

Discretization of Model

While studying the behavior of fluid, you will find that there are infinite number of particles with different physical properties in the fluid stream. These infinite particles will generate infinite equations to solve if mathematical model is to be solved directly which is not possible. So, we convert the mathematical model into finite number of elements and then we can solve equations for each of the element. In case of CFD, elements are called cells. The process of converting a mathematical model into finite element equations is called discretization. In simple words, the partial differential and integral equations are converted into algebraic equations of **A.x = B**.

There are many methods to perform discretization of mathematical model like Finite Difference Method, Finite Volume Method, and Finite Element Method which use mesh type structuring of the model. There are also methods which do not use mesh like Smooth Particle Hydrodynamics (SPH) and Finite Pointset Method (FPM). Note that there is again loss of information at the stage of discretization.

Analyzing with Numerical Schemes

Numerical scheme is the complete setup of equations with all the parameters and boundary conditions defined as needed. The scheme also includes numerical methods by which these equations will be solved and limiting points upto which the equations will be solved. When you start analyzing the problem with specified numerical scheme, there is no input required from your side in system; everything is automatic by computer. Every numerical scheme need to satisfy some basic requirements which are consistency, stability, convergence, and accuracy.

Solution

At this stage, we get the solution of equations as different basic variables like pressure, speed, volume, and so on. The resulting flow variables are obtained at each grid point/mesh point whether the scheme is time dependent (transient) or steady.

Visualization (Post-processing)

At this stage, everything is based on interpretation of designer. Various desired parameters are derived from basic calculated parameters at this stage. You also need to check whether results of CFD analysis are realistic or not.

Now, one question left here to understand in more detail is discretization. Discretization can be performed by various methods like FDM, FVM, FEM, and so on as discussed earlier. Out of these methods, Finite Difference Method and Finite Volume Methods are the most used methods for CFD. Here, we will discuss FDM in detail.

FINITE DIFFERENCE METHOD

At discretization stage, the model need to be converted into numerical grid with different cells defined by nodes; refer to Figure-1. For a 2D problem, i and j will be used for direction in horizontal and vertical directions.

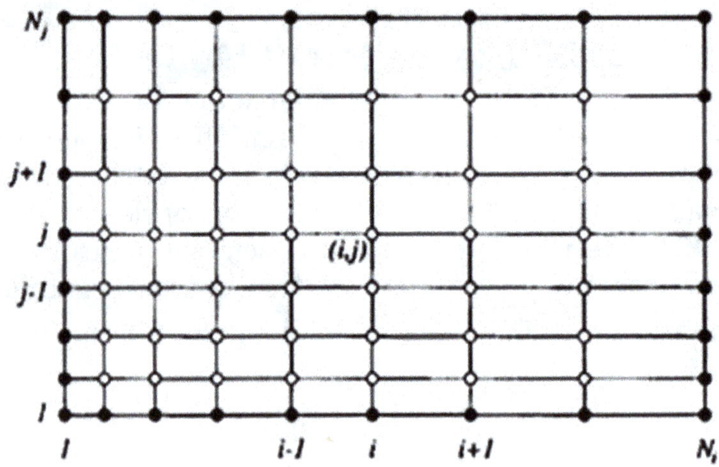

Figure-1. Grid with cartesian coordinates of nodes

If you recall the definition of derivative: $\left(\frac{\partial u}{\partial x}\right)_{x_i} = \lim_{\Delta x \to 0} \frac{u(x_i + \Delta x) - u(x_i)}{\Delta x}$

The equation represents slope of tangent to a curve u(x) which can be geometrically represented as shown in Figure-2.

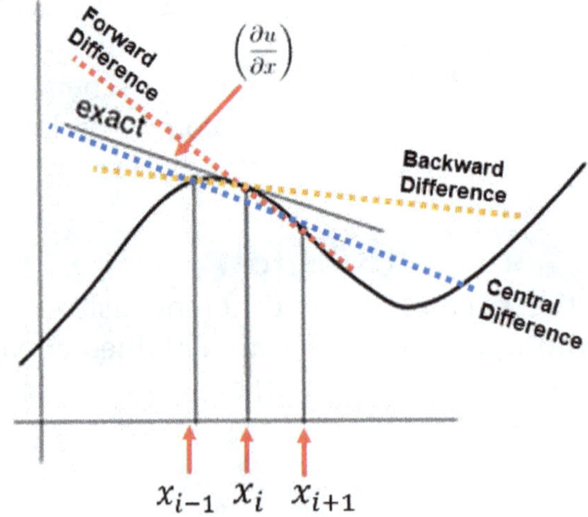

Figure-2. Geometrical interpretation of derivative

Now, assume that we have value of x_i, x_{i+1}, and x_{i-1} then there are three ways in which we can find the slope of curve.

Backward Difference which uses x_i and x_{i-1}.
Forward Difference which uses x_i and x_{i+1}.
Central Difference which uses x_{i+1} and x_{i-1}.

If we assume that Δx is difference between two consecutive nodes then lower the value of Δx more accurate we will get the value of slope.

For a uniform grid where Δx is same for each node then based on

Backward Difference : $\frac{du}{dx} \cong \frac{u_i - u_{i-1}}{\Delta x}$

Forward Difference : $\frac{du}{dx} \cong \frac{u_{i+1} - u_i}{\Delta x}$

Central Difference : $\frac{du}{dx} \cong \frac{u_{i+1} - u_{i-1}}{2\Delta x}$

Now, if you expand the backward difference derivative using Taylor's series then

Consider a function u(x) and its derivative at point x,

$$\left(\frac{\partial u}{\partial x}\right)_{x_i} = \lim_{\Delta x \to 0} \frac{u(x_i + \Delta x) - u(x_i)}{\Delta x}$$

If u(x + Δx) is expanded in Taylor series about u(x), we obtain

$$u(x + \Delta x) = u(x) + \Delta x \frac{\partial u(x)}{\partial x} + \frac{(\Delta x)^2}{2} \frac{\partial^2 u(x)}{\partial x^2} + \frac{(\Delta x)^3}{3!} \frac{\partial^3 u(x)}{\partial x^3} + \cdots$$

If we substitute this value in derivative equation then

$$\frac{\partial u(x)}{\partial x} = \lim_{\Delta x \to 0} \left(\frac{\partial u(x)}{\partial x} + \frac{\Delta x}{2} \frac{\partial^2 u(x)}{\partial x^2} + \cdots \right)$$

Based on this equation, we can write u in Taylor Series at i+1 and i-1 as:

$$u_{i+1} = u_i + \Delta x \left(\frac{\partial u}{\partial x}\right)_i + \frac{\Delta x^2}{2} \left(\frac{\partial^2 u}{\partial x^2}\right)_i + \frac{\Delta x^3}{3!} \left(\frac{\partial^3 u}{\partial x^3}\right)_i + \frac{\Delta x^4}{4!} \left(\frac{\partial^4 u}{\partial x^4}\right)_i + \cdots$$

$$u_{i-1} = u_i - \Delta x \left(\frac{\partial u}{\partial x}\right)_i + \frac{\Delta x^2}{2} \left(\frac{\partial^2 u}{\partial x^2}\right)_i - \frac{\Delta x^3}{3!} \left(\frac{\partial^3 u}{\partial x^3}\right)_i + \frac{\Delta x^4}{4!} \left(\frac{\partial^4 u}{\partial x^4}\right)_i + \cdots$$

Re-writing the above two equations, we get:

Forward difference : $\left(\frac{\partial u}{\partial x}\right)_i = \frac{u_{i+1} - u_i}{\Delta x} + O(\Delta x)$

Backward Difference : $\left(\frac{\partial u}{\partial x}\right)_i = \frac{u_i - u_{i-1}}{\Delta x} + O(\Delta x)$

Central Difference : $\left(\frac{\partial u}{\partial x}\right)_i = \frac{u_{i+1} - u_{i-1}}{2\Delta x} + O(\Delta x^2)$

Note that central difference takes second derivative of delta x so the value of error is lesser than compared to forward or backward difference. For example if forward or backward difference gives an error of 0.001 then second derivative will be 0.001 x 0.001 which gives 0.000001. This error value is a lot lesser.

You can learn more about these methods in books dedicated to Finite Difference Method and solution techniques.

FOR STUDENT NOTES

Chapter 7

Practical and Practice

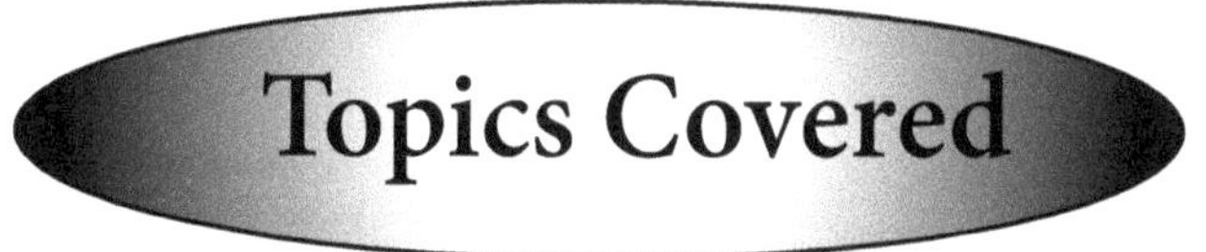

The major topics covered in this chapter are:

- ***Introduction***
- ***Practical 1 Air Flow on Fin***
- ***Practical 2 Flow of Two Gases in a Chamber***
- ***Practical 3 Air Flow by Fan***
- ***Practice 1 Air Flow through Ball Valve***

INTRODUCTION

In this chapter, we will perform CFD analysis on various actual design projects. You are free to modify the model and analysis the effects of flow in different situations. The practical and practice problems are given next.

PRACTICAL 1 Air Flow on Fin

In this practical, we will analysis the flow around a fin that can be used as spoiler on the car. We will also find out the forces acting on various faces of the model due to air flow. The boundary conditions for the problem are shown in Figure-1. Note that the model file is available in the resource kit folder of the book. The procedure to perform this analysis is given next.

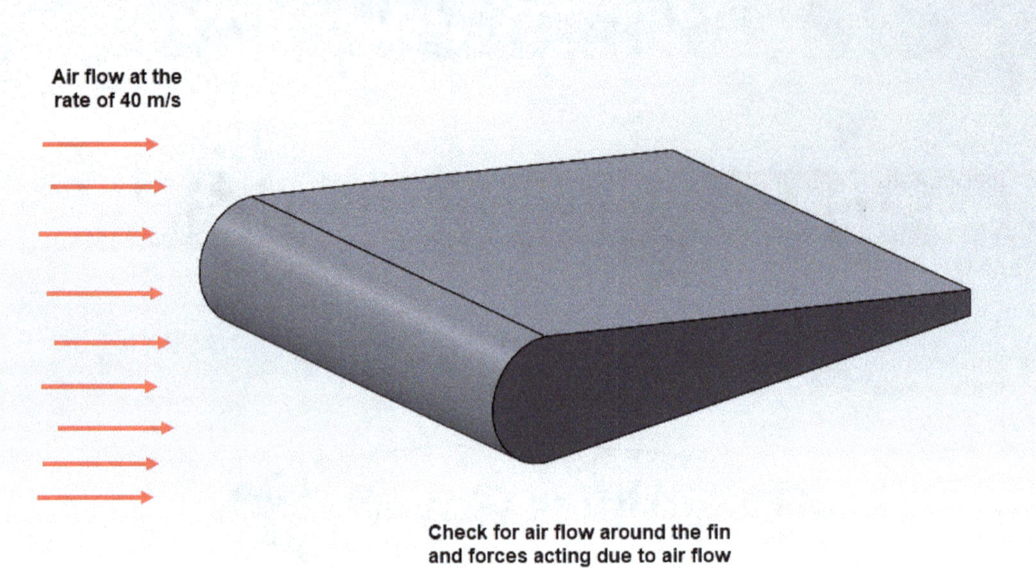

Figure-1. Boundary_conditions_for_Practical_1.png

Opening Model and Starting Project

- Start SolidWorks using the desktop icon or **Start** menu.
- Click on the **Open** button from the **Quick Access Toolbar** or **Open** option from the **File** menu. The **Open** dialog box will be displayed.
- Select the part file for this practical and click on the **Open** button. The model for file will be displayed; refer to Figure-2.
- Click on the **Add-Ins** option from the **Options** drop-down in the **Quick Access Toolbar**. The **Add-Ins** dialog box will be displayed. Select the check boxes for SOLIDWORKS Flow Simulation from the dialog box and click on the **OK** button to activate the add-in if not done yet.
- Click on the **Wizard** tool from the **Flow Simulation CommandManager** in the **Ribbon**. The **Wizard** dialog box will be displayed.

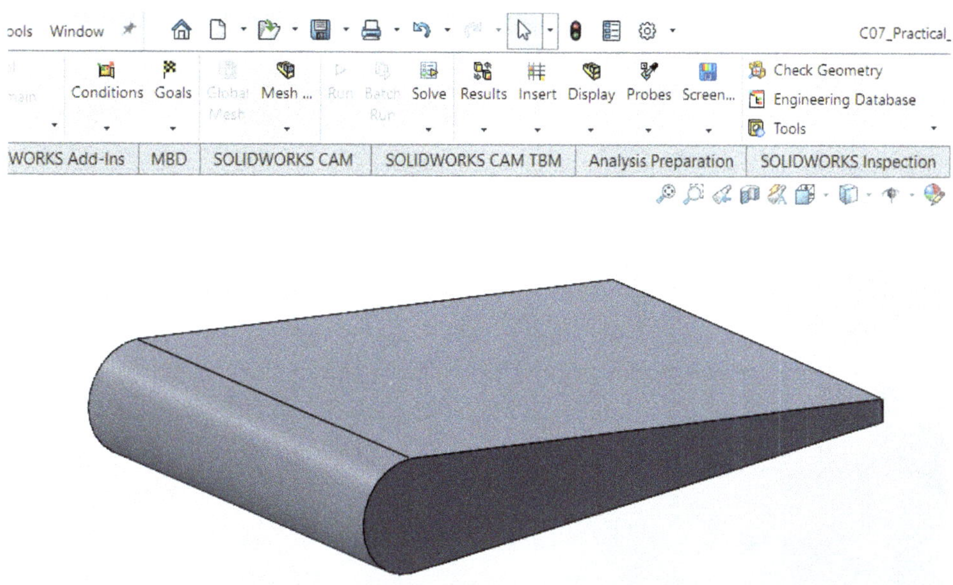

Figure-2. Model_for_Practical_1.png

- Specify desired name of the project in the **Project name** edit box and click on the **Next** button. The **Unit System** page of the dialog box will be displayed; refer to Figure-3.

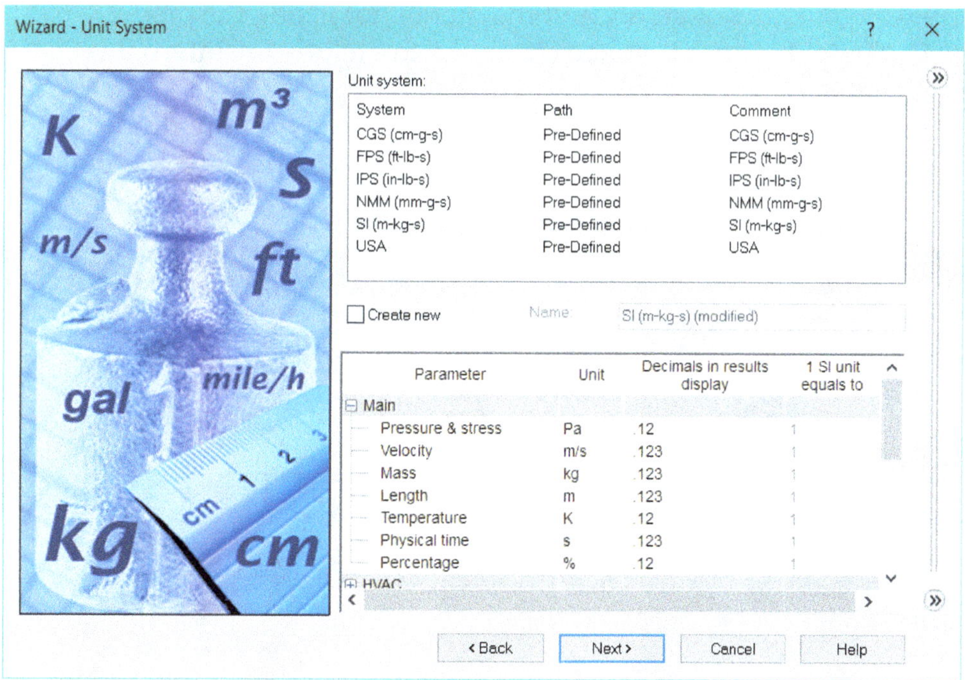

Figure-3. Unit_System_page.png

- Select the **NMM (mm-g-s)** unit system from the **Unit system** area of the dialog box and click on the **Next** button. The **Analysis Type** page will be displayed in the dialog box; refer to Figure-4.

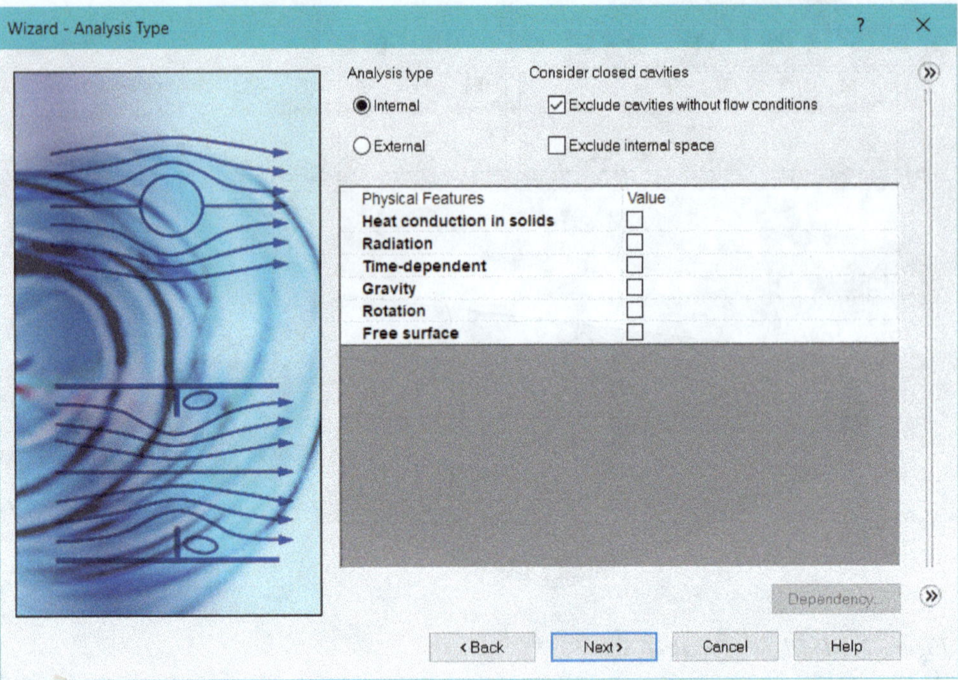

Figure-4. Analysis_Type_page.png

- Select the **External** radio button from the **Analysis type** area of the dialog box because we are interested in flow of air around the model not inside the model.
- Select the **Exclude internal space** check box to exclude the internal voids in the model if any from analysis.
- Select the **Gravity** check box to include effect of gravity on the fluid flow. Click on the **Next** button from the dialog box. The **Default Fluid** page of the dialog box will be displayed; refer to Figure-5.

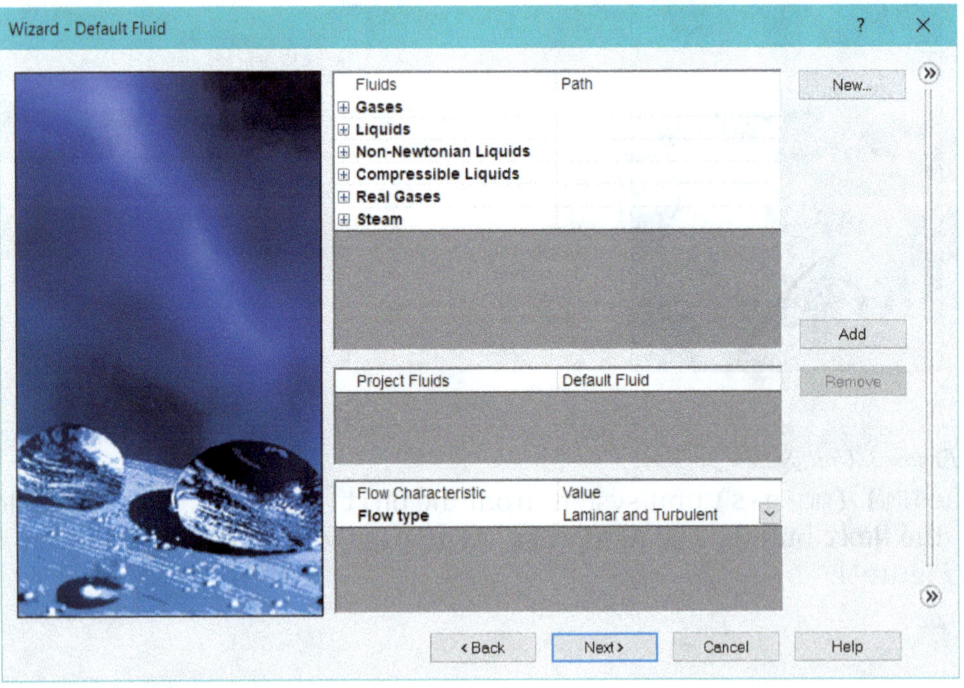

Figure-5. Default_Fluid_page.png

- Expand the **Gases** category and double-click on air to select as default fluid.
- Select the **Humidity** check box if you want to include humidity in analysis. After selecting the fluid, click on the **Next** button from the dialog box. The **Wall Conditions** page will be displayed.

- Set the wall thermal condition to default value which is **Adiabatic wall** and surface roughness as **18** micrometer.
- After setting desired parameters, click on the **Next** button. The **Initial and Ambient Conditions** dialog box will be displayed; refer to Figure-6.

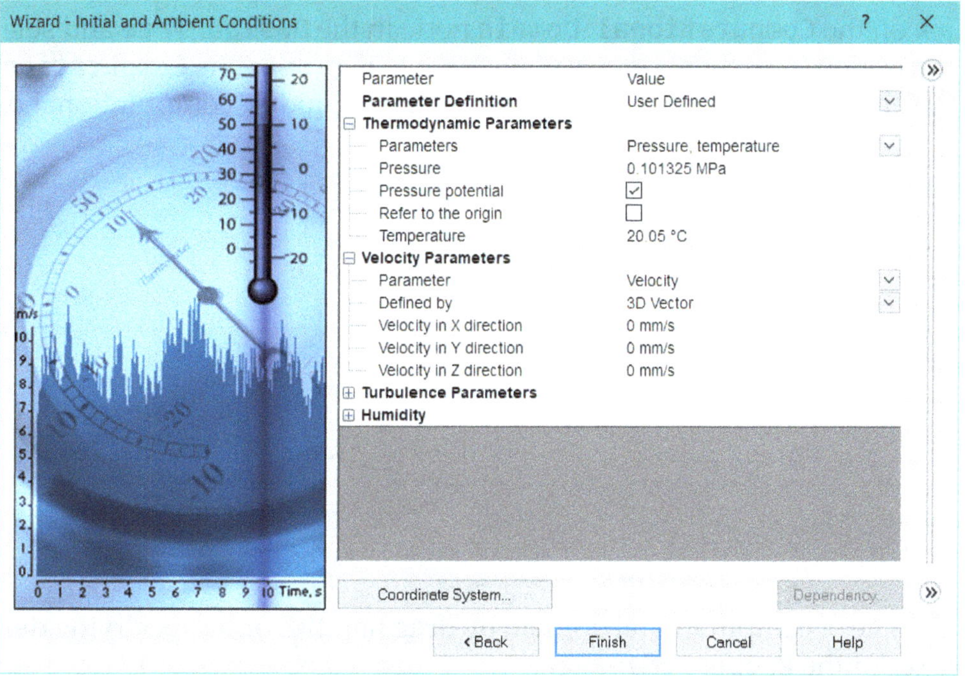

Figure-6. Initial_and_Ambient_Conditions_page.png

- Specify the velocity value as **40000 mm/s** in the **Velocity in X direction** edit box.
- Specify the value in **Temperature** edit box as 25 °C.
- Set the other parameters as desired and click on the **Finish** button from the dialog box. The model will be displayed with computational domain; refer to Figure-7.

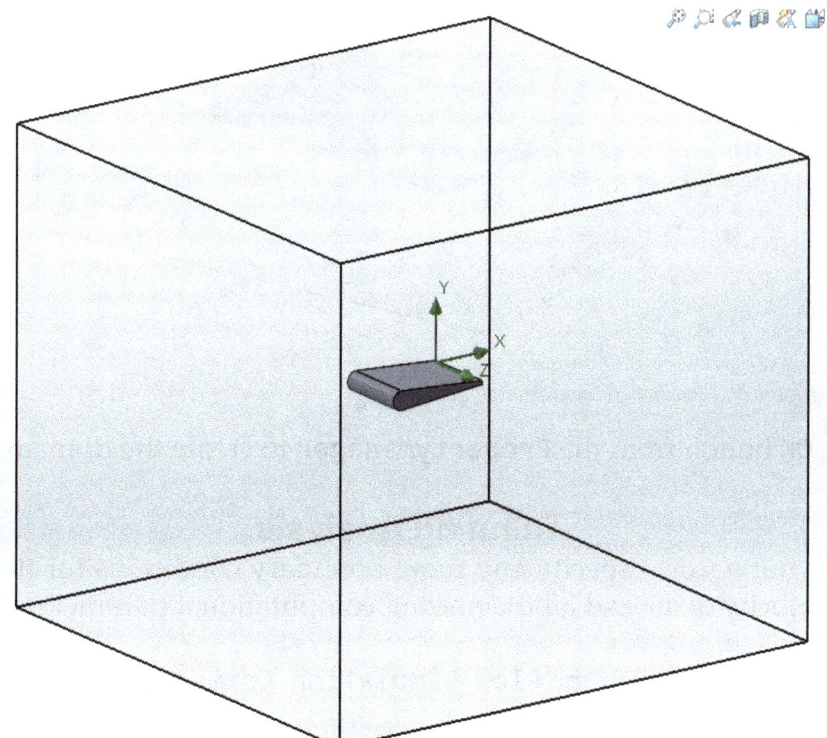

Figure-7. Model_with_computational_domain.png

Modifying Computational Domain

By default, the computational domain of analysis is very large as compared to model. A large computational domain takes more time to perform analysis. It is better to keep the computational domain minimum surrounding the model.

- Right-click on the **Computational Domain** node in the **Design Tree** and select the **Edit Definition** option from the shortcut menu; refer to Figure-8. The **Computational Domain PropertyManager** will be displayed with options to modify computational domain.

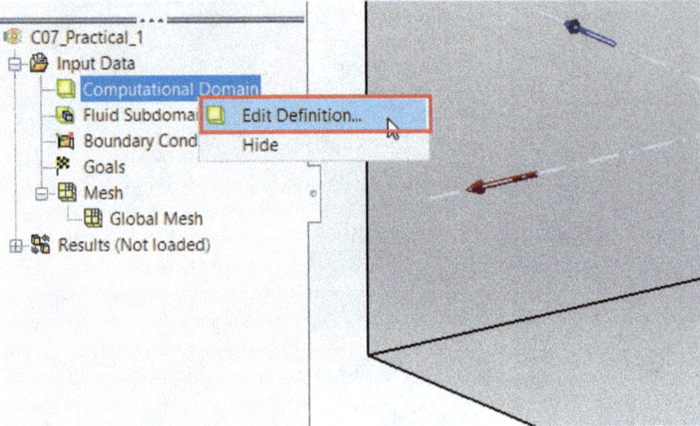

Figure-8. Edit_Definition_option.png

- Modify the size of computational domain by drag handles or by specifying desired values in edit boxes of the **PropertyManager**.

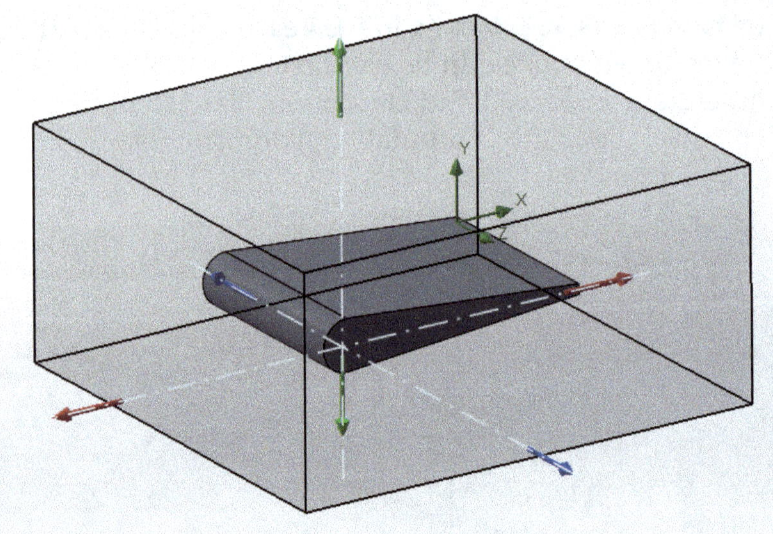

Figure-9. Modified_computational_domain.png

- Click on the **OK** button from the **PropertyManager** to create the domain.

Running Analysis

Note that we do not need to specify any more boundary conditions for this analysis as we have specified velocity of air and all the needed computational parameters.

- Click on the **Run** button from the **Flow Simulation CommandManager** in the **Ribbon**. The **Run** dialog box will be displayed; refer to Figure-10.

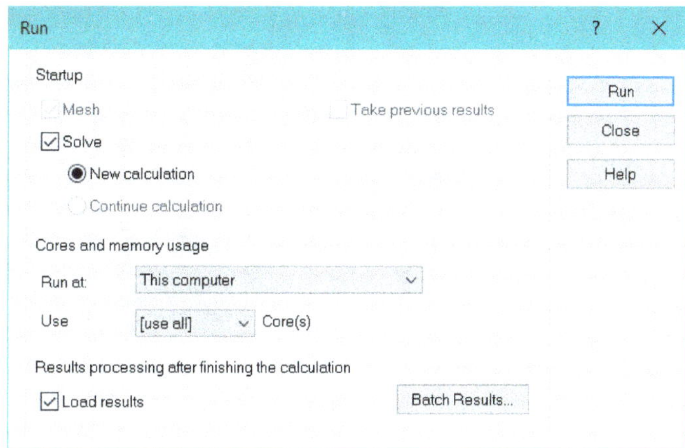

Figure-10. Run_dialog_box

- Set the parameters as shown in figure and click on the **Run** button. The Solver dialog box will be displayed showing progress of analysis. Once the analysis is complete, close the **Solver** dialog box by using the **x** button at the top right corner of the dialog box.

Generating Results

- Expand the **Results** node and right-click on the **Flow Trajectories** result from the **Design Tree**. The shortcut menu for result will be displayed.
- Select the **Insert** option from the shortcut menu. The **Flow Trajectories PropertyManager** will be displayed.
- Select a plane perpendicular to model face and set the offset value to define plane from where flow will start; refer to Figure-11.

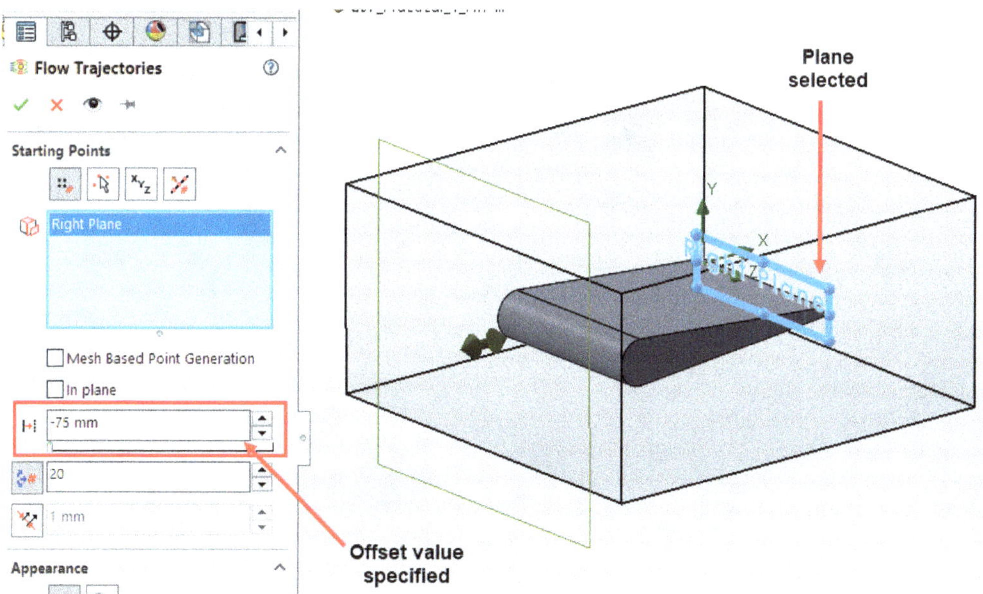

Figure-11. Plane_selected_for_flow_trajectories.png

- Click on the **Detailed Preview** button from the **PropertyManager** to check preview of flow trajectories; refer to Figure-12.
- Set the desired parameters like number of points, size of arrows, and so on.
- Click on the **OK** button to generate the flow trajectory result.

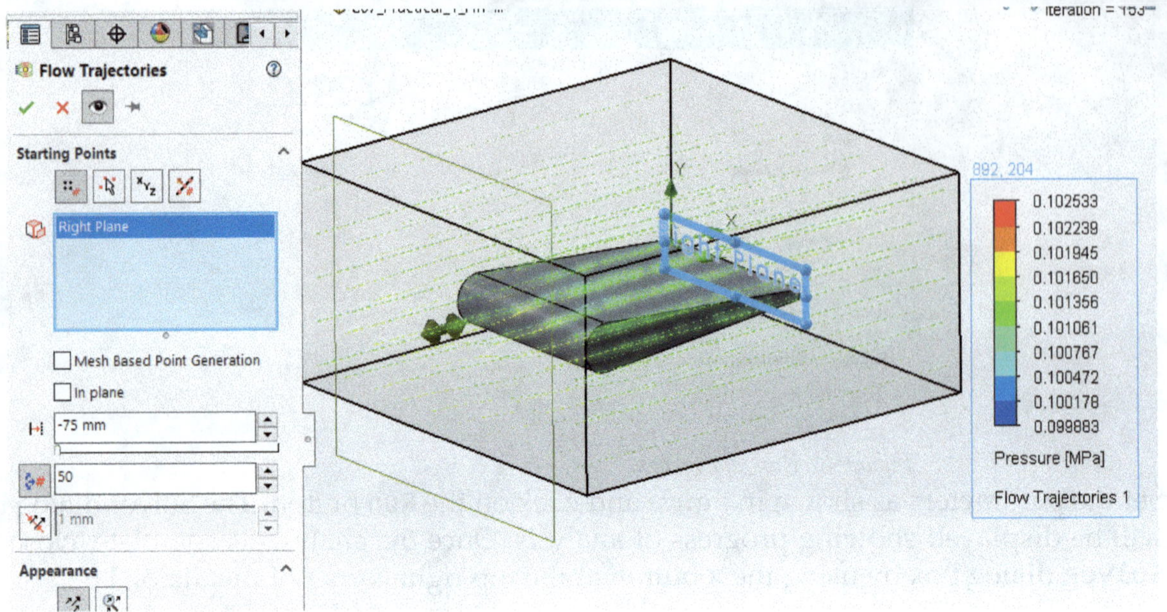

Figure-12. Preview_of_flow_trajectories.png

Now, we will check the results for forces acting on the faces of model due to air flow.

- Right-click on the **Surface Parameters** result node from the **Design Tree** and select the **Insert** option from the shortcut menu. The **Surface Parameters PropertyManager** will be displayed; refer to Figure-13.

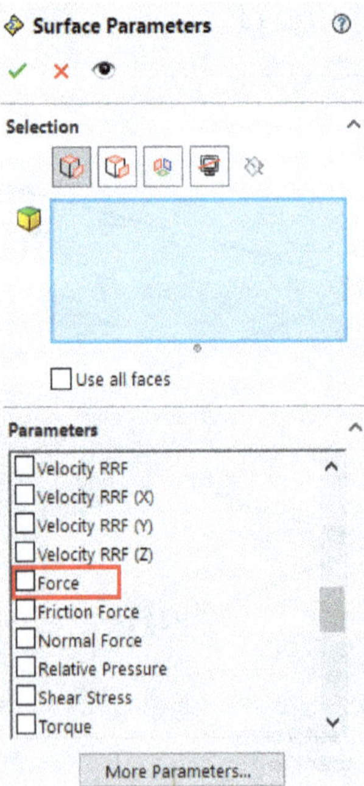

Figure-13. Surface Parameters PropertyManager

- Select the **Force** check box from the **Parameters** rollout to check forces on selected surfaces.
- Click on the **Detailed Preview** button from the top in the **PropertyManager** and select the face of model on which you want to check the force parameters; refer to Figure-14.

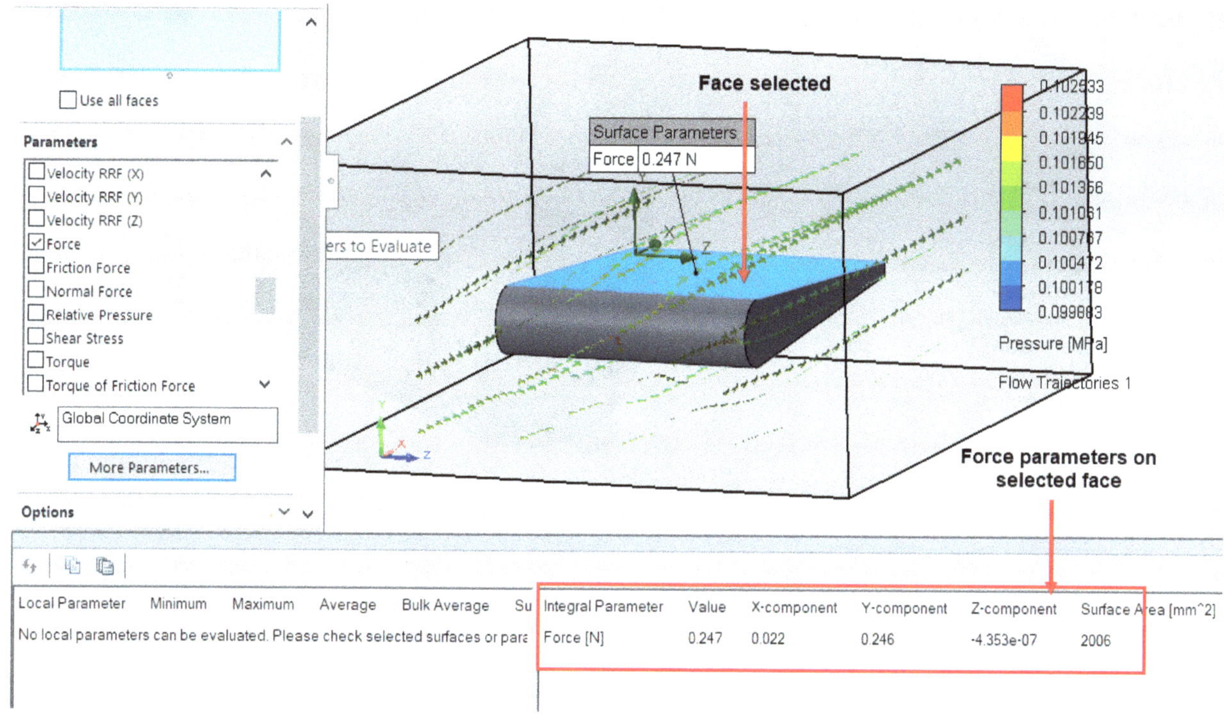

Figure-14. Force_result_on_selected_face.png

- Click on the **OK** button from the **PropertyManager** to create the results.

PRACTICAL 2 FLOW OF TWO GASES IN A CHAMBER

In this practical, we will check the mixed flow of two gases in a chamber through different entry points. The model for this practical is shown in Figure-15 with boundary conditions.

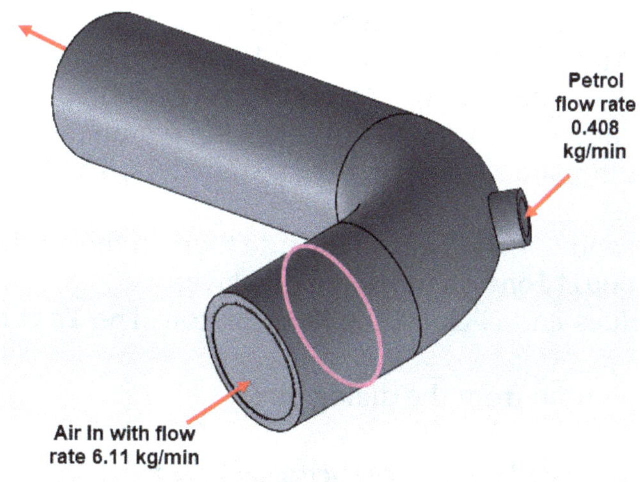

Figure-15. Model_for_Practical_2.png

Opening Model and Starting Project

- Start SolidWorks using the desktop icon or **Start** menu.
- Click on the **Open** button from the **Quick Access Toolbar** or **Open** option from the **File** menu. The **Open** dialog box will be displayed.
- Select the part file for this practical and click on the **Open** button. The model for file will be displayed; refer to Figure-16.

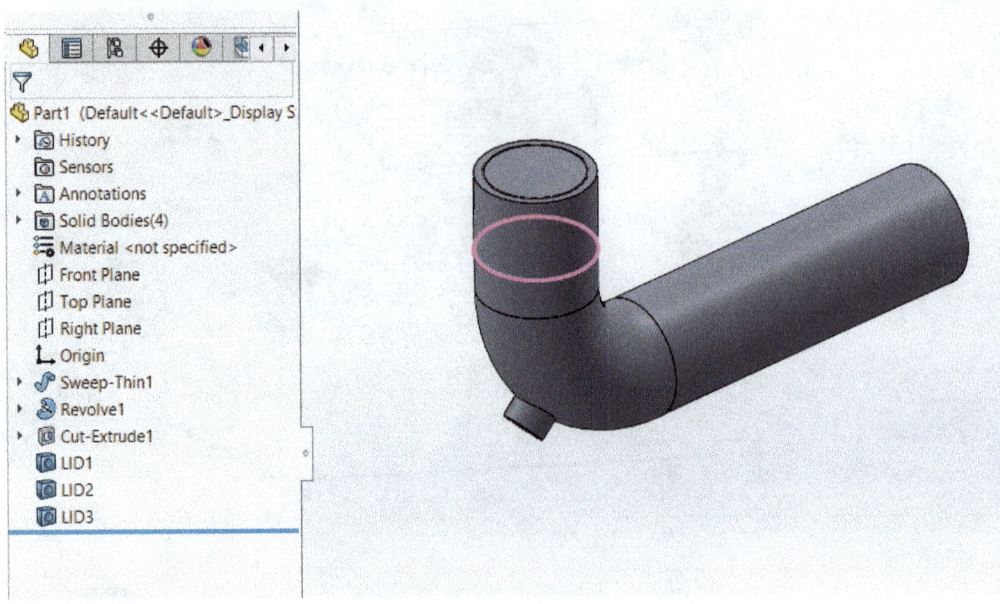

Figure-16. Model_for_Practical_2_.png

- Click on the **Add-Ins** option from the **Options** drop-down in the **Quick Access Toolbar**. The **Add-Ins** dialog box will be displayed. Select the check boxes for SOLIDWORKS Flow Simulation from the dialog box and click on the **OK** button to activate the add-in if not done yet.
- Click on the **Wizard** tool from the **Flow Simulation CommandManager** in the **Ribbon**. The **Wizard** dialog box will be displayed.
- Specify desired name for the project and click on the **Next** button. The **Unit System** page of the dialog box will be displayed.
- Select the **SI (m-kg-s)** unit system from the page and click on the **Next** button. The **Analysis Type** page of dialog box will be displayed.
- Select the **Internal** radio button from the **Analysis type** area of the dialog box. Select the **Exclude cavities without flow conditions** check box. Since, we concerned about only the flow of fuel and air so we will not select the parameters related to heat transfer.
- Click on the **Next** button after specifying the parameters. The **Default Fluid** page will be displayed.
- Double-click on **Air** and **Ethanol** gas to include them in the study and click on the **Next** button. The **Wall Conditions** page will be displayed.
- Keep the default values and click on the **Next** button. The **Initial Conditions** page will be displayed.
- Click on the **Finish** button from the dialog box.

Applying Boundary Conditions

- Click on the **Section View** button from the **Heads-up View Toolbar**; refer to Figure-17. The **Section View PropertyManager** will be displayed with preview of section; refer to Figure-18.

Figure-17. Section View button

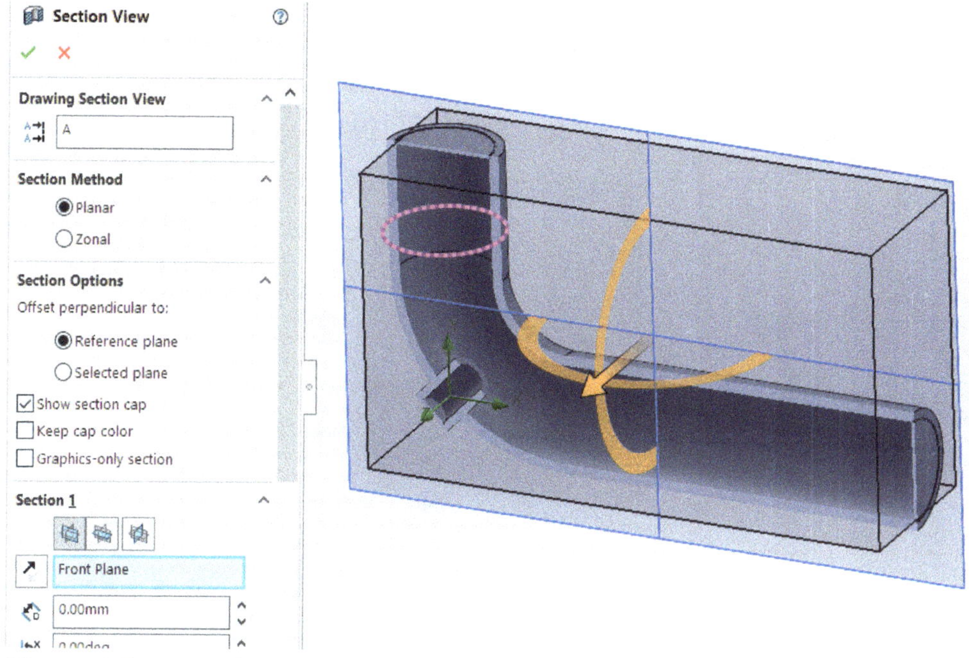

Figure-18. Preview_of_section.png

- Make sure **Front Plane** is selected and model section is created from half. After setting desired parameters, click on the **OK** button from the **PropertyManager**. The section view will be displayed; refer to Figure-19.

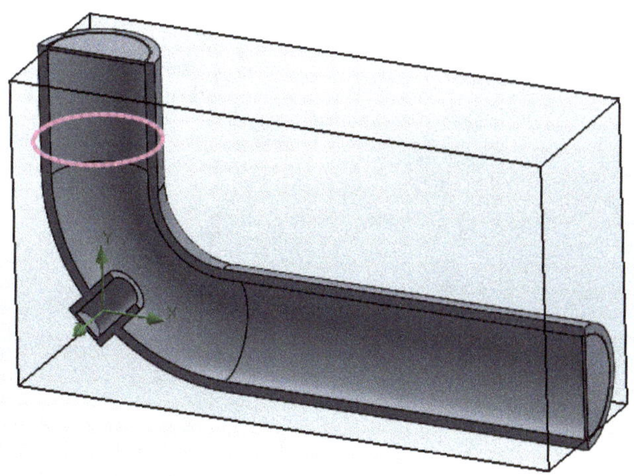

Figure-19. Section_view_of_model.png

- Right-click on the **Boundary Conditions** node from the **Design Tree** and select the **Insert Boundary Condition** option from the shortcut menu displayed. The **Boundary Condition PropertyManager** will be displayed.
- Select the **Flow Openings** button from the **Type** rollout and **Inlet Mass Flow** option from the list.
- Specify **6.11 kg/min** value in the **Mass Flow Rate** edit box and select the top inner face of the model for air flow as shown in Figure-20.

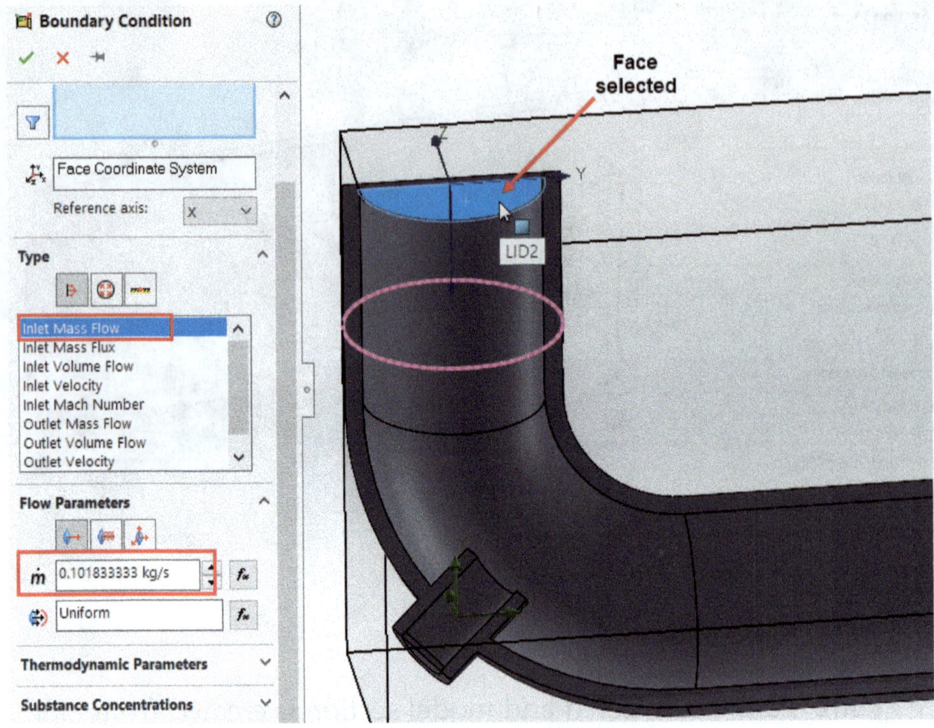

Figure-20. Air_inlet_boundary_condition.png

- Expand the **Substance Concentrations** rollout and make sure **Air** concentration is set to **1** and **Ethanol** concentration is set to **0**; refer to Figure-21.

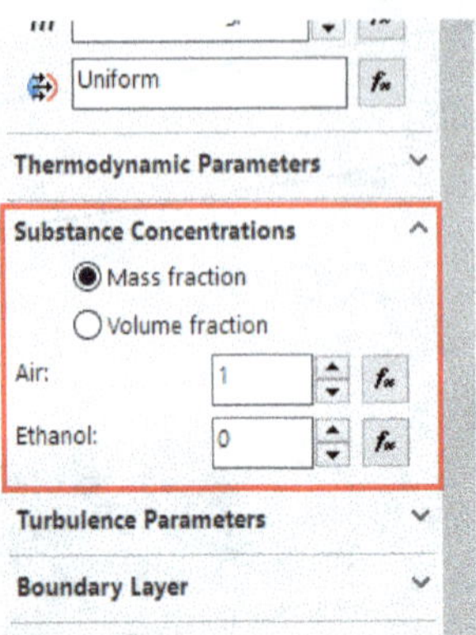

Figure-21. Setting air concentration

- After setting desired parameters, click on the **OK** button from the **PropertyManager**.
- Right-click again on the **Boundary Conditions** node and select the **Insert Boundary Condition** option from the shortcut menu displayed to specify flow rate of ethanol in the chamber. The **Boundary Condition PropertyManager** will be displayed.
- Specify inlet mass flow rate as **0.408 kg/min** in the **Mass Flow Rate** edit box of **Flow Parameters** rollout in the **PropertyManager** as discussed earlier; refer to Figure-22.
- Select the face of small opening to be used for inlet of ethanol and specify the ethanol concentration as 1; refer to Figure-23.

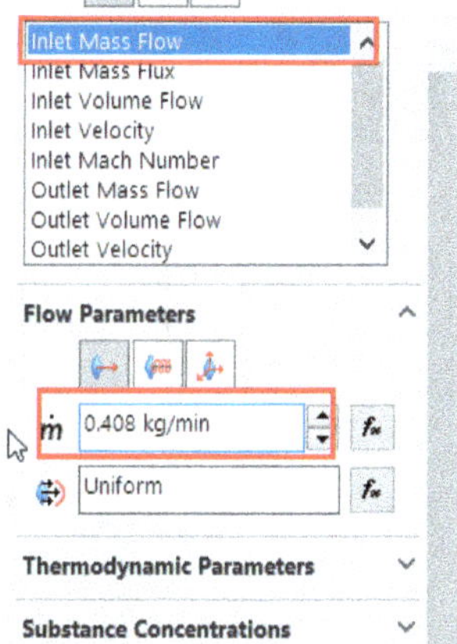

Figure-22. Flow_rate_for_ethanol

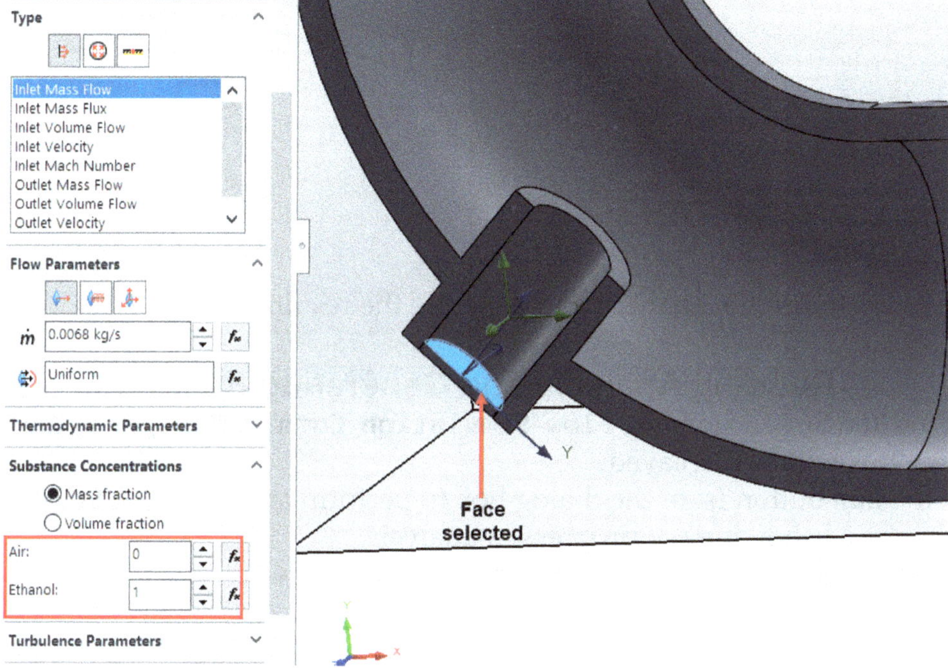

Figure-23. Face_selected_for_ethanol_inlet.png

- After setting the parameters, click on the **OK** button from the **PropertyManager**.
- Right-click again on the **Boundary Conditions** node and select the **Insert Boundary Condition** option from the shortcut menu displayed to specify pressure and temperature conditions at outlet. The **Boundary Condition PropertyManager** will be displayed.
- Select the **Pressure Openings** button from the **Type** rollout and select the **Total Pressure** option from the list.
- Specify the pressure value as **34500 Pa** and temperature as **500 K** in the respective edit boxes.
- Select the inner face of model lid as shown in Figure-24 and then click on the **OK** button from the **PropertyManager** to apply the parameters.

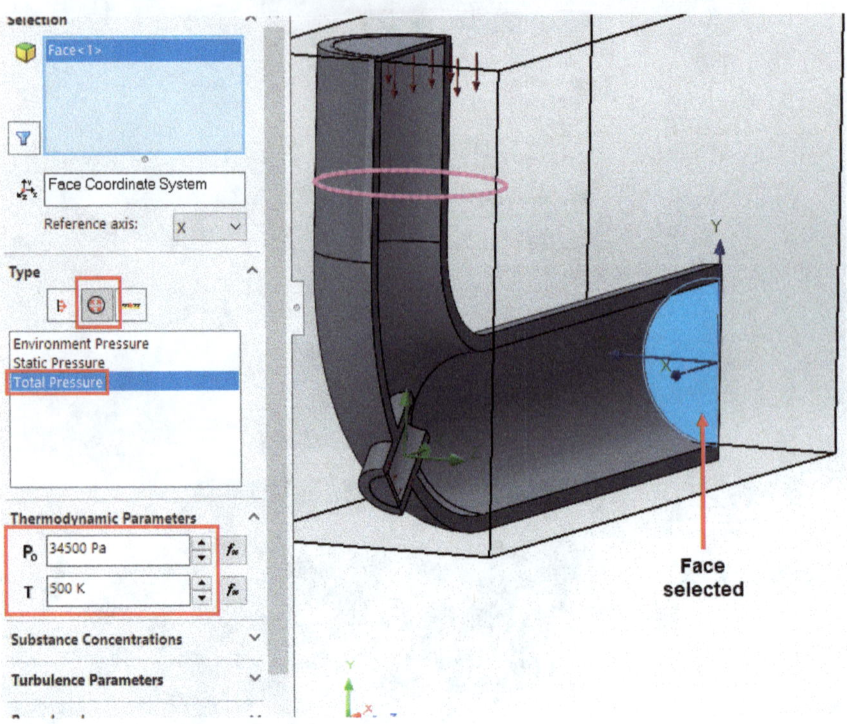

Figure-24. Pressure_parameters_at_outlet.png

We are now ready to run the analysis and generate the results.

Running Analysis and Generating Results

- Click on the **Run** button from the **Flow Simulation CommandManager** in the **Ribbon**. The Run dialog box will be displayed.
- Click on the **Run** button from the dialog box to perform analysis. The **Solver** dialog box will be displayed showing the progress in analysis.
- Once the analysis is finished, close the **Solver** dialog box.
- You can generate the flow trajectories and other results as discussed in previous practical.

Practical 3 Air Flow by Fan

In this practical, we will check the flow of air passing through a fan running at an rpm of 150; refer to Figure-25. Note that in this practical, we will use rotary region to perform analysis. The steps to perform this analysis are given next.

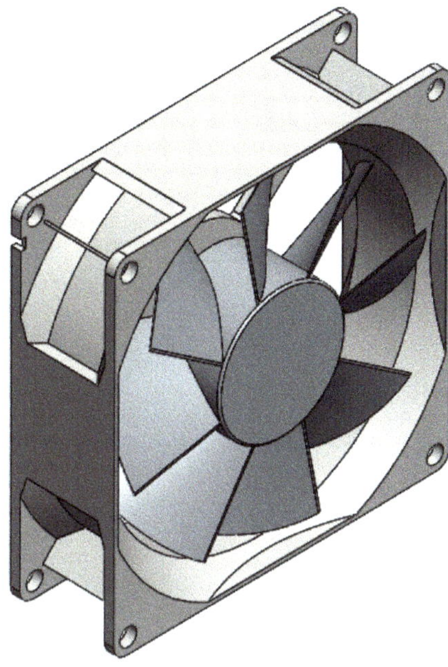

Figure-25. Model_for_Practical_3.png

Opening Model and Starting Project

- Start SolidWorks using the desktop icon or **Start** menu.
- Click on the **Open** button from the **Quick Access Toolbar** or **Open** option from the **File** menu. The **Open** dialog box will be displayed.
- Select the part file for this practical and click on the **Open** button. The model for file will be displayed in the drawing area.
- Click on the **Add-Ins** option from the **Options** drop-down in the **Quick Access Toolbar**. The **Add-Ins** dialog box will be displayed. Select the check boxes for **SOLIDWORKS Flow Simulation** from the dialog box and click on the **OK** button to activate the add-in if not done yet.
- Click on the **Wizard** tool from the **Flow Simulation CommandManager** in the **Ribbon**. The **Wizard** dialog box will be displayed.
- Specify desired name for the project and click on the **Next** button. The **Unit System** page of the dialog box will be displayed.
- Select the **SI (m-kg-s)** unit system from the page.
- Expand the **Loads&Motion** node from the bottom list and select the **RPM** unit for **Angular Velocity**; refer to Figure-26.

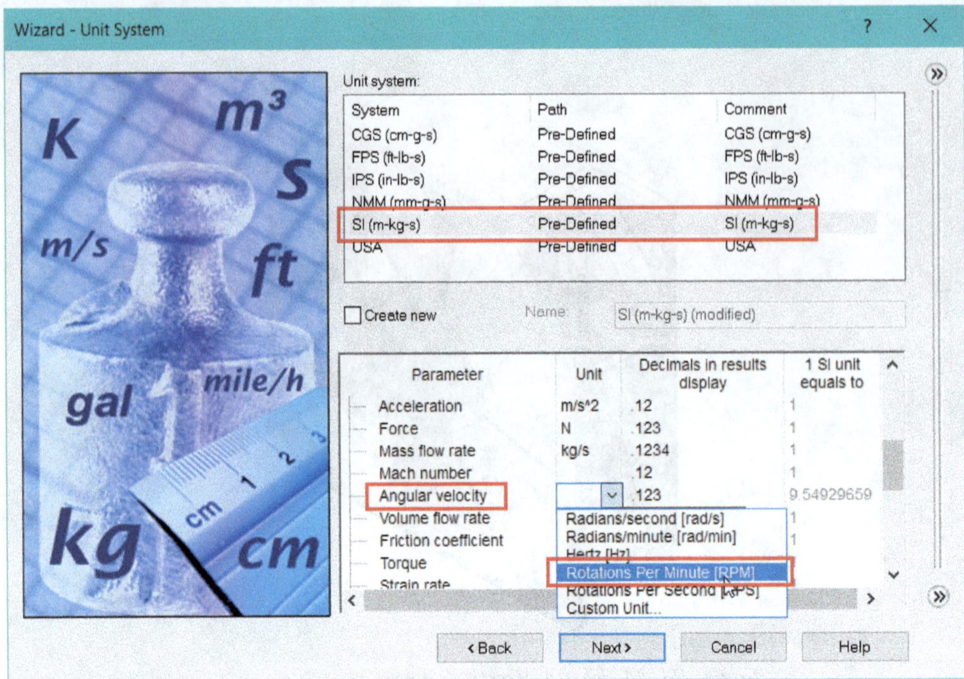

Figure-26. RPM_option_for_angular_velocity.png

- Click on the **Next** button. The **Analysis Type** page of dialog box will be displayed.
- Select the **External** radio button from the **Analysis type** area of the dialog box. Select the **Exclude cavities without flow conditions** check box.
- Select the **Rotation** check box from the list box and select the **Local region(s) (Averaging)** option from the **Type** drop-down list for rotation; refer to Figure-27.

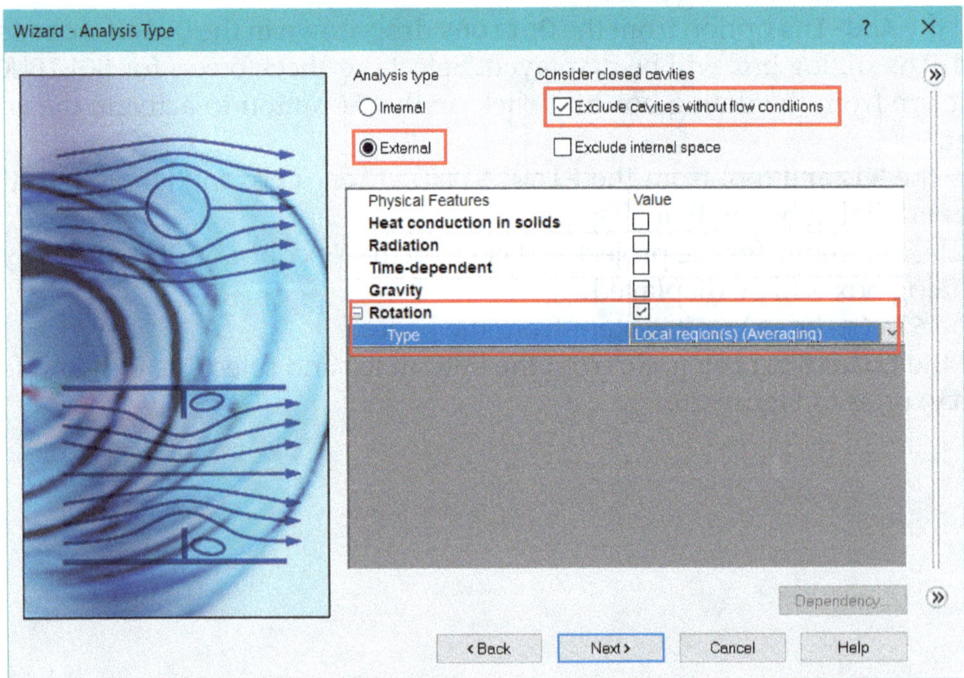

Figure-27. Defining_Rotation_options.png

- Click on the **Next** button after specifying the parameters. The **Default Fluid** page will be displayed.
- Double-click on **Air** gas to include it in the study and click on the **Next** button. The **Wall Conditions** page will be displayed.
- Keep the default values and click on the **Next** button. The **Initial Conditions** page will be displayed.

- Click on the **Coordinate System** button from the bottom in this page. The **Coordinate system** dialog box will be displayed; refer to Figure-28. Make sure the **X** option is selected in the **Reference axis** drop-down so that the fan rotates about X axis.

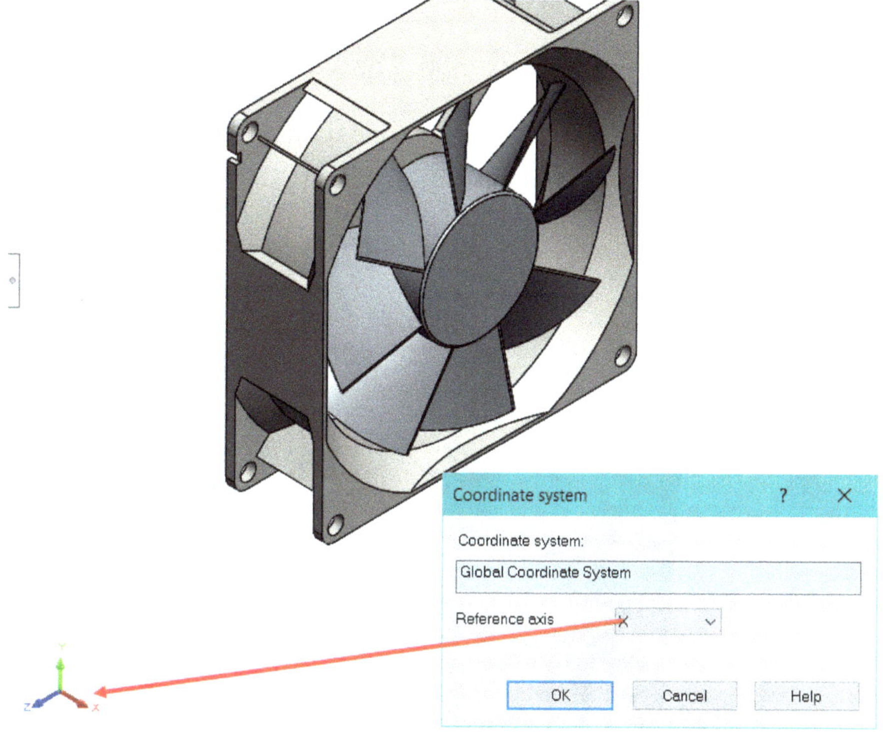

Figure-28. Coordinate_system_dialog_box

- Click on the **OK** button from the **Coordinate system** dialog box and then click on the **Finish** button from the dialog box. The computational domain of the analysis will be displayed.
- Select **Computation Domain** option from the **Design Tree** and then reduce the size of computation domain using the arrows displayed on the model; refer to Figure-29.

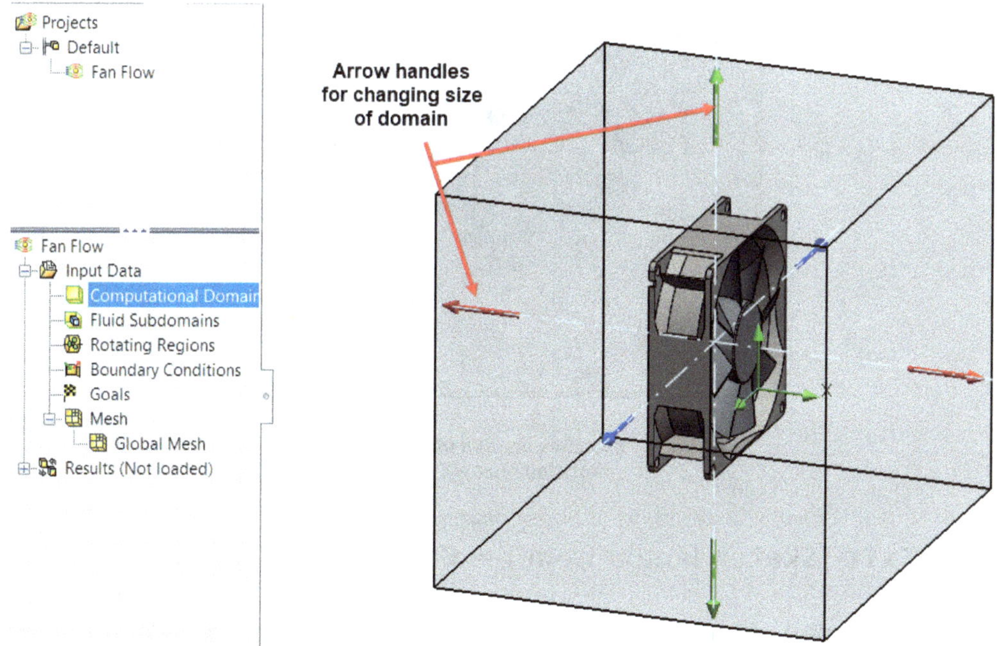

Figure-29. Changing_size_of_computational_domain.png

Defining Rotating Region

- To use a rotating region, we will need a cylinder shape component. So, we will first create a new component. Click on the **New Part** tool from the **Insert Components** drop-down in the **Assembly CommandManager** in the **Ribbon**. You will be asked to select face/plane for positioning new part.
- Select the face of model as shown in Figure-30. The sketching mode will be activated and the model will display transparent.

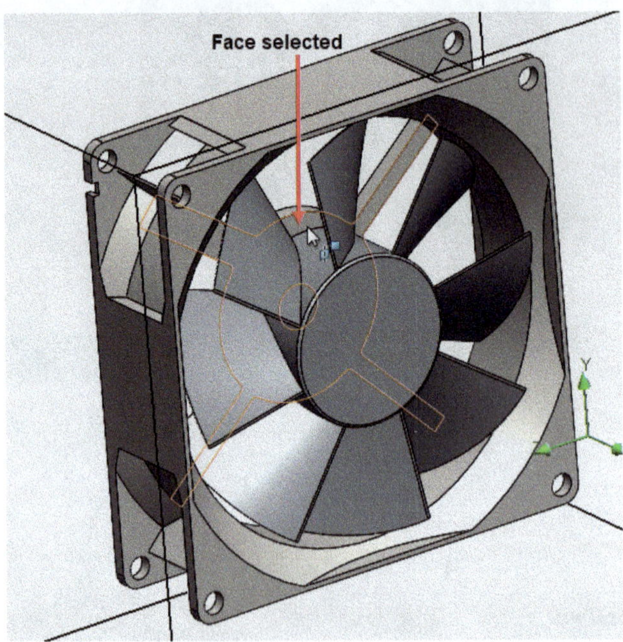

Figure-30. Face_selected_for_positioning.png

- Create two concentric circles at the origin as shown in Figure-31 to define the rotating region boundaries.

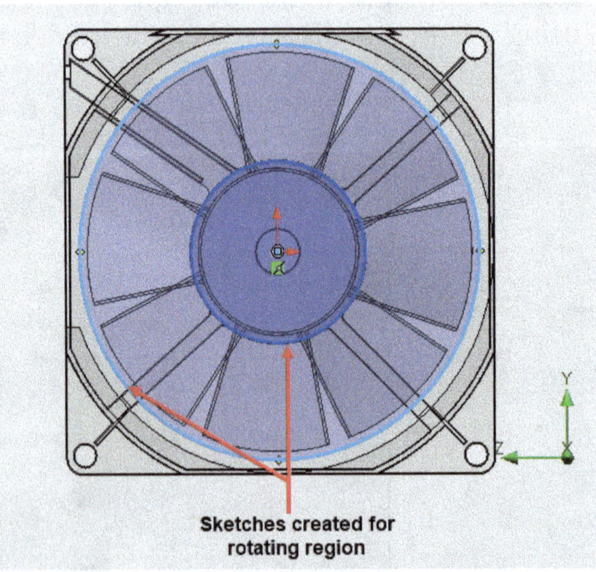

Figure-31. Sketch_for_defining_rotating_region.png

- Click on the **Exit Sketch** button from the **CommandManager** to exit the sketching environment.
- Click on the **Extruded Boss/Base** tool from the **Features CommandManager** in the **Ribbon**. The **Boss-Extrude PropertyManager** will be displayed.

- Select the two circles to be extruded. Preview of the extrude feature will be displayed; refer to Figure-32.

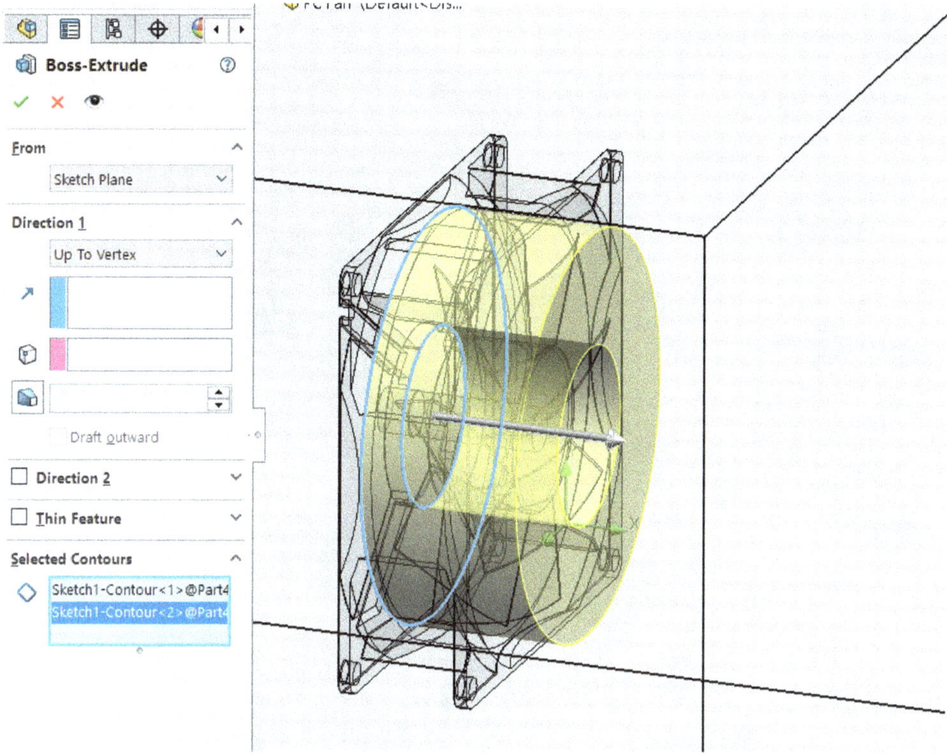

Figure-32. Preview_of_extrude_feature.png

- Select the **Up To Surface** option from the drop-down in **Direction 1** rollout and select the top surface fan; refer to Figure-33.

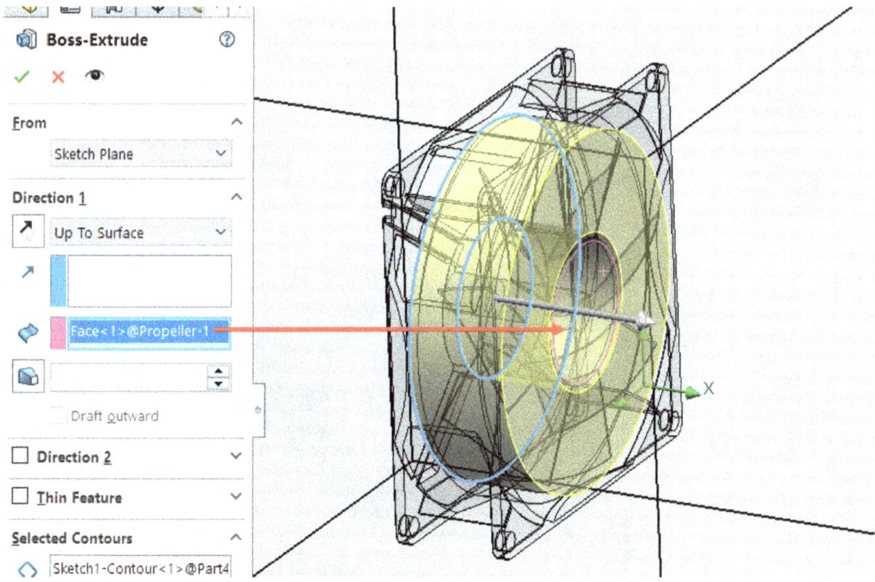

Figure-33. Face_selected_for_defining_extrude_extent.png

- Click on the **OK** button from the **PropertyManager** to create the feature.
- Click on the **Edit Component** toggle button from the **Features CommandManager** to exit component editing mode. The solid extrude body will be displayed as shown in Figure-34.
- Right-click on the new body and **Change Transparency** button from the shortcut menu; refer to Figure-35. The new part will become transparent; refer to Figure-36.

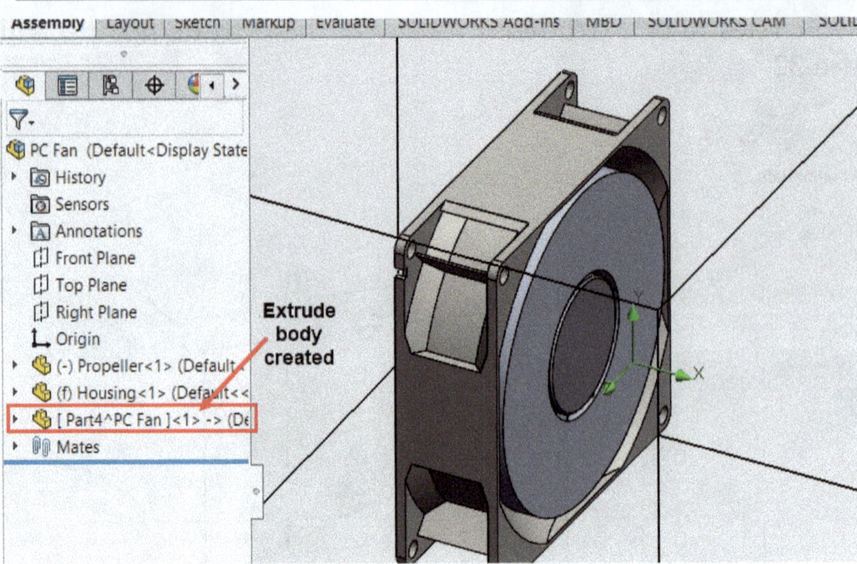

Figure-34. Extrude_body_created.png

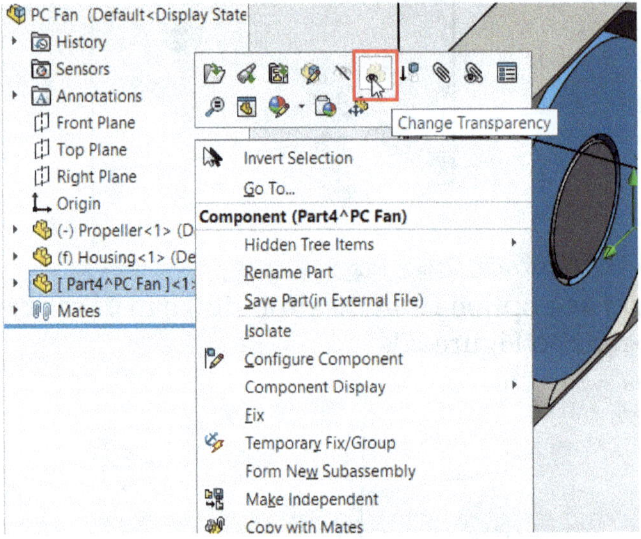

Figure-35. Change_transparency_button.png

Figure-36. Setting_transparency.png

- Open the **Flow Simulation Analysis Manager** in the **Design Tree** and right-click on **Rotating Regions** option from the **Design Tree**. A shortcut menu will be displayed; refer to Figure-37.

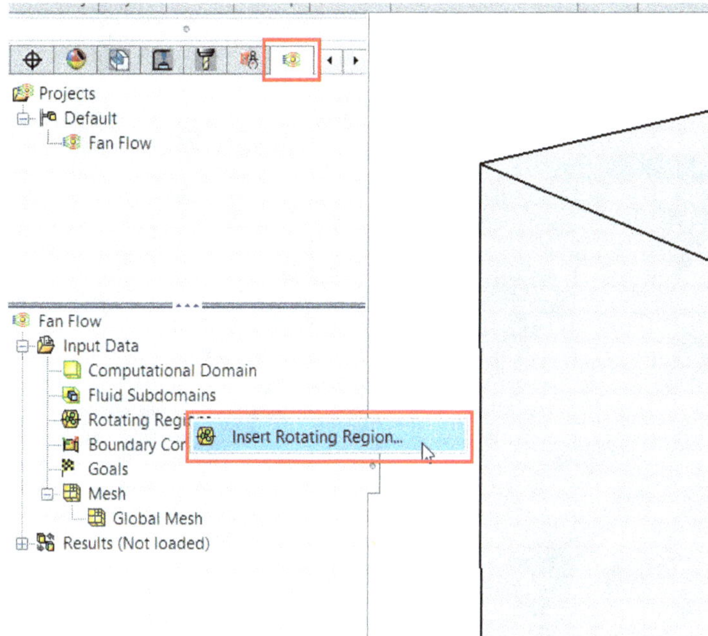

Figure-37. Shortcut_menu_for_rotating_region.png

- Select the **Insert Rotating Region** option from the shortcut menu. The **Rotating Region PropertyManager** will be displayed; refer to Figure-38.

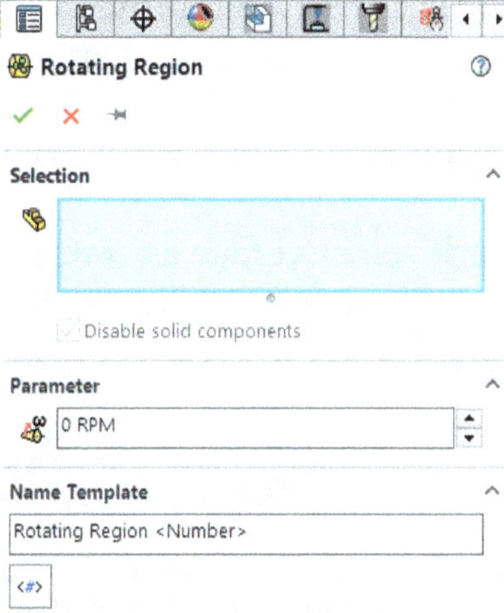

Figure-38. Rotating_Region_PropertyManager

- Select the new part created for rotating region and specify angular velocity as **150 RPM** in the **Parameter** rollout; refer to Figure-39.

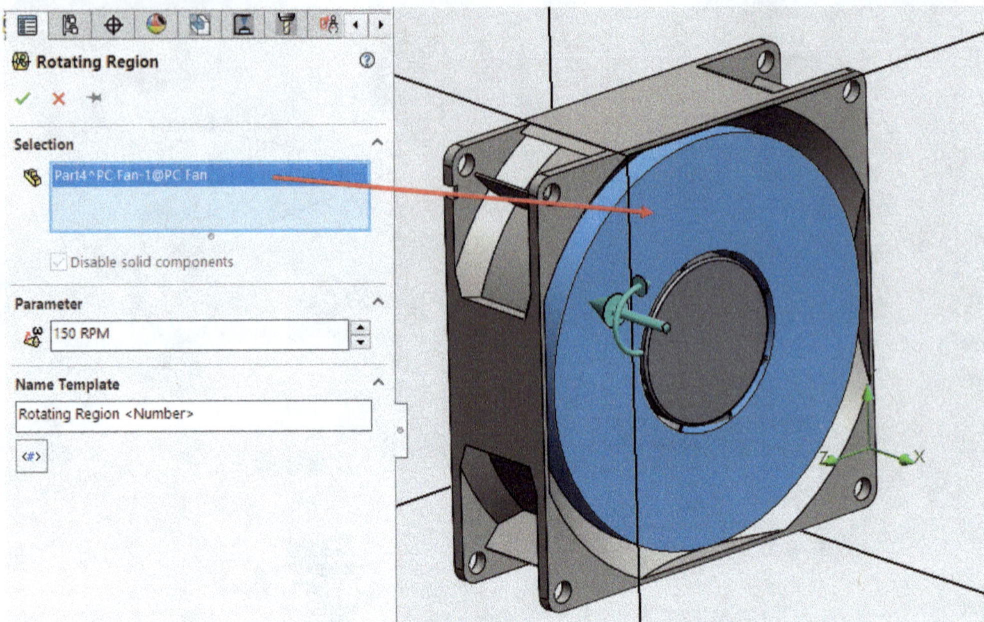

Figure-39. Setting_parameters_for_rotating_region.png

- Click on the **OK** button from the **PropertyManager** to create the region.

Defining Stator for Fan Motion

- Right-click on the **Boundary Conditions** node in the **Design Tree** and select **Insert Boundary Condition** option from the shortcut menu displayed. The **Boundary Condition PropertyManager** will be displayed.
- Select the **Wall** button from the **Type** rollout and **Real Wall** option from the list in the rollout.
- Select the **Stator** check box to mark selected face as stator for rotating region.
- Select the cylindrical face of housing as shown in Figure-40.

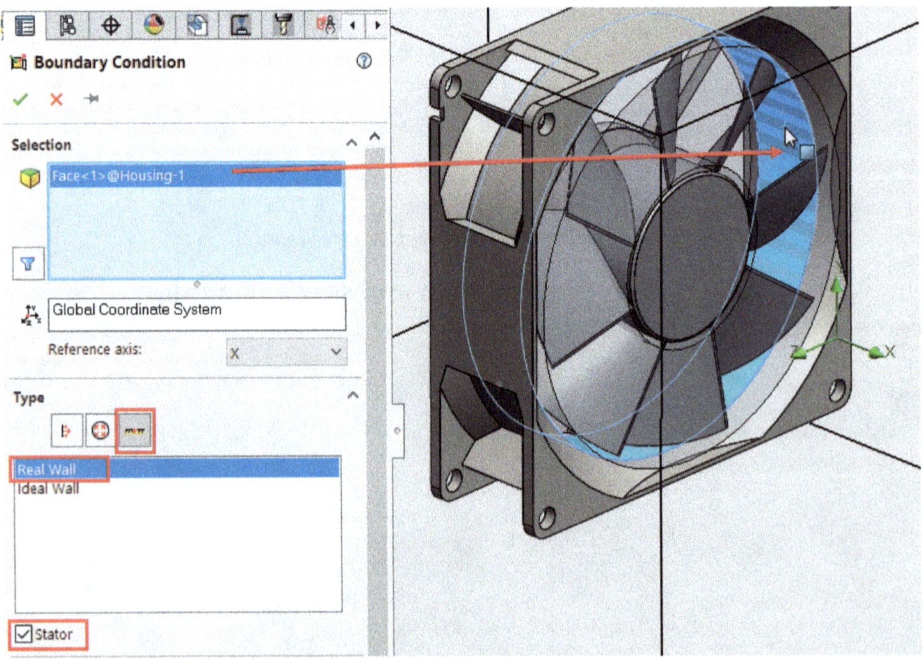

Figure-40. Face_selected_for_stator.png

- Click on the **OK** button from the **PropertyManager**.

Now we are ready to run the analysis and check the results.

Running Analysis and Checking Results

- Click on the **Run** button from the **Flow Simulation CommandManager** in the **Ribbon**. The Run dialog box will be displayed.
- Click on the **Run** button from the dialog box. The **Solver** dialog box will be displayed.
- Once the analysis is complete without warning, close the **Solver** dialog box.
- Expand the **Results** node and right-click on **Flow Trajectories** option. A shortcut menu will be displayed.
- Select the **Insert** option from the shortcut menu. The **Flow Trajectories PropertyManager** will be displayed.
- Select the Right plane and offset it up to the boundary of computational domain to define start point for flow of air; refer to Figure-41.

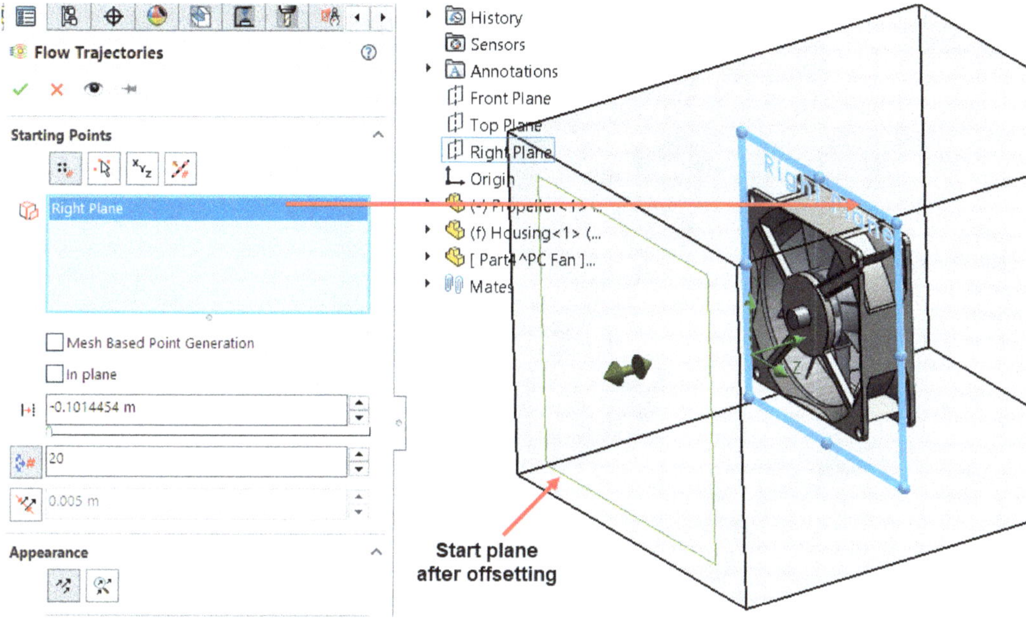

Figure-41. Setting_start_point_for_flow.png

- Set the desired parameters as discussed earlier and click on the **OK** button. The flow trajectory results will be displayed; refer to Figure-42.

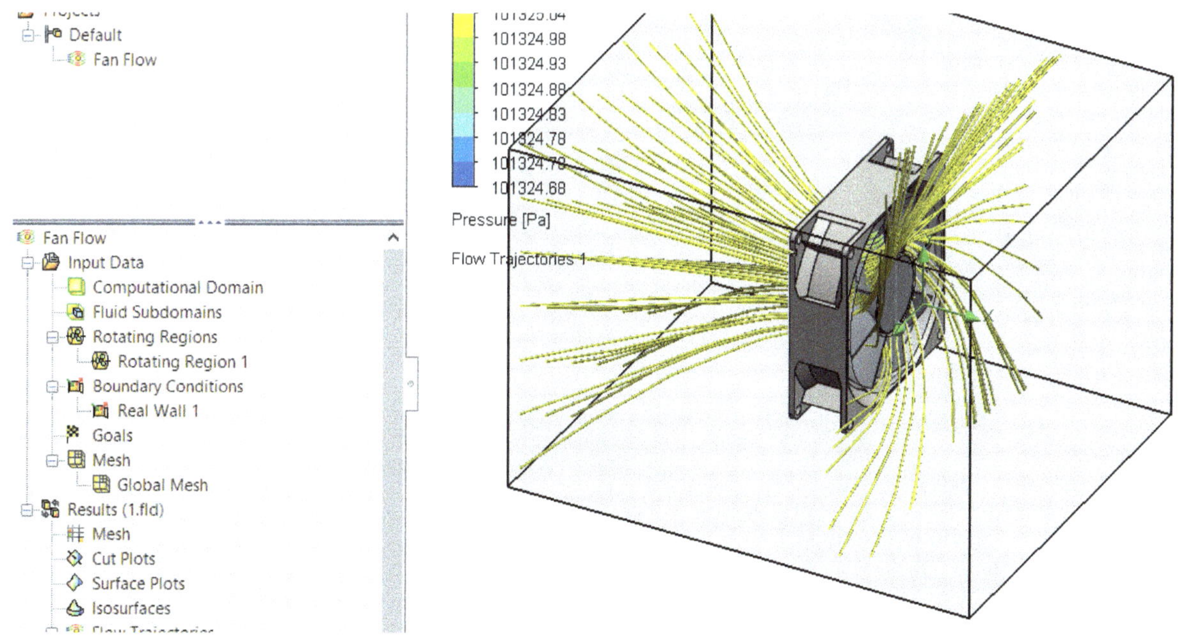

Figure-42. Flow_trajectory_results.png

You can check different results by modifying shape of fan blade and increasing/decreasing the RPM of fan.

PRACTICE 1

Using the model shown in Figure-43 check the flow of water in the valve with different opening orientation of valve.

Figure-43. Model_for_Practice_1.png

Index

Ethics of an Engineer

- Engineers shall hold paramount the safety, health and welfare of the public and shall strive to comply with the principles of sustainable development in the performance of their professional duties.

- Engineers shall perform services only in areas of their competence.

- Engineers shall issue public statements only in an objective and truthful manner.

- Engineers shall act in professional manners for each employer or client as faithful agents or trustees, and shall avoid conflicts of interest.

- Engineers shall build their professional reputation on the merit of their services and shall not compete unfairly with others.

- Engineers shall act in such a manner as to uphold and enhance the honor, integrity, and dignity of the engineering profession and shall act with zero-tolerance for bribery, fraud, and corruption.

- Engineers shall continue their professional development throughout their careers, and shall provide opportunities for the professional development of those engineers under their supervision.

OTHER BOOKS BY CADCAMCAE WORKS

Autodesk Inventor 2021 Black Book

Autodesk Revit 2021 Black Book

Autodesk Fusion 360 Black Book (V 2.0.6508)

AutoCAD Electrical 2021 Black Book
AutoCAD Electrical 2020 Black Book

SolidWorks 2021 Black Book
SolidWorks 2020 Black Book
SolidWorks 2019 Black Book

SolidWorks Simulation 2021 Black Book
SolidWorks Simulation 2020 Black Book

SolidWorks Flow Simulation 2021 Black Book
SolidWorks Flow Simulation 2020 Black Book

SolidWorks Electrical 2021 Black Book
SolidWorks Electrical 2020 Black Book

Mastercam X7 for SolidWorks 2014 Black Book
Mastercam 2017 for SolidWorks Black Book

Creo Parametric 7.0 Black Book
Creo Parametric 6.0 Black Book
Creo Parametric 5.0 Black Book

Creo Manufacturing 4.0 Black Book

Autodesk CFD 2018 Black Book

Basics of Autodesk Inventor Nastran 2020
Basics of Autodesk Inventor Nastran 2021

ETABS 2016 Black Book
ETABS 2018 Black Book

www.ingramcontent.com/pod-product-compliance
Lightning Source LLC
Chambersburg PA
CBHW060534060326
40690CB00017B/3490
* 9 7 8 1 7 7 4 5 9 0 0 8 9 *